Introduction to Asymptotic Methods

CRC Series: Modern Mechanics and Mathematics

Series Editors: David Gao and Ray W. Ogden

PUBLISHED TITLES

Beyond Perturbation: Introduction to The Homotopy Analysis Method
by Shijun Liao

Mechanics of Elastic Composites
by Nicolaie Dan Cristescu, Eduard-Marius Craciun, and Eugen Soós

Continuum Mechanics and Plasticity
by Han-Chin Wu

Hybrid and Incompatible Finite Element Methods
by Theodore H.H. Pian and Chang-Chun Wu

Introduction to Asymptotic Methods
by Jan Awrejcewicz and Vadim A. Krysko

FORTHCOMING TITLE

Microstructural Randomness in Mechanics of Materials
by Martin Ostroja Starzewski

Introduction to Asymptotic Methods

Jan Awrejcewicz
Vadim A. Krysko

CRC Press
Taylor & Francis Group
Boca Raton London New York

CRC Press is an imprint of the
Taylor & Francis Group, an **informa** business
A CHAPMAN & HALL BOOK

CRC Press
Taylor & Francis Group
6000 Broken Sound Parkway NW, Suite 300
Boca Raton, FL 33487-2742

First issued in paperback 2019

ISBN-13: 978-1-58488-677-8 (hbk)
ISBN-13: 978-0-367-39090-7 (pbk)

Library of Congress Card Number 2006042615

Library of Congress Cataloging-in-Publication Data

Awrejcewicz, J. (Jan)
Introduction to asymptotic methods / Jan Awrejcewicz, Vadim A. Krysko.
p. cm. -- (Modern mechanics and mathematics ; 5)
Includes bibliographical references and index.
ISBN 1-58488-677-3 (alk. paper)
1. Singular perturbations (Mathematics) 2. Differential equations--Asymptotic theory.
1. Krys'ko, V.A. (Vadim Anatol'evich), 1937- II. Title. III. CRC series--modern mechanics and mathematics ; 5.

QA372.A973 2006
515'.392--dc22 2006042615

Visit the Taylor & Francis Web site at
http://www.taylorandfrancis.com

and the CRC Press Web site at
http://www.crcpress.com

Contents

Introduction xv

1 Elements of mathematical modeling **1**

1.1 Structure of a mathematical model 1

1.2 Examples of reducing problems to a dimensionless form . . . 3

1.3 Mathematical model adequacy and properties. Regular and singular perturbations 7

2 Expansion of functions and mathematical methods **9**

2.1 Expansions of elementary functions into power series 9

2.1.1 Newton's binomial 9

2.1.2 Taylor and Maclaurin series 10

2.1.3 Estimation of approximating functions values 13

2.1.4 Estimation of approximating values of definite integrals 14

2.1.5 Approximating solution of Cauchy problem for ordinary differential equations 14

2.2 Mathematical methods of perturbations 16

2.2.1 Perturbation along a parameter and coordinates. Classical method of a small parameter 16

2.2.2 Comparison of infinitely small and infinitely large functions. Scaling functions. Magnitude order quantities . 17

2.2.3 Asymptotical sequences and decompositions 22

2.2.4 Nonuniform asymptotic decompositions 25

2.2.5 Operations on asymptotic decompositions 26

2.2.6 Asymptotic series. Comparison of asymptotic and convergent series. Advantages of application of asymptotic series and decompositions 29

Exercises 32

3 Regular and singular perturbations **35**

3.1 Introduction. Asymptotic approximation with respect to a parameter 36

3.2 Nonuniformities of a classical perturbation approach 40

3.3 Method of "elongated" parameters 43

3.4 Method of deformed variables 47

3.5 Method of scaling and full approximation 52

3.5.1 Deformation of one independent variable 53

3.5.2 Deformation of two independent variables 55
3.5.3 Method of full approximation 57
3.6 Multiple scale methods 58
3.6.1 Introduction 58
3.6.2 Derivative decomposition along one and two variables 62
3.6.3 Application to the problems of vibrations 63
3.7 Variations of arbitrary constants 68
3.8 Averaging methods 72
3.8.1 Methods of Van der Pol and Krylov-Bogolubov-Mitropolskiy (KBM) 72
3.8.2 Duffing's problem and the averaging procedure 75
3.9 Matching asymptotic decompositions 77
3.9.1 Fundamental notions and terminology 77
3.9.2 Example with a boundary layer 82
3.9.3 Fundamental rules and order of matching 84
3.9.4 Construction of matched asymptotic expansion 87
3.9.5 Example with a singularity 88
3.9.6 On the choice of internal variables 90
3.10 On the sources of nonuniformities 92
3.11 On the influence of initial conditions 94
3.12 Analysis of strongly nonlinear dynamical problems 97
3.13 A few perturbation parameters 108
Exercises 111

4 Wave-impact processes **115**
4.1 Definition of a cylinder-like piston wave 115
4.1.1 Defining the problem, its solution and analysis 115
4.1.2 Nonlinear solution in the vicinity of the piston 118
4.1.3 Nonlinear solution in the vicinity of the front of the impact wave 119
4.1.4 Methods of strained coordinates and renormalization . 124
4.1.5 Effectiveness of various asymptotic methods 130
4.2 One-dimensional nonstationary nonlinear waves 130
4.2.1 Formulation of the problem and its solution 130
4.2.2 Renormalization method and singularities 133
4.2.3 Analytical method of characteristics 134
4.2.4 Multiple scales method 137

5 Padé approximations **141**
5.1 Determination and characteristics of Padé approximations . 141
5.2 Application of Padé approximations 144
5.2.1 Simple examples 144
5.2.2 Supersonic flow round a thin cone in circumsonic regime 145
5.2.3 Damping of the ball-shaped waves of pressure in a free space and in a tube 146

5.2.4 Analysis of the "blow-up" phenomenon 149
5.2.5 Homoclinic orbits 158
5.2.6 Vibrations of nonlinear system with nonlinearity close to sign (x) 160
Exercises 163

6 Averaging of ribbed plates **165**
6.1 Averaging in the theory of ribbed plates 165
6.2 Kantorovich-Vlasov-type methods 169
6.2.1 Kantorowich-Vlasov method (KVM) 170
6.2.2 Vindiner method (VM) 170
6.2.3 Method of variational iterations (MVI) 171
6.2.4 Agranowsky-Baglay-Smirnov method (ABSM) 171
6.2.5 Combined method (CM) 172
6.2.6 Kantorovich-Vlasov method with the amendment . . . 173
6.2.7 Vindiner method with the amendment 173
6.2.8 Vindiner method and variational iterations 173
6.3 Transverse vibrations of rectangular plates 174
6.4 Deflections of rectangular plates 187

7 Chaos foresight **191**
7.1 The analyzed system 192
7.2 Melnikov-Gruendler's approach 194
7.3 Melnikov-Gruendler function 195
7.4 Numerical results 199

8 Continuous approximation of discontinuous systems **203**
8.1 An illustrative example 203
8.2 Higher dimensional systems 208

9 Nonlinear dynamics of a swinging oscillator **217**
9.1 Parametrical form of canonical transformations 218
9.2 Function derivative 218
9.3 Invariant normalization of Hamiltonians 220
9.4 Algorithm of invariant normalization with the help of parametric transformations 221
9.5 Algorithm of invariant normalization for asymptotical determination of the Poincaré series 224
9.6 Examples of asymptotical solutions 225
9.7 A swinging oscillator 228
9.8 Normal form 230
9.9 Normal form integral 231

References **233**

Index 243

List of Figures

1.1 Free vibrations of one-degree-of-freedom system. 3

2.1 Dependence of the sum of elements of convergent series A on the number of its elements. 31
2.2 Sum of divergent series B. 32

3.1 Graphs of solutions $u_0(x)$, $u_\varepsilon(x)$. 38
3.2 Integral curves of equation (3.59). 50
3.3 Exact solution and its approximations through different numbers of series (3.107) terms. 60
3.4 Exact solution and its approximations through different numbers of series (3.108) terms. 60
3.5 Variations in time of functions $u(t)$, $a(t)$, $\beta(t)$. 71
3.6 External y^o, internal y^i and composite y^c decompositions. . 83
3.7 Comparison of results obtained through different analytical methods: a) $\omega_0 = 1$, $\varepsilon = 0.1$; b) $\omega_0 = \varepsilon = 1$. 95
3.8 Period of vibrations of a mathematical pendulum vs initial deflection θ_0. 96
3.9 Numerical solution x_0 and x and x yielded by equation (3.270), and x obtained from equation (3.272) for $\gamma = 0.1$, $\omega = 1$, $\varepsilon = 0.1$, $x(0) = 0$, $x'(0) = 1$ and for different values of n: a) 3; b) 7; c) 21. 99
3.10 Numerical solution x_0 and x obtained from equation (3.270), and x obtained from equation (3.272) for $\gamma = 0.1$, $\omega = 1$, $\varepsilon = 1$, $x(0) = 0$, $x'(0) = 1$ and for different values of n: a) 3, b) 7, c) 21. 100
3.11 Numerical solution x_0 and x yielded by equation (3.270), and x obtained from (3.272) for $\gamma = 0.1$, $\omega = 1$, $\varepsilon = 1$, $x(0) = 0$, $x'(0) = 10$ and for different values of n: a) 3, b) 7, c) 21. . . 101
3.12 Approximation of $\Theta/\sin\Theta$ for various values of n: a) $n = 2$; b) $n = 3$; c) $n = 6$; d) $n = 11$. 104
3.13 Comparison of solutions of equations (3.299) (dashed curve), and equations (3.302) and (3.304) for various values of n: a) $n = 3$, b) $n = 7$, c) $n = 11$, d) $n = 21$. 108

5.1 Comparison of the results of numerical computations, the application of PA and the experimental research (see the text). 147

5.2 Results of computations obtained from the approximation of constant shock waves (curve 1), numerical computations (curve 2) and the experiment (circles) – see text. 149
5.3 The numerical solution of equation (5.20) for various initial conditions: a) $\varepsilon = 0.1$; b) $\varepsilon = 0.01$; c) $\varepsilon = 0.001$. 153
5.4 The numerical, asymptotic, and PA solution for various initial conditions: a) $\varepsilon = 0.1$; b) $\varepsilon = 0.01$; c) $\varepsilon = 0.001$ ($AP[2,1]$). . 155
5.5 The numerical, asymptotic, and PA solution PA for various initial conditions: a) $\varepsilon = 0.1$; b) $\varepsilon = 0.01$; c) $\varepsilon = 0.001$ ($AP[3,1]$). 156
5.6 The numerical, asymptotic, and PA solution for various initial conditions: a) $\varepsilon = 0.1$; b) $\varepsilon = 0.01$; c) $\varepsilon = 0.001$ ($AP[2,2]$). . 157
5.7 The solution of Cauchy problem (5.68), (5.69) for $A = 1$ with the use of Runge-Kutta method (——), and the approximations: (5.79) (------); (5.80) (- - - - - -); (5.81) (– – – –) for different values of n. 162

6.1 A homogeneous beam supported discretely with parallel bearings. 166
6.2 Dependencies w, w_0, w_1 on y. 166
6.3 Relations between the accurate solution and the averaged one. 168
6.4 Computational scheme and the support of the analyzed plate. 175
6.5 Results of computations M_2^* with respect to the wave number. 186
6.6 Numerical computations of relation $M_1^*(x_1)$. 188
6.7 Numerical computations of relation $M_2^*(x_1)$. 189
6.8 Numerical computations of relation $W^{II}(x_1)$. 189

7.1 The analyzed system. 193
7.2 The threshold curve. 199
7.3 Phase portraits and Poincaré maps ($\Gamma' = 0.98, w' = 0.1$). . . 200
7.4 Phase portraits and Poincaré maps ($\Gamma' = 1.02, w' = 0.1$). . . 200
7.5 Phase portraits and Poincaré maps ($\Gamma' = 1.1, w' = 0.1$). . . . 201

8.1 The considered one-degree-of-freedom mechanical system with friction. 204
8.2 Vector field of (2.3) for (a) $w_0 = -0.3$, $y_{0i} \in (-1.2, 2)$ for $i = 0 \ldots 32$, (b) $w_0 = -0.2$, $y_{0i} \in (-1.3, 1.3)$ for $i = 0 \ldots 26$, (c) $w_0 = 1.2$, $y_{0i} \in (-1.5, 1.5)$ for $i = 0 \ldots 30$, (d) $w_0 = 1.0$, $y_{0i} \in (-1.6, 1.6)$ for $i = 0 \ldots 32$, (e) $y_0 = 1.08803$, $w_{0i} \in (-0.4, 2.2)$ for $i = 0 \ldots 26$, (f) $y_0 = -1.0$, $w_{0i} \in (-0.4, 2.3)$ for $i = 0 \ldots 27$, (g) $y_0 = 0.321889$, $w_{0i} \in (-0.4, 2.3)$ for $i = 0 \ldots 27$, (h) $w_{0k} \in (-0.2, 2.1)$ for $k = 0 \ldots 23$, $y_{0i} \in (-1.3, 1.3)$ for $i = 0 \ldots 26$. 206
8.3 Stable solution to the (3.16) ($x_0 = 0$): (a) $x(t)$, $y_1(t)$, and (b) $x(y_1)$ for $\bar{y}_0 = [0.321889, 0, 0]$, $\delta = 0.1$. 212

8.4 Unstable solution to the (3.16) ($x_0 = 0$): (a) $x(t)$, $\bar{y}(t)$, and (b) $x(y_1)$ for $\bar{y}_0 = [1.08803, 0.01, 0.01]$, $\delta = 0.1$; $x(t)$ for $\bar{y}_0 = [0.321889, 0.1, 0.1]$ if (c) $\delta = 0.2$ and (d) $\delta = 0.07$; $y_2(t)$ for $\bar{y}_0 = [0.321889, 0.1, 0.1]$ if (e) $\delta = 0.5$ and (f) $\delta = 0.2$. 213

9.1 Scheme of a swinging oscillator. 228

List of Tables

3.1 The estimation of errors of values of A_1 and A_2 104

5.1 Exact solutions and relative errors. 148
5.2 Numerical values t^* obtained with the use of PA. 154
5.3 The values t^* and t_0 obtained with the use of PA. 158
5.4 Further iterations. 159
5.5 The results obtained with the use of Padé approximations. . . 160
5.6 The numerical estimation $1/4T_i$, $i = 1, 2, 3$ where: T_1, T_2, T_3 are the periods associated with adequate approximations (5.79), (5.80) and (5.81). 163

9.1 Properties of algorithms 223

Introduction

Learning about various phenomena and processes occurring in the universe around us, building durable and resistant elements of machines in different branches of industry, including engineering, car, aviation, and shipbuilding industry and astronautics, results in the rapid development of technology. On the other hand, it also leads to brand new questions and problems that science has to face.

Among the theoretical (numerical and analytical) methods of solving many problems of applied mathematics, physics and technology, *asymptotical methods* are those that deserve special attention [6, 29, 118]. It is worth noticing that their application is not limited to introducing one *"small" perturbation parameter*, but it can be much wider.

The application of asymptotical methods very often results in discovering the essential characteristics of analyzed processes. For some ranges of parameter changes and of the changes of reference inputs of the processes under consideration, the obtained results are quantitative. Moreover, the results very often serve as the verification and are used in tests, which leads to obtaining more effective algorithms of numerical evaluation.

The main aim of this book is to introduce the reader to the mathematical methods of perturbation theory and to review the most important methods of singular perturbations within the scope of application of differential equations. Many other aspects of perturbation methods and their development are also described in monographs [23, 29, 65, 121, 166].

The authors of this book attempt to emphasize the dynamics of the development of perturbation methods and to present the development of ideas associated with this field.

Let us consider how the solution of differential equation depends on parameter ε (and on the coordinates and time) and let us assume that, for the small ε, we know the formula describing the "deviation" from the known boundary solution obtained for $\varepsilon = 0$ (a so-called *generating solution*). Since both the analyzed differential equation and boundary conditions depend on ε, and since the deviation mentioned above is determined by the application of analytical transformations, we will assume that it depends analytically also on ε. Due to that, the perturbance will be always searched in the form of asymptotic solutions with regarding the small parameter ε.

The simplest example of *perturbation analysis* is the classical direct application of *the method of small parameter*. However, it turns out that in most of the cases of irregular or singular problems, the results obtained from the classical approach are not satisfactory and may lead to false conclusions. It is worth noticing that in the solutions associated with linear theory, which is the direct consequence of applying the classical method of small parameter, from the very beginning the basic assumption of this theory is violated, namely the one that says that the demanded function and its derivatives should be small (limited) in the whole range of its definiteness.

The desire for solving this essential problem through decades has led many scientists to study various asymptotic methods of singular perturbations, which allow to take into account nonlinearity of the analyzed problems. According to the monograph [166], the characteristic feature of the problem of singular excitations is the fact that none of the asymptotic series is uniformly adapted for application in the whole researched range of solutions.

Moreover, the problem of singular perturbations occurs also when [166] not only the first approximation, but also approximations of higher order have the mentioned characteristic. In the latter case, it is obvious that it leads both to difficulties in evaluation and to application of the expected characteristics of a solution.

In the first chapter of the book, we show the possibility of adequate, and at the same time formally simple, structure of mathematical model of the analyzed processes. We also emphasize the fundamental elements of the process of mathematical modeling. Two simple examples exhibit the process of transition from the dimension quantities to the dimensionless ones through the perturbation parameter ε, characterizing the analyzed problem.

The second chapter deals with mathematical background of asymptotic approaches. First, the distributions of the majority of elementary functions into power series are presented together with the possibility of their application in various approximate evaluations. Secondly, perturbations along a parameter and coordinates as well as the classical method of a small parameter are described. Then, a comparison of infinitely small and infinitely large functions is carried out, and symbols of estimation of magnitude order are given and illustrated. In addition, asymptotic sequences and decompositions are defined and illustrative examples are included. Nonuniform asymptotic decompositions are introduced, and basic operators acting on them are explained. Advantages of application of asymptotic sequences and decompositions are also addressed. Seven exercises to be solved by a reader finish this chapter.

Regular and singular perturbations are discussed in Chapter 3. First, nonuniformities of a classical perturbation method using an example of a Duffing oscillator are discussed. Various approaches suitable to study either regular or singular problems are described: the method of "elongated" parameters, the method of deformed variables, the method of scaling and full approximation, the multiple scale method, the derivative decomposition along one and two variables, the method of variation of arbitrary constants, averaging (Van der Pol and Krylov-Bogolubov-Mitropolskiy methods). Special attention is paid to matching of asymptotic decompositions, where after the introduction of basic notations and definitions an example with a boundary layer is studied. Then a construction of matched asymptotic expansion either using additive or mulitiplicative version is rigorously stated. A choice of internal variables is briefly discussed, and the sources of the occurrence of nonuniformities of asymptotic decompositions are described. The problems related to asymptotic analysis to strongly nonlinear dynamical systems are addressed. Finally, an application of a few perturbation parameters is demonstrated. Six exercises are also attached.

Asymptotic methods may be useful for investigating even complex mathematical problems occurring during mathematical modeling of some impacting wave processes in fluids and gases, which is shown in Chapter 4. For didactic purposes, further solutions are limited to first approximations, and the comparison of the efficiency of results is carried out with the use of various methods of the theory of singular perturbations.

First, the definition of a cylinder-like piston wave is introduced, and the vicinity of both piston and front of the impact wave is outlined. It is illustrated, how an application of elongated coordinates and renormalization methods yields highly accurate results (note that up to now there is no efficient numerical approach to solve this problem appropriately). In the second part of Chapter 4, one-dimensional nonstationary nonlinear waves are analyzed. After mathematical formulation of the problem, irregularity of the solution is illustrated and discussed. Then it is shown how an application of the methods of renormalization, characteristics and multiple scales allows for overcoming the singularities to achieve the uniformly suitable solutions.

Chapter 5 deals with Padé approximation and their applications in the analysis of various problems of applied mathematics. It should be emphasized that recently an increase of interest in applying Padé approximations to mechanical problems is observed. First, basic relations and characteristics of Padé approximations are introduced, and then a few simple examples are followed by the analysis of a supersonic flow around a thin cone in a circumsonic regime. The example provided emphasizes important features and advantages of applying Padé approximations, especially for practical use. It is worth noticing that modeling of burning of the fuel in a combustion engine, various biological models, buckling of constructions, etc. are all associated with an occurrence of the sudden blow-up phenomenon. This problem is addressed among others, and solved efficiently with the use of Padé approximations. Six exercises illustrating these considerations are added.

Chapter 6 contains examples of application of averaging to analyze ribbed plates. Averaging procedure is illustrated, as well as Kantorovich-Vlasov method and its modifications are discussed, among others. Examples include applications of the described methods to both static and dynamic problems.

In Chapter 7, two coupled oscillators with negative Duffing type stiffness that are self (due to friction) and externally (harmonically) excited are studied. Fundamental solutions of the homoclinic orbit are constructed. Then, Melnikov-Gruendler asymptotic approach is used to define Melnikov's function, including smooth and stick-slip chaotic behavior. Theoretical considerations are supported by numerical examples.

In Chapter 8, differential equations with discontinuous nonlinearities are analyzed. Those nonlinearities are continuously approximated by using one-parametric families of continuous functions. To study the dynamics of the approximated equation, we split it for variables near and far from the discontinuities. We scale the variables near discontinuities to get singular differential equations. Then, we use results from the theory of singularly perturbed differential equations, like Tichonov theorem. Finally, we combine dynamics of singularly perturbed and normal differential equations to get the dynamics of the original approximated differential equations. We use this method for the study of persistence and stability of a periodic solution of the discontinuous systems under the continuous approximation. Summarizing, the method is based on construction of Poincaré maps along periodic solutions of discontinuous systems and their continuous approximations. Then Tichonov theorem is applied to study the relationship between those Poincaré maps. Some transversal assumptions are needed to derive those Poincaré maps. A simple model of one-degree of freedom mechanical system to illustrate the main idea of the used method is presented. Then this method is extended for higher dimensional general discontinuous systems.

In Chapter 9, a method of construction of canonical variables transformation in the parametric form is proposed. The introduced definition of a normal form does not require any partition to either autonomous – nonautonomous, or resonance – nonresonance cases, since a parametrized guiding function is used. In the first part of this chapter the parametric form of canonical transformation and the method of normalization are introduced and illustrated. In the second part, dynamics of a swinging oscillator is analyzed.

Both authors greatly appreciate the help of Mr. W. Dziubinski and Mrs. A. Debska with the final manuscript preparation.

The first author (J.A.) wishes to acknowledge the financial support by the Polish Ministry of Education and Science under the grant No. 4 T07A 031 28.

Chapter 1

Elements of mathematical modeling

In this chapter we show the possibility of adequate, and at the same time formally simple, structure of mathematical model of the analyzed processes. We also emphasize the fundamental elements of the process of mathematical modeling. Two simple examples exhibit the process of transition from the dimension quantities to the dimensionless ones through the perturbation parameter ε, characterizing the analyzed problem.

1.1 Structure of a mathematical model

An ideal scientific theory should include a set of minimal axioms (elementary rules and notions that are taken without proof) on the basis of which any problem can be solved with the use of formal, i.e., mathematical logic. It turns out, however, that the complexity of phenomena and processes surrounding us, together with limited creative capabilities of man effectively prevent strict application of a scientific theory.

Processes taking place in the real world are investigated with the use of appropriate mathematical devices and basing on *mathematical models* built for these processes. Although such models can only be approximate and can never claim to be fully adequate to the processes they describe, it is obvious that for the descriptions of phenomena and processes, the construction of mathematical models is of essential importance. For a mathematical model to be constructed, it is necessary that the most characteristic features of the process under analysis be identified and taken into account. On the other hand, an accepted mathematical model should be relatively simple but at the same time provide all necessary information about the process under investigation. Consequently, certain characteristics are fully taken into account, others are considered to a certain extent only, while others become entirely neglected. To a high degree, this procedure, known as an *idealization process*, is responsible for the final success of the investigation.

In order to enable practical exploitation of mathematical tools, a problem under consideration needs to be simplified, while the introduced assumptions, which should always be not only mathematically but also physically verifiable, ought to be experimentally confirmed. It is not required to carry out experiments related to problems that do not exhibit deeper differences, but qualitatively new problems demand new experiments since introduced simplifications can frequently lead to conclusions that are not realized physically.

Idealization process requires certain ordering of its different elements, which is achieved by comparison of the elements with one another and to particular characteristic quantities chosen earlier. For example, if one of the system elements is 1 cm in length, a natural question arises whether this quantity is big or small. This question can be decided depending on the initial formulation of the problem considered. It is obvious that when a satellite movement on the circum-earth orbit is examined, 1 cm can be treated as a negligibly small quantity. On the other hand, if the intermolecular distance is in the focus of attention, 1 cm becomes a quantity of gigantic value. Let us illustrate the point with one more example. It is generally known that air is compressible. Is it, however, always necessary to take the air compressibility into account? Again, it depends on the formulation of the original problem. If an object under investigation moves in air at small velocity V, the mathematical model construction needs not account for compressibility. However, if the object's speed is very high, close to the acoustic velocity or higher, then compressibility must not be neglected. In this case it is convenient to introduce a dimensionless quantity $M = V/a$ called Mach number, which plays an important role in aerodynamics. When $M \ll 1$, an idealized mathematical model of incompressible gas can be applied, while for greater values of Mach number, air compressibility should be taken into consideration. A similar situation occurs when a mathematical model is constructed in the domains of science and technology where other characteristic dimensionless numbers are significant, usually built as combinations of three-dimensional quantities, i.e., length L, time T, and mass M. For the sake of convenience it is assumed that the dimension of combination FT^2/ML equals 1 (where F stands for force). In other words, one of the quantities F, T, L, and M may be chosen to be independent.

To conclude, the first step in the discussed procedure should include reduction of all variables of a process under examination to dimensionless quantities. This can be achieved through assigning of all initial variables of the process to certain characteristic quantities (corresponding to the variables in terms of the dimension) or their combinations: length L, velocity V, friction coefficient μ, spring stiffness k, dynamic viscosity coefficient ν, and the like. This is why application of a dimensional theory to a problem of any branch of science or technology always leads to the determination of a set of characteristic dimensionless quantities (similarity parameters) whose values characterize qualitatively the essence of examined processes (process similarity rules). The numbers of Mach, Nusselt, Reynolds, Strouhal, Froude, Biot, together with many others, very big or very small, can serve as examples here [62].

The two examples discussed below provide an illustration of how a problem is reduced to its dimensionless form and a small parameter appears.

1.2 Examples of reducing problems to a dimensionless form

Consider the motion of a rigid body of mass m attached to a frame by means of a massless linear spring with stiffness coefficient k in a medium characterized by viscous damping μ^* (Figure 1.1).

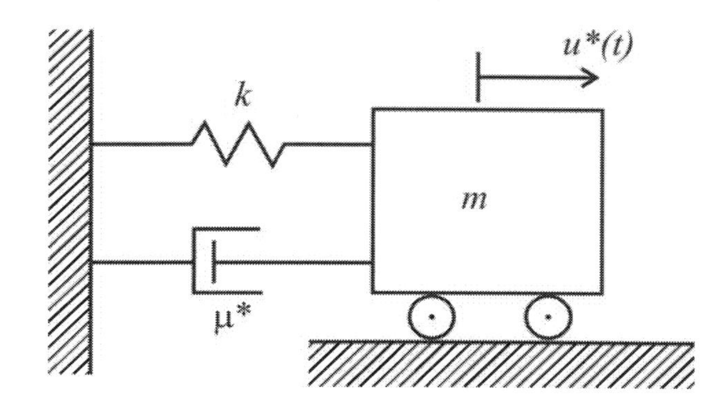

FIGURE 1.1: Free vibrations of one-degree-of-freedom system.

This is a classic system of free vibrations (with damping) often referred to as *the problem of linear oscillator with damping.* From Newton's second law of dynamics, the equation

$$m\left(d^2u^*/dt^{*2}\right) + \mu^*\left(du^*/dt^*\right) + ku^* = 0 \tag{1.1}$$

is obtained, where dependent variables $u^*\ (t^*)$ describe displacement of the examined body from a certain initial position u_0^*, and time t^* is an independent variable. Let us assume that at an initial moment $t^* = 0$ the body is in an initial position a^* and the initial speed of its motion is zero, i.e.,

$$u_0^* = u^*(0) = a^*, \qquad du^*/dt^*(0) = 0\ . \tag{1.2}$$

To reduce problem (1.1), (1.2) to a dimensionless form, characteristic values should be first assumed. Let them be: u_0^* – a characteristic length value, and $\sqrt{k/m} = \omega_0$ – the eigenfrequency of the considered body (for $\mu^* = 0$ in the differential equation). From (1.1), we obtain

$$d^2u^*/dt{*}^2 + 2h\left(du^*/dt\right) + \omega_0^2u^* = 0\ . \tag{1.3}$$

The characteristic equation corresponding to it assumes the form of

$$\lambda^2 + 2h\lambda + \omega^2 = 0\ ,$$

and it yields two roots $\lambda_{1,2} = -h \pm i\sqrt{\lambda}$, where $\lambda = \sqrt{\omega^2 - h^2}$, $2h = \mu^*/m$, $i^2 = -1$.

In the case of $h = 0$, since the real part of the complex conjugated roots equals zero, the general solution of differential equation (1.3) has the form

$$u^*(t, c_1, c_2) = c_1 \cos \omega_0 t^* + c_2 \sin \omega_0 t^* \ ,$$

where c_1 i c_2 are arbitrary constant numbers. From the initial conditions, we determine $c_1 = a^*$, $c_2 = 0$ and

$$u^*(t) = a^* \cos \omega_0 t^* \ ,$$

which means that $\omega_0 = \sqrt{k/m}$ is the eigenfrequency, and a^* is the amplitude of vibrations.

Introducing dimensionless quantities t and $u(t)$ in accordance with the formulas

$$t = \omega_0 t^* \ , \qquad u = u^*/u_0^* \ , \tag{1.4}$$

and then transforming (1.1) and (1.2), and making use of both (1.4) and the composite function differentiation rule, we arrive at

$$\frac{du^*}{dt^*} = \frac{du^*}{dt}\frac{dt}{dt^*} = \frac{d(u_0^* u)}{dt}\frac{dt}{d(t/\omega_0)} = \omega_0 u_0^* \frac{du}{dt} \ . \tag{1.5}$$

Analogically, we obtain

$$\frac{d^2u^*}{dt^{*2}} = \omega_0^2 u_0^* \frac{d^2u}{dt^2} \ . \tag{1.6}$$

Substituting (1.4)–(1.6) to (1.1), (1.2), we have:

$$m\omega_0^2 u_0^* \frac{d^2u}{dt^2} + \mu^* \omega_0 u_0^* \frac{du}{dt} + k u_0^* u = 0 \ ,$$

$$u^*(0) = u_0^* u(0) = a^* \ , \qquad \omega_0 u_0^*(0) = 0 \ ,$$

which finally gives

$$m\omega_0^2 \frac{d^2u}{dt^2} + \mu^* \omega_0 \frac{du}{dt} + ku = 0$$

or

$$\frac{d^2u}{dt^2} + \mu^* \frac{du/dt}{m\omega_0} + k\frac{u}{m\omega_0^2} = 0 \ ,$$

$$u(0) = \frac{a^*}{u_0^*} \ , \qquad \frac{du}{dt(0)} = 0 \ .$$

Introducing a dimensionless viscosity coefficient

$$\mu = h\omega_0^{-1} \tag{1.7}$$

and selecting for example $u_0^* = a^*$, we arrive at the following dimensionless form of a mathematical problem connected with *a damped linear oscillator analysis*:

$$\frac{d^2u}{dt^2} + 2\mu\frac{du}{dt} + u = 0 \ , \tag{1.8}$$

$$u(0) = 1 \ , \quad \frac{du}{dt(0)} = 0 \ . \tag{1.9}$$

Formulas (1.8), (1.9) define the initial problem (*Cauchy problem*) for the linear differential equation (1.8) of the second order with constant coefficients. Its solution depends on one dimensionless parameter μ, being the ratio of the forces of resistance and inertia (or, in other words, of damping and elasticity) and it has the form

$$u = e^{-\mu t} \cos\sqrt{\omega_0^2 - \mu^2 t} \ . \tag{1.10}$$

If μ is small, it is said that the oscillator is weakly damped. If $\mu = 0$, then harmonic eigenfrequency $\omega_0 = \sqrt{k/m}$ occurs. It is worth noticing that exploiting the solution to problem (1.8), (1.9), and using (1.4), (1.7) for calculations, it is possible to obtain a solution of the initial problem (1.1), (1.2) for different sets of values m, k, μ^*, a^*. This is where the very essence of the transition to a dimensionless model before the solving procedure starts should be searched. The solution of problem (1.8), (1.9) is discussed in Chapter 3.

Analogous procedure can be applied in transition to dimensionless quantities while analyzing other more complex processes.

Consider a widely analyzed *model for a nonlinear problem of one-dimensional vibrations* of a mass suspended by a nonlinear spring. Let, at a time instant $t = 0$, the mass be displaced from its equilibrium position by value a^*. If it is released with no initial speed, it will vibrate around its equilibrium position. The displacement of the mass from its initial position (vibration amplitude) at an arbitrary time instant t^* will be denoted by $u^*(t^*)$.

If the resistance of the medium in which the vibrations take place is neglected, then the vibration process is described by the differential equation

$$\frac{d^2u^*}{dt^{*2}} + f(u^*) = 0 \ , \tag{1.11}$$

where d^2u^*/dt^{*2} is acceleration of the system, $f(u^*)$ is a nonlinear force defined by properties of the spring. Let $u^* = 0$ be an equilibrium position of the system, then $f(0) = 0$. Assume that $f(u^*)$ is analytical at point $u^* = 0$. Then, in the neighborhood of this point, and remembering that $f(0) = 0$, function $f(u^*)$ can be decomposed into *the Maclaurin series*

$$f(u^*) = u^*\frac{df}{du^*(0)} + \frac{u^{*2}}{2!}\frac{d^2f}{du^{*2}(0)} + \frac{u^{*3}}{3!}\frac{d^3f}{du^{*3}(0)} + \ldots , \tag{1.12}$$

where $df/du^*(0) > 0$.

Substituting (1.12) to (1.11), we obtain a differential equation with respect to function u^*. Let us consider its most widely used form

$$\frac{d^2u^*}{dt^{*2}} + u^* \frac{df}{du^*(0)} + \frac{u^{*3}}{3!} \frac{d^3 f}{du^{*3}(0)} = 0 , \tag{1.13}$$

where $df/du^*(0) > 0$.

Equation (1.13) is called the *Duffing equation* [51].

Let us reduce (1.13) to a dimensionless form. For this purpose, scales of length l and time T are introduced and the following dimensionless variables are taken:

$$u = \frac{u^*}{l} , \qquad t = \frac{t^*}{T} . \tag{1.14}$$

In accordance with the differentiation rule we obtain

$$\frac{d}{dt^*} = \frac{d}{dt}\frac{dt}{dt^*} = \frac{1}{T}\frac{d}{dt} , \qquad \frac{d^2}{dt^{*2}} = \frac{1}{T^2}\frac{d^2}{dt^2} . \tag{1.15}$$

With (1.13) and (1.14) taken into account (1.15) takes the form

$$\frac{l}{T^2}\frac{d^2u}{dt^2} + lu\frac{df}{du^*(0)} + \frac{l^3u^3}{3!}\frac{d^3 f}{du^{*3}(0)} = 0$$

or

$$\frac{d^2u}{dt^2} + uT^2\frac{df}{du^*(0)} + \frac{u^3l^2T^2}{3!}\frac{d^3 f}{du^{*3}(0)} = 0 .$$

Let the characteristic timescale T be introduced so that

$$T^2\frac{df}{du^*(0)} = 1 ,$$

and assume that

$$\frac{l^2T^2}{6}\frac{d^3 f}{du^{*3}(0)} = \frac{l^2}{6}\frac{d^2 f}{du^{*2}(0)} = \varepsilon . \tag{1.16}$$

Having performed these operations on Duffing equation (1.13), for the dimensionless amplitude $u\ (t;\varepsilon)$ we arrive at the following form

$$\frac{d^2u}{dt^2} + u + \varepsilon u^3 = 0 , \quad 0 \le t < \infty . \tag{1.17}$$

Introducing a dimensionless displacement of the mass $a = a^*/l$ at the initial time instant $t = 0$, as the initial conditions we have

$$u(0) = a , \qquad \frac{du}{dt(0)} = 0 . \tag{1.18}$$

As it can be observed, contrary to problem (1.8), (1.9) which is linear, the initial problem (1.17), (1.18)) is nonlinear. While finding a solution to (1.8),

(1.9) as well as analytical analysis of this solution does not cause significant troubles, the construction and the analytical analysis of a solution to Duffing problem (1.17), (1.18) cannot be claimed to be a simple task [119, 121].

In the two examples discussed above the behavior of the solutions depended on one dimensionless parameter μ or ε. Numerous mathematical models of various vibration processes are described in [18]. In more complex cases, the number of parameters can be greater.

1.3 Mathematical model adequacy and properties. Regular and singular perturbations

Assume that a mathematical model of a process is given, i.e., a certain boundary problem (initial, with boundary conditions, or mixed-hybrid) for an equation (normal, partial, differential-integral or combined of any of these types) has been formulated. The question concerning correctness of the problem formulation can be answered in different ways.

It is said that a boundary problem is formulated correctly if a solution to it exists, is unique, and its dependence on the boundary conditions is continuous. The sense of the first two properties is obvious. The third property means that small changes in the boundary conditions should result in small changes in the solutions. For instance, the boundary (initial) problems (1.8), (1.9) or (1.17), (1.18) formulated earlier are well-posed because there is a solution to them and the solution is only one, which is shown in Chapter 3. It is obvious that small changes in initial conditions (1.9) and (1.18) lead to small changes in their solutions.

Another essential question is how the quality of an accepted mathematical model can be assessed?

As it has already been mentioned, an idealized mathematical model is always of approximate nature. That is why, when a solution to this model has been constructed and is subject to analysis, the question arises about the adequacy (correspondence) of an idealized mathematical model to the real model under examination, i.e., the question concerning the degree of consistency of the object's behavior described on the basis of the model. As the model adequacy criteria, one can apply comparison of the results obtained basing on the model with the data resulting from experiments or theoretical estimations and/or the data obtained using other mathematical models.

If there is no consistency or some results need modification, then it is necessary to correct the model and consider the possibility of taking into account certain results that have been neglected so far.

To this purpose a more complex (extended) model of the process is generally constructed so that small factors neglected in a simplified model can be taken into account. Then, closeness of the solutions received basing on the simplified model and the extended one is analyzed.

This procedure is continued until the desired results are obtained.

It should be pointed out that to construct a "good" mathematical model of a complex process is a difficult task that requires extensive professional knowledge and good intuition from the researcher. The mathematical modeling steps described above constitute only fundamental elements of the procedure. Mathematical modeling can frequently involve the theory of dimensions and the similarity theory, which, however, is a separate research area dealt with in selected positions in the literature (see for instance [151]).

Finally, let us give some thought to so-called perturbations introduced in the process of an extended mathematical model construction.

When small results, which have been neglected so far, are taken into account, in the basic model there appear additional terms with small multipliers accentuating the smallness of the additional factors. It is these small multipliers that are referred to as *small parameters*. Equation terms involving small parameters are called *perturbations*. The initial equation with no perturbations is said to be *unperturbated*, whereas an "extended" equation, corresponding to an extended model, is called *a perturbated equation* or *an equation with perturbations*.

For example, if in Duffing equation introduced earlier (1.17), parameter ε, that characterizes the nonlinearity degree of a vibrating oscillator is taken to be small, then the equation $d^2u/dt^2 + u = 0$ is unperturbated, the term εu^3 is the perturbation, and equation (1.17) is a perturbated equation.

Perturbations that occur in different problems are conventionally divided into two classes: regular and singular.

A small perturbation that causes a small change in the solution of an unperturbated equation is said to be regular. *A singular perturbation* on the other hand, although also small, leads to essential qualitative changes in the solution of an unperturbated equation. For instance, in Duffing problem described above (1.17), (1.18) a small singular forcing εu^3 is responsible for a change in the frequency and amplitude of vibrations of the oscillator. A more detailed description of singular perturbations is given in Section 3.1.

Since in practice most problems involve *singular forcing*, further chapters 3–5 concentrate on mathematical tools used in the theory of *singular perturbations* and on applications of the theory, with a special emphasis put on the impact of singular perturbations on parameters of various processes.

Chapter 2

Expansion of functions and mathematical methods

In this chapter we deal with mathematical background of asymptotic approaches. First, expansions of the majority of elementary functions into power series are presented together with the possibility of their application in various approximate evaluations. Secondly, perturbations along a parameter and coordinates as well as the classical method of a small parameter are described. Then, a comparison of infinitely small and infinitely large functions is carried out, and symbols of estimation of magnitude order are given and illustrated. In addition, asymptotic sequences and decompositions are defined and illustrative examples are included. Nonuniform asymptotic decompositions are introduced, and basic operators acting on them are explained. Advantages of application of asymptotic sequences and decompositions are also addressed.

2.1 Expansions of elementary functions into power series

While constructing approximate solutions to algebraic, differential and integral equations and their systems, as well as during the estimation of various integrals, *asymptotic series* are widely applied with respect either to a parameter or to an independent variable. The mentioned expansions into power series are usually constructed using both the Taylor and Maclaurin series or using other special tools.

2.1.1 Newton's binomial

The following formulas for abbreviated products are often used

$$(a+b)^2 = a^2 + 2ab + b^2 \ , \quad (a+b)^3 = a^3 + 3a^2b + 3ab^2 + b^3 \ .$$

Recall that the generalization of these formulas into an arbitrary case of integer n is Newton's binomial of the form

$$(a+b)^n = \sum_{m=0}^{n} C_n^m a^{n-m} b^m = \sum_{m=0}^{n} \frac{n!}{m!\,(n-m)!} a^{n-m} b^m =$$

$$=a^n + na^{n-1}b + \frac{n\,(n-1)}{2!}a^{n-2}b^2 + \ldots + \frac{n\,(n-1)}{2!}a^2b^{n-2} + nab^{n-1} + b^n\ , \quad (2.1)$$

where $C_n^m = \binom{n}{m}$ is the amount of m-elementary combinations without replications of n elements; $n! = 1\cdot 2\cdot \ldots \cdot (n-1)\cdot n$; $0! = 1$. Observe that Newton's binomial (2.1) consists of $(n+1)$ components. In the case of arbitrary positive or negative n, in order to approximate $(a+b)^n$, the following series consisting of the infinite number of terms is used:

$$(a+b)^n = a^n + na^{n-1}b + \frac{n\,(n-1)}{2!}a^{n-2}b^2 + \frac{n\,(n-1)\,(n-2)}{3!}a^{n-3}b^3 + \cdots =$$

$$= \sum_{m=0}^{\infty} C_n^m a^{n-m} b^m\ . \qquad (2.2)$$

The question arises: Is the series (2.2) always convergent? Observe that

$$\lim_{m\to\infty} \frac{m\text{-th term}}{(m-1)\text{-th term}} = \lim_{m\to\infty} \frac{c_n^m a^{n-m} b^m}{c_n^{m-1} a^{n-m+1} b^{m-1}} =$$

$$= \frac{b}{a} \lim_{m\to\infty} \frac{n!\,(m-1)!\,(n-m+1)!}{m!\,(n-m)!\,n!} = \frac{b}{a} \lim_{m\to\infty} \frac{(n-m+1)}{m} = -\frac{b}{a}\ .$$

According to d'Alembert's principle, the series (2.2) is convergent only when the condition $|b/a| < 1$ is satisfied. In applications, very often the following particular form of series (2.2) is used

$$(1+x)^n = 1 + nx + \frac{n\,(n-1)}{2!}b^2 + \cdots + \frac{n\,(n-1)\cdot \ldots \cdot (n-m+1)}{m!}x^m + \cdots\ . \qquad (2.3)$$

In this case $m = 1, 2, \ldots$, and the condition of convergence for the series (2.3) has the form $|x| < 1$.

Another special case of series (2.3) is represented by infinite and decreasing geometrical series of the form

$$\frac{1}{1-x} = (1-x)^{-1} = 1 + x + x^2 + \cdots + x^n + \cdots = \sum_{n=0}^{\infty} x^n, \quad |x| < 1\ . \quad (2.4)$$

2.1.2 Taylor and Maclaurin series

If a function $f(x)$ is infinitely times differentiable in a point $x = x_0$, then in the vicinity of this point it can be approximated through *Taylor's series* of the form

$$f\,(x) = f\,(x_0) + f'\,(x_0)\,(x - x_0) + \frac{f''\,(x_0)}{2!}\,(x - x_0)^2 + \cdots =$$

$$= \sum_{n=0}^{\infty} \frac{f^{(n)}\,(x_0)}{n!}\,(x - x_0)^n\ . \qquad (2.5)$$

Taking in (2.5) $x_0 = 0$, the following *Maclaurin's series* is obtained

$$f(x) = f(0) + xf'(0) + \frac{x^2 f''(0)}{2!} + \cdots = \sum_{n=0}^{\infty} \frac{f^{(n)}(0)}{n!} x^n . \tag{2.6}$$

In addition, if in the series (2.5) and (2.6) only a bounded number of terms is taken, then Taylor's and Maclaurin's formulas are yielded. In order to estimate an error introduced by neglecting the omitted terms, Lagrange's formula is often applied [126, 127].

According to (2.6), one may obtain an expansion of elementary functions into series with respect to x. Furthermore, one may estimate the values of x, for which the series are convergent, which is illustrated by the examples below.

Example 2.1. Let $f(x) = e^x$. Then $f'(x) = f''(x) = \cdots = f^{(n)}(x) = = e^x$; $f(0) = f'(0) = f''(0) = \cdots = f^{(n)}(0) = e^0 = 1$ and the series (2.6) yields

$$e^x = 1 + x + \frac{x^2}{2!} + \frac{x^3}{3!} + \cdots = \sum_{n=0}^{\infty} \frac{x^n}{n!}, \quad |x| < \infty . \tag{2.7}$$

Applying formula (2.7) and employing the properties of a logarithmic function, one gets

$$a^x = e^{x \ln a} = 1 + x \ln a + \frac{(x \ln a)^2}{2!} + \cdots = \sum_{n=0}^{\infty} \frac{(x \ln a)^n}{n!}, \quad |x| < \infty . \tag{2.8}$$

Example 2.2. Let $f(x) = \sin x$. Then $f'(x) = \cos x$, $f''(x) = -\sin x$, $f'''(x) = -\cos x$, and so on, $f(0) = 0$, $f'(0) = 1$, $f''(0) = 0$, $f'''(0) = -1$, and so on, the formula (2.6) yields

$$\sin x = x - \frac{x^3}{3!} + \frac{x^5}{5!} - \frac{x^7}{7!} + \cdots = \sum_{n=0}^{\infty} \frac{(-1)^n x^{2n+1}}{(2n+1)!}, \quad |x| < \infty . \tag{2.9}$$

Example 2.3. Let $f(x) = \cos x$. This example is analogical to the previous one, and one gets

$$\cos x = 1 - \frac{x^2}{2!} + \frac{x^4}{4!} - \frac{x^6}{6!} + \cdots = \sum_{n=0}^{\infty} \frac{(-1)^n x^{2n}}{(2n)!} = \frac{d}{dx}(\sin x) \quad |x| < \infty . \tag{2.10}$$

Example 2.4. Let $f(x) = \ln(1+x)$. Then $f'(x) = 1/(1+x)$, $f''(x) = -1/(1+x)^2$, $f'''(x) = 2/(1+x)^3$, and so on, $f(0) = 0$, $f'(0) = 1$, $f''(0) = -1$,

$f'''(0) = 2$, and so on, and formula (2.6) yields

$$\ln(1+x) = x - \frac{x^2}{2} + \frac{x^3}{3} - \frac{x^4}{4} + \cdots = \sum_{n=1}^{\infty} (-1)^{n+1} x^n/n \,, \quad |x| \le 1 \,, \quad x \ne -1 \,. \tag{2.11}$$

Introducing the exchange of x na $-x$ in the formula (2.11), one gets

$$\ln(1-x) = -x - \frac{x^2}{2} - \frac{x^3}{3} - \frac{x^4}{4} - \cdots = -\sum_{n=1}^{\infty} x^n/n \,, \quad |x| < 1 \,, \tag{2.12}$$

whereas formulas (2.11) and (2.12) yield

$$\ln\frac{1+x}{1-x} = \ln(1+x) - \ln(1-x) = 2\left(x + \frac{x^3}{3} + \frac{x^5}{5} + \cdots\right) = 2\sum_{n=0}^{\infty} \frac{x^{2n+1}}{2n+1} \,,$$

$$|x| < 1 \,. \tag{2.13}$$

Example 2.5. Let $f(x) = \tan(x)$. Since $\tan x = \sin x/\cos x$, dividing (2.9) by (2.10) and using formula (2.47) or formula (2.6), one gets

$$\tan x = x + \frac{x^3}{3} + \frac{2x^5}{15} + \frac{17x^7}{315} + \cdots \,, \quad |x| < \pi/2 \,. \tag{2.14}$$

Repeating the earlier steps, one finds

$$\mathrm{ctan}x = \frac{1}{x} - \frac{x}{3} - \frac{x^3}{45} - \frac{2x^5}{945} + \ldots \,, \quad |x| < \pi \,. \tag{2.15}$$

Observe that (2.14) and (2.15) can be obtained by applying Maclaurin's formula (2.6). According to (2.6), one gets

$$\arcsin(x) = x + \frac{1}{2}\frac{x^3}{3} + \frac{1}{2}\frac{3}{4}\frac{x^5}{5} + \frac{1}{2}\frac{3}{4}\frac{5}{6}\frac{x^7}{7} + \ldots = x + \frac{x^3}{6} + \frac{3x^5}{40} + \frac{5x^7}{112} + \ldots \,, \; |x| < 1 \,. \tag{2.16}$$

Observe that sometimes it is worthy to apply other methods. For example, one may find the derivatives of functions first, then distribute the obtained result into series (2.3), and finally carry out the integration. For example, let

$$d(\arctan(x)) = \frac{1}{1+x^2}dx = (1+x^2)^{-1}dx = (1 - x^{-2} + x^4 - \ldots)dx$$

and after integration one gets

$$\arctan(x) = x - \frac{x^3}{3} + \frac{x^5}{5} - \frac{x^7}{7} + \ldots = \sum_{n=1}^{\infty} (-1)^{n-1} x^{2n-1}/(2n-1) \,, \quad |x| < 1 \,. \tag{2.17}$$

In what follows, we give formulas of distribution of hyperbolic functions into power series [178], i.e.:

$$\mathrm{sh}\,(x) = x + \frac{x^3}{3!} + \frac{x^5}{5!} + \ldots = \sum_{n=0}^{\infty} \frac{x^{2n+1}}{(2n+1)!}\,, \quad \mathrm{ch}\,(x) = (\mathrm{sh}\,(x))'\,, \quad |x| < \infty\,, \tag{2.18}$$

$$\mathrm{tgh}(x) = x - \frac{x^3}{3} + \frac{2x^5}{15} - \frac{17x^7}{315} + \ldots\,, \quad |x| < \frac{\pi}{2}\,, \tag{2.19}$$

$$\mathrm{cth}\,(x) = \frac{1}{x} + \frac{x}{3} - \frac{x^3}{45} + \frac{2x^5}{945} - \ldots\,, \quad |x| < \pi\,, \tag{2.20}$$

$$\mathrm{arsh}\,(x) = x - \frac{x^3}{6} + \frac{3x^5}{40} + \frac{5x^7}{112} + \ldots\,, \quad |x| < 1\,, \tag{2.21}$$

$$\mathrm{arth}\,(x) = x + \frac{x^3}{3} + \frac{x^5}{5} + \frac{x^7}{7} + \ldots = \sum_{n=1}^{\infty} \frac{x^{2n-1}}{2n-1}\,, \quad |x| < 1\,. \tag{2.22}$$

It is worth noticing that some of given formulas (2.3)–(2.22) are used further in our book.

2.1.3 Estimation of approximating functions values

In what follows, we exhibit the advantages of the earlier introduced functions distribution through two examples.

Example 2.6. Find a value of $\sin(10^\circ)$ with the accuracy of 0.0001 without tables.

Recall that $10^\circ = \pi/18$, and employing formula (2.9), one gets

$$\sin(10^\circ) = \sin\frac{\pi}{18} = \frac{\pi}{18} - \frac{1}{6}\left(\frac{\pi}{18}\right)^3 + \frac{1}{120}\left(\frac{\pi}{18}\right)^5 - \ldots\,.$$

The obtained sign-changeable series is convergent, according to Leibniz criterion. Now, we apply a well-known theorem saying that if a series is convergent with respect to Leibniz criterion, then an error occurring during estimation of its absolute sum value does not exceed the value of the first term of all neglected terms.

Therefore, in order to find a sum of this series with the accuracy of 0.0001 it is sufficient to account for only its two first terms to get:

$$\frac{1}{6}\left(\frac{\pi}{18}\right)^3 < 0.001\,,$$

$$\frac{1}{120}\left(\frac{\pi}{18}\right)^3 < 0.0001\,.$$

All computations should be carried out only with the accuracy of $\pi \approx 3.14159$, $\pi^3 \approx 31.00624$. As a result, one gets $\sin(10^\circ) \approx 0.173$, which overlaps satisfactorily with the value included in mathematical tables.

Example 2.7. Find an approximate value of arcsin(1/3) using first four terms (different from zero) of a series approximating a given function.

Applying (2.17), one gets

$$\arcsin\frac{1}{3} = \frac{1}{3} + \frac{1}{6}\left(\frac{1}{3}\right)^3 + \frac{3}{40}\left(\frac{1}{3}\right)^5 + \frac{5}{112}\left(\frac{1}{3}\right)^7 + \ldots \approx 19.4712\,.$$

2.1.4 Estimation of approximating values of definite integrals

In the case when a primary function cannot be approximated in the form of combinations of elementary functions, one may apply formulas given in the item 2.1.2. An integrand can be distributed into a power series, which next can be easily integrated.

Example 2.8. Find a value of Laplace function

$$\Phi(x) = \frac{1}{\sqrt{2\pi}}\int\limits_0^x e^{-\frac{z^2}{2}}dz$$

for different values of x.

In accordance with the theory of probability [109], Laplace function defines the probability of a normalized random quantity x in the range $(0, x)$. Using both formulas (2.7), we carry out the distribution of the integrand

$$\Phi(x) = \frac{1}{\sqrt{2\pi}}\int\limits_0^x \left(1 - \frac{z^2}{2} + \frac{z^4}{8} - \frac{z^6}{48} + \ldots\right)dz =$$

$$= \frac{1}{\sqrt{2\pi}}\left(z - \frac{z^3}{6} + \frac{z^5}{40} - \frac{z^7}{336} + \ldots\right)\Bigg|_0^x = \frac{1}{\sqrt{2\pi}}\left(x - \frac{x^3}{6} + \frac{x^5}{40} - \frac{x^7}{336} + \ldots\right).$$

We obtain, for example, $\Phi(0.5) \approx 0.1915$; $\Phi(2) \approx 0.4772$, which is in agreement with the values given in mathematical tables.

2.1.5 Approximating solution of Cauchy problem for ordinary differential equations

Consider an ordinary first order differential equation with a given initial condition (*Cauchy problem*). In the case when an integration of this equation is not straightforward, one may approximate its solution in a power series with unknown coefficients.

The first coefficient is defined through the initial conditions, whereas next ones are defined by the initial differential equation and its derivatives (*the method of successive differentiation*).

Example 2.9. Find four first and different from zero terms of Maclaurin series of the following Cauchy problem

$$y' = e^x + y \ , \qquad y(0) = 5 \ . \tag{2.23}$$

Let us distribute the solution $y(x)$ into Maclaurin series

$$y(x) = y(0) + xy'(0) + \frac{x^2}{2!}y''(0) + \frac{x^3}{3!}y'''(0) + \dots \ .$$

One has to find $y(0)$, $y'(0)$, $y''(0)$, and so on.

From the initial condition (2.23) we have $y(0) = 5$, and the differential equation yields

$$y'(0) = e^0 + y(0) = 1 + 5 = 6 \ .$$

In order to find $y''(0)$, $y'''(0)$, and so on, one has to successively differentiate the equation (2.23) and use the initial condition. As a result, one gets

$$y''(x) = e^x + y'(x) \ , \quad y''(0) = e^0 + y'(0) = 1 + 6 = 7 \ ,$$

$$y'''(x) = e^x + y''(x) \ , \quad y'''(0) = e^0 + y''(0) = 1 + 7 = 8 \quad \text{and so on.}$$

The presented process can be continued. By substituting the found values into Maclaurin series, the solution to Cauchy problem (2.23) is finally obtained. Recall that equation (2.23) is the linear differential equation of first order of the form $y' + P(x)y = Q(x)$, and its general solution reads

$$y(x, c) = \frac{1}{\mu(x)} \left(c + \int Q(x)\mu(x)dx \right) \ , \qquad c = const \ ,$$

where the integral multiplier is $\mu(x) = e^{\int P(x)dx}$. In our case $P(x) = -1$, $Q(x) = e^x$, which gives

$$y(x, c) = e^x \left(c + \int e^x e^{-x} dx \right) = e^x (c + x) \ .$$

Satisfying the initial condition (2.23) one gets $c = 5$ and the exact solution to the Cauchy problem reads

$$y(x) = (5 + x)e^x \ .$$

By distributing the above solution into a series in accordance with (2.7), one obtains

$$y(x) = (5 + x) \left(1 + x + \frac{x^2}{2!} + \frac{x^3}{3!} + \dots \right) = 5 + 6x + \frac{7}{2}x^2 + \frac{4}{3}x^3 + \dots \ ,$$

which coincides with the earlier obtained approximated solution.

Eventually, the following three essential remarks follow.

1. The methods of searching for the solution to Cauchy problem using both power series and successive differentiation of a differential equation can be successfully applied to solve differential equations of a higher order and systems of differential equations.

2. The presented method is particularly efficient when finding an analytical solution of a differential equation is either impossible or very troublesome.

3. A question of conditions for which the solution to the Cauchy problem in a form of power series exists is not addressed here, neither is a question related to the accuracy of the obtained solutions discussed. (In the case of Cauchy problem 2.23 a comparison of both exact and approximate solutions indicated their conciseness).

2.2 Mathematical methods of perturbations

Currently, various *methods of the theory of perturbations* exist and are applied in different branches of applied mathematics, physics, engineering, and even economy and social sciences. In what follows, some of them will be illustrated and discussed (see also [11]).

2.2.1 Perturbation along a parameter and coordinates. Classical method of a small parameter

Note that the mathematical formulation of many problems of applied mathematics, physics, etc., can be presented in the following general manner: Find a function $u(x;\varepsilon)$, satisfying both a certain differential equation $N(u,x;\varepsilon)=0$ and the attached initial and/or boundary conditions (or initial-boundary ones); x denotes a scalar or vector independent variable; N – is a known differential operator; $\varepsilon>0$ is a small (perturbation) parameter. It is well known that generally the stated problem cannot be solved analytically owing to potential nonlinearities and complexity of equations and/or complexity of initial or boundary value problems. However, if there is a value of the parameter $\varepsilon=\varepsilon_0$, for which the stated problem can be solved in an exact manner (in fact, after rescaling the problem is reduced to that of $\varepsilon_0=0$), then for a small ε, one may look for a solution in the following form

$$u(x;\varepsilon)=u_0(x)+\varepsilon u_1(x)+\varepsilon^2 u_2(x)+\dots\ . \tag{2.24}$$

In the above, u_i $(i=0,1,\dots)$ do not depend on ε; $u_0(x)$ is the solution to the problem for $\varepsilon=0$ (the so-called unperturbed solution).

A series of (2.24) type will be referred as *a perturbation along a parameter*.

On the other hand, series can be also constructed along coordinates (physical and time ones) for their large or small values. In the last case they are referred to as *perturbations along coordinates* of a certain known solution $u_0(x_0;\varepsilon) = \lim_{x\to x_0} u(x;\varepsilon)$, where ε has a certain fixed value.

A value x_0, through a special choice, can be taken either as 0 or ∞. Furthermore, one may find a deviation $u(x;\varepsilon)$ from $u_0(x_0;\varepsilon)$ for x, in the vicinity of $x_0 = 0$ (*a direct decomposition along a coordinate*) or for decreased values of x for $x_0 = \infty$ (*an inverse decomposition of a coordinate*).

Note that a key feature of decomposition along coordinates is characterized through a computation of the solution with respect to only one parameter value, regardless of the knowledge of other values.

In what follows in chapters 3–5 many examples addressing the stated questions will be studied.

Since in a series with respect to the parameter (2.24) it is assumed that the functions are bounded, each of the further introduced corrections in the series (2.24) is essentially smaller from the previous ones due to $\varepsilon \ll 1$. In other words, the introduced correction improves solution $u(x;\varepsilon)$ obtained in a previous step. This is the basic idea of the method of small parameters.

Formally, *a classical method of a small parameter* consists of the following steps: Series (2.24) is first introduced both to the equation $N(u,x;\varepsilon) = 0$ and to the boundary (initial) conditions. The equation is then developed into series for small ε, and terms standing by the same powers of ε are grouped in each of these series. Since equations should be satisfied for all values of ε, and since a power series with respect to ε is linearly independent, all of the coefficients standing by powers of ε should be equal to zero.

When equating the coefficients standing by ε^0 (the so-called process of linearization of original nonlinear problem) to zero, the boundary value problems occur for the first approximation $u_1(x)$, second approximation $u_2(x)$, etc. As a result, linear but nonhomogeneous problems with respect to u_i $(i = 1, 2, \ldots)$ occur, and they can be successively differentiated. The obtained solution consists of the information obtained in the previous steps of computations. It should be noticed, however, that perturbation equations with respect to a small parameter rarely possess so simple form as the series (2.24).

In order to analyze more adequately certain particular problems related to the application of a classical method of small parameter, some additional notions will be introduced further.

2.2.2 Comparison of infinitely small and infinitely large functions. Scaling functions. Magnitude order quantities

Consider a function $f(\varepsilon)$ with real ε and let us analyze its limit for $\varepsilon \to \varepsilon_0$, where ε_0 is a certain characteristic value. The following three different cases will be further analyzed: $\varepsilon_0 = 0$, $\varepsilon_0 = \infty$, $\varepsilon_0 = c$, $0 < |c| < \infty$. Assuming that $\varepsilon_0 = 0$, two other considered cases can be analyzed in a similar way.

First, consider a case when $\lim_{\varepsilon \to 0} f(\varepsilon)$. Note that the limit may depend on the approach, i.e., if it is achieved from right-hand side or from left-hand side ($\varepsilon \to -0$).

One may easily recall a figure of e^x, $-\infty < x < \infty$, where $\lim_{\varepsilon \to +0} e^{-1/\varepsilon} = 0$, and $\lim_{\varepsilon \to -0} e^{-1/\varepsilon} = \infty$.

It is further assumed that $\varepsilon \geq 0$.

If $\lim_{\varepsilon \to 0} f(\varepsilon)$ exists, then a limit of the function $f(\varepsilon)$ can be described qualitatively, since one of the following three possibilities takes place

$$\left.\begin{array}{l} f(\varepsilon) \to 0 \\ f(\varepsilon) \to A \\ f(\varepsilon) \to \infty \end{array}\right\} \qquad \text{for} \qquad \varepsilon \to 0 \ , \quad 0 < |A| < \infty \ .$$

However, an estimation basing on first and third cases seems not to be the appropriate one, because there is an infinite set of functions approaching zero for $\varepsilon \to 0$. For example, in accordance with formulas (2.7), (2.10), (2.11) and (2.14) one gets

$$\lim_{\varepsilon \to 0} \sin \varepsilon = 0 \ , \quad \lim_{\varepsilon \to 0} \sin \varepsilon^n = 0 \quad \text{for} \ \ n > 0 \ ,$$

$$\lim_{\varepsilon \to 0} \tan 2\varepsilon = 0 \ , \quad \lim_{\varepsilon \to 0} (1 - \cos \varepsilon) = 0 \ ,$$

$$\lim_{\varepsilon \to 0} [\ln(1+\varepsilon)]^5 = 0 \ , \quad \text{and so on.}$$

Similarly, there are infinitely many functions tending to infinity for $\varepsilon \to 0$. For example, according to (2.9) and (2.10), one gets

$$\lim_{\varepsilon \to 0} \frac{1}{\sin \varepsilon} = \infty \ ,$$

$$\lim_{\varepsilon \to 0} -\text{ctan}\varepsilon = \infty \ ,$$

$$\lim_{\varepsilon \to 0} \varepsilon^n = \infty \quad \text{for } n < 0 \ , \quad \lim_{\varepsilon \to 0} \frac{1}{1 - \cos 3\varepsilon} = \infty \ .$$

In the cases considered above, one deals with so-called infinitely small and infinitely large function.

Definition 2.1. A function $f(\varepsilon)$ is called *infinitely small* for $\varepsilon \to 0$, if $\lim_{\varepsilon \to 0} f(\varepsilon) = 0$.

Definition 2.2. A function $f(\varepsilon)$ is called *infinitely large* for $\varepsilon \to 0$, if $\lim_{\varepsilon \to 0} f(\varepsilon) = \infty$.

In order to describe the character of variations of infinitely small and infinitely large functions more adequately, one has to include velocities of their approaches to either zero or infinity. Usually, in order to estimate the mentioned velocities, a comparison with the velocities approaching zero or infinity of some known functions is made. Such functions are called *comparison functions* or *scaling functions.*

The following natural powers of the parameter ε belong to the simplest and most widely used functions: $1, \varepsilon, \varepsilon^2, \ldots$, as well as the inversed powers of the form $\varepsilon^{-1}, \varepsilon^{-2}, \ldots$. It is clear that for small ε, the following inequality chain holds $\ldots > \varepsilon^{-2} > \varepsilon^{-1} > 1 > \varepsilon > \varepsilon^2 > \ldots$

However, the introduced set of scaling functions is called *deficient.* It cannot be used to describe the function $\ln \frac{1}{\varepsilon}$, which for $\varepsilon \to 0$ tends to infinity, but slower than any ε, also slower than $\varepsilon^{-\alpha}$, where α is an arbitrary small positive number. Applying de l'Hospital formula, one gets

$$\lim_{\varepsilon\to 0} \frac{\ln \frac{1}{\varepsilon}}{\varepsilon^{-\alpha}} = \lim_{\varepsilon\to 0} \frac{\ln \frac{1}{\varepsilon}}{\left(\frac{1}{\varepsilon}\right)^{\alpha}} = \lim_{x\to\infty} \frac{\ln x}{x^{\alpha}} = \lim_{x\to\infty} \frac{1}{x\alpha x^{\alpha-1}} = \frac{1}{\alpha} \lim_{x\to\infty} \frac{1}{x^{\alpha}} = 0 \ .$$

The role of scaling function may be also fulfilled by the function $e^{-1/\varepsilon}$, because for $\varepsilon \to 0$ it tends to zero faster than each of powers of ε

$$\lim_{\varepsilon\to 0} \frac{e^{-1/\varepsilon}}{e^{n}} = \lim_{\varepsilon\to 0} \frac{\varepsilon^{-n}}{e^{1/\varepsilon}} = \lim_{\varepsilon\to 0} \frac{\left(\frac{1}{\varepsilon}\right)^{n}}{e^{1/\varepsilon}} = \lim_{x\to 0} \frac{x^{n}}{e^{x}} = \lim_{x\to\infty} \frac{n!}{e^{x}} = 0 \ .$$

In this case, in order to identify the indeterminacy of $\frac{\infty}{\infty}$ type one has to apply de l'Hospital rule n times. Terms proportional to $e^{-1/\varepsilon}$ are called *exponentially* (or *transcendentally*) small.

Note that an inversed function $e^{1/\varepsilon}$ tends to infinity faster than any other arbitrary power ε, because

$$\lim_{\varepsilon\to 0} \frac{e^{1/\varepsilon}}{e^{-n}} = \lim_{\varepsilon\to 0} \frac{e^{1/\varepsilon}}{\left(\frac{1}{\varepsilon}\right)^{n}} = \lim_{x\to\infty} \frac{e^{x}}{x^{n}} = \lim_{x\to\infty} \frac{e^{x}}{n!} = \infty \ .$$

In this case, like in the previous one, de l'Hospital's rule is applied also n times.

It is not difficult to show that $\lim\limits_{\varepsilon\to 0} \varepsilon^{n} \ln \varepsilon = 0$ for $n > 0$. According to de l'Hospital's rule

$$\lim_{\varepsilon\to 0} \frac{\ln \varepsilon}{\varepsilon^{-n}} = \lim_{\varepsilon\to 0} \frac{\frac{1}{\varepsilon}}{-n\varepsilon^{-n-1}} = -\frac{1}{n} \lim_{\varepsilon\to 0} \frac{1}{\varepsilon^{-n}} = -\frac{1}{n} \lim_{\varepsilon\to 0} \varepsilon^{n} = 0 \ .$$

Note that a deeper estimation of the investigated function is sometimes required, and hence other scaling functions should be applied.

Let a chosen scaling function have the form $q(\varepsilon)$. In order to compare the investigated function $f(\varepsilon)$ for $\varepsilon \to 0$ with scaling one $q(\varepsilon)$, one of the following two *symbols of estimation of magnitude order*, i.e., either O ("O capital") or o ("o small"), is used.

Definition 2.3. It is said that $f(\varepsilon) = O(q(\varepsilon))$ for $\varepsilon \to 0$, if

$$\lim_{\varepsilon\to 0} \frac{f(\varepsilon)}{q(\varepsilon)} = A \ , \quad 0 < |A| < \infty \ . \tag{2.25}$$

In addition, it is said that the functions $f(\varepsilon)$ and $q(\varepsilon)$ have same *order of magnitude*.

Symbol O can also be introduced in a different way. Namely, $f(\varepsilon) = O(q(\varepsilon))$ for $\varepsilon \to 0$, if there exists a positive number A (coefficient of proportionality), $0 < A < \infty$, independent from ε, and $\varepsilon_0 > 0$, such that $|f(\varepsilon)| \leq A|q(\varepsilon)|$ holds for all $\varepsilon \leq \varepsilon_0$. Observe that in accordance with this definition, symbol O practically coincides with an upper estimation of a magnitude order of the function $f(\varepsilon)$. Some examples follow.

According to the formula (2.9), one has:

$$\lim_{\varepsilon \to 0} \frac{\sin \varepsilon}{\varepsilon} = \lim_{\varepsilon \to 0} \left(1 - \frac{\varepsilon^2}{3!} + \frac{\varepsilon^4}{5!} - \ldots \right) = 1 ,$$

whereas according to the definition (2.25), one gets $\sin \varepsilon = O(\varepsilon)$ for $\varepsilon \to 0$. Furthermore, due to the introduced definition one gets $\sin 5\varepsilon = O(\varepsilon)$, $\sin \varepsilon^3 = O(\varepsilon^3)$ for $\varepsilon \to 0$.

Using (2.3), (2.7), (2.10), (2.11), (2.14)–(2.17), and applying the definition (2.25), one has:

$$\sqrt[n]{1 + a\varepsilon} = O(1) , \quad e^{\varepsilon} = O(1) ,$$

$$\cos \varepsilon = O(1) , \quad \tan\varepsilon = O(\varepsilon) , \quad \text{ctan}\varepsilon = O(1/\varepsilon) ,$$

$$\arcsin \varepsilon = O(\varepsilon) , \quad \arctan\varepsilon = O(\varepsilon) , \quad \ln(1 + \varepsilon) = O(\varepsilon) .$$

One deals with the more complex problem when both functions f and q depend on both parameter ε and x. We say that $f(x;\varepsilon) = O(q(x;\varepsilon))$ for $\varepsilon \to 0$, if there is a number $A > 0$, independent from ε, and for the number $\varepsilon_0 > 0$ the following inequality holds

$$|f(x;\varepsilon)| \leq A\,|q(x;\varepsilon)| \quad \text{for all } \ \varepsilon \leq \varepsilon_0 . \tag{2.26}$$

If A and ε_0 do not depend on x, it is said that the inequality estimation (2.26) holds alike with respect to variable x. For example, the estimation $\sin(x + \varepsilon) = O(1)$ is uniform with respect to x for $\varepsilon \to 0$, but the estimation $\varepsilon/x = O(\varepsilon)$ for $\varepsilon \to 0$ is not uniform. One may easily show using formula (2.7), that

$$f(x;\varepsilon) = e^{\varepsilon\, x} - 1 = \varepsilon x + \frac{\varepsilon^2 x^2}{2!} + \frac{\varepsilon^3 x^3}{3!} + \ldots ,$$

and hence the estimation $e^{\varepsilon x} - 1 = O(\varepsilon)$ is not true for large values of x, for instance for $x = O(1/\varepsilon)$.

Very often the following series is used $f(x;\varepsilon) = x^{\varepsilon} \equiv e^{\varepsilon \ln x} = 1 + \varepsilon \ln x + \frac{1}{2}\varepsilon^2 \ln^2 x + \ldots$ which is not uniform for $x = 0$ or $x = \infty$.

In many cases, information with respect to an investigated function $f(\varepsilon)$ is not sufficient to define its rate of approaching a limit for $\varepsilon \to 0$. However, this information allows for the estimation if this rate is larger than the velocity of variations of an associated scaling function $q(\varepsilon)$. For this purpose, the estimating symbol o ("o small") is applied.

Definition 2.4. It is said that $f(\varepsilon) = o(q(\varepsilon))$ for $\varepsilon \to 0$, if

$$\lim_{\varepsilon \to 0} \frac{f(\varepsilon)}{q(\varepsilon)} = 0 \ . \tag{2.27}$$

In other words, it means that $f(\varepsilon)$ has higher *order magnitude* of "smallness" than function $q(\varepsilon)$.

2.2.2.1 Examples

Using (2.7)–(2.11), (2.14) and the definition (2.27) one may get the following relations for $\varepsilon \to 0$:

$$\sin\varepsilon = o(1) \ , \quad 1 - \cos 7\varepsilon = o(\varepsilon) \ ,$$

$$\tan\varepsilon^2 = o(\varepsilon) \ , \quad e^{\varepsilon} - 1 = o\left(\varepsilon^{1/2}\right) \ ,$$

$$\ln(1+\varepsilon) = o\left(\varepsilon^{1/3}\right) \ , \qquad \text{and so on.}$$

It appears that for above functions $f(\varepsilon)$, the following estimations hold:

$$e^{\varepsilon - 1} = o\left(\varepsilon^{2/3}\right) \ ,$$

$$1 - \cos 7\varepsilon = o\left(\varepsilon^{1/3}\right) \ ,$$

$$\ln(1+\varepsilon) = o\left(\varepsilon^{4/5}\right) \ ,$$

$$\sin\varepsilon = o\left(\varepsilon^{1/2}\right) \ ,$$

$$\tan\varepsilon^2 = o\left(\varepsilon^{3/2}\right) \ , \qquad \text{and so on.}$$

Obtaining different results associated with the estimation of magnitude order of function $f(\varepsilon)$ with the use of the symbol "o small" proves that the application of the symbol "O capital" gives more exact information with respect to behavior of function $f(\varepsilon)$ for $\varepsilon \to 0$.

Mathematical estimation of magnitude order through symbols O and o does not overlap formally with a physical intuitive estimation of this quantity, because the quantitative value of a proportionality coefficient does not appear in a mathematical definition. According to the definition (2.25), we have $A\varepsilon = O(\varepsilon)$ for $0 < A < \infty$. On the other hand, in the majority of physical problems, the proportionality coefficient A is of order 1, which means that mathematical and physical estimations of a magnitude of considered quantities coincide.

Let us finally outline some obvious rules of application of symbols o and O:

- The order of a product (division) of quantities is equal to a product (division) of orders of individual terms of the operation;
- The order of a sum (difference) of terms is equal to the order of a dominating element (for instance, it can be a term ε^m with the smallest value of m);
- A sum of a finite number of infinitely small terms of the same order possesses same order of magnitude.

Symbols of order magnitude can be integrated with respect to ε, as well as with respect to other variables; differentiation of order magnitude, in general, is not permitted. However, in many branches of applied mathematics, physics and engineering analyzed with the use of asymptotical approaches, it is assumed that derivatives are of the same orders as differentiated functions.

More information on the mentioned symbols can be found in reference [5].

2.2.3 Asymptotical sequences and decompositions

An asymptotical sequence is defined as the sequence of scaling functions satisfying a condition that each successive term of asymptotical sequence should be of an order smaller than the previous sequence term. In the item 2.2.2, the analysis of scaling functions is carried out. It is shown, among others, that scaling function powers of ε, as well as logarithmic, exponential, and other functions can be used. Let us denote the n-th term of an arbitrary sequence of scaling functions by $\delta_n(\varepsilon)(n = 0, 1, 2, \ldots)$.

Definition 2.5. A sequence of scaling functions given in general form $\delta_n(\varepsilon)$ is called *an asymptotical sequence*, if

$$\delta_{n+1}(\varepsilon) = o(\delta_n(\varepsilon)) \quad \text{for} \quad \varepsilon \to 0 \; ; \quad n = 0, 1, 2, \ldots \tag{2.28}$$

or if

$$\lim_{\varepsilon \to 0} \frac{\delta_{n+1}}{\delta_n(\varepsilon)} = 0 \; , \qquad n = 0, 1, 2, \ldots \tag{2.29}$$

The last definition coincides with the earlier introduced (2.27) symbol of "o small."

As an example, let us mention some of the scaling functions considered in item 2.2.2 with respect to their magnitude order:

$$e^{1/\varepsilon}, \; \varepsilon^{-n}, \; \varepsilon^{-2}, \; \varepsilon^{-1}, \; \varepsilon^{-\alpha}, \; \ln 1/\varepsilon, \; 1, \; \varepsilon^{\alpha}, \; \varepsilon \ln \varepsilon, \; \varepsilon, \; \varepsilon^2 \ln \varepsilon, \; \varepsilon^2, \; \varepsilon^n, \; e^{-1/\varepsilon}.$$

In the above, it is assumed that $n > 2$, $0 < \alpha < 1$. Observe that between two arbitrarily chosen members of an asymptotical sequence, one may put an infinite set of other scaling functions without violating the condition (2.29).

In practice, during solving various problems through asymptotic methods, two main approaches are applied. In the first case, an asymptotic sequence is taken a priori, which may result in either omitting the required terms or in the occurrence of nonrequired terms. In the second case, an asymptotic sequence is defined through the successive definition of its terms during successive steps of asymptotic construction. It occurs that the second approach is more effective.

The introduced definition of an asymptotic sequence allows defining an asymptotic decomposition.

Definition 2.6. Sum of the form $\sum_{n=0}^{\infty} a_n \delta_n(\varepsilon)$, where a_n does not depend on ε, and $\delta_n(\varepsilon)$ is an asymptotic sequence, is called *an asymptotic decomposition of a function* $f(\varepsilon)$, if

$$f(\varepsilon) = \sum_{n=0}^{N} a_n \delta_n(\varepsilon) + o(\delta_N(\varepsilon)) \quad \text{for } \varepsilon \to 0 \tag{2.30}$$

or

$$f(\varepsilon) = \sum_{n=0}^{N} a_n \delta_n(\varepsilon) + O(\delta_{N+1}(\varepsilon)) \quad \text{for } \varepsilon \to 0\,. \tag{2.31}$$

An asymptotic decomposition of the function $f(\varepsilon)$ with respect to an asymptotic sequence $\delta_n(\varepsilon)$ is defined in the following way

$$f(\varepsilon) \sim \sum_{n=0}^{\infty} a_n \delta_n(\varepsilon) \quad \text{for } \varepsilon \to 0\,. \tag{2.32}$$

Symbol $\sim$ occurring in (2.32) denotes that the sum $\sum_{n=0}^{\infty} a_n \delta_n(\varepsilon)$ for limiting transition $\varepsilon \to 0$ is asymptotically equal to the function $f(\varepsilon)$.

Notice that the given function $f(\varepsilon)$ can be defined through the infinite number of asymptotic decompositions of the form (2.32), since there is the infinite number of applied sequences $\delta_n(\varepsilon)$. For example

$$\sin 2\varepsilon \sim 2\varepsilon - \frac{4}{3}\varepsilon^3 + \frac{4}{15}\varepsilon^5 + \ldots \sim 2\tan\varepsilon - 2\tan^3\varepsilon - 2\tan^5\varepsilon + \ldots \sim$$
$$\sim 2\ln(1+\varepsilon) + \ln(1+\varepsilon^2) - 2\ln(1+\varepsilon^3) + \ldots \sim$$
$$\sim 6\left(\frac{\varepsilon}{3+2\varepsilon^2}\right) - \frac{756}{5}\left(\frac{\varepsilon}{3+2\varepsilon^2}\right)^5 + \ldots$$

In the given example, it is rather difficult to guess the form of two last terms, because their magnitude order and the order of terms of the first sequence do not coincide. In spite of that, all of the mentioned scaling functions governed by formulas (2.9), (2.14), (2.11) and (2.2) yield the equivalent results.

One may prove that if a chosen asymptotic sequence is $\delta_n(\varepsilon)$, then there is only one asymptotic decomposition of the given function $f(\varepsilon)$ (with respect to this sequence).

Proof. Let the following approximation holds for $\varepsilon \to 0$

$$f(\varepsilon) \sim \sum_{n=0}^{\infty} a_n \delta_n(\varepsilon) = a_0\delta_0(\varepsilon) + a_1\delta_1(\varepsilon) + a_2\delta_2(\varepsilon) + \dots, \tag{2.33}$$

where scaling functions $\delta_n(\varepsilon)$ are known, and where $\delta_{n+1} = o(\delta_n(\varepsilon))$, $n = 0, 1, \dots$. Dividing (2.33) by $\delta_0(\varepsilon)$, one gets

$$\frac{f(\varepsilon)}{\delta_0(\varepsilon)} \sim a_0 + \frac{a_1\delta_1(\varepsilon)}{\delta_0(\varepsilon)} + \dots ,$$

and from $\varepsilon \to 0$, one obtains

$$a_0 = \lim_{\varepsilon\to 0}\left(\frac{f(\varepsilon)}{\delta_0(\varepsilon)} - a_1\frac{\delta_1(\varepsilon)}{\delta_0(\varepsilon)} + \dots\right),$$

and, consequntly,

$$a_0 = \lim_{\varepsilon\to 0}\frac{f(\varepsilon)}{\delta_0(\varepsilon)}, \quad \text{because} \quad \lim_{\varepsilon\to 0}\frac{\delta_n(\varepsilon)}{\delta_0(\varepsilon)} = 0 \quad \text{for} \quad n \geq 1 .$$

Let us move the term $a_0\delta_0(\varepsilon)$ into the left-hand side of (2.33), and let us divide it by $\delta_1(\varepsilon)$ to get

$$\frac{f(\varepsilon) - a_0\delta_0(\varepsilon)}{\delta_1(\varepsilon)} \sim a_1 + a_2\frac{\delta_2(\varepsilon)}{\delta_1(\varepsilon)} + \dots$$

where for $\varepsilon \to 0$, one gets

$$a_1 = \lim_{\varepsilon\to 0}\frac{f(\varepsilon) - a_0\delta_0(\varepsilon)}{\delta_1(\varepsilon)} .$$

Proceeding in an analogical way, the following formulas are obtained

$$a_0 = \lim_{\varepsilon\to 0}\frac{f(\varepsilon)}{\delta_0(\varepsilon)}, \quad a_n = \lim_{\varepsilon\to 0}\frac{f(\varepsilon) - \sum\limits_{m=0}^{n-1} a_m\delta_m(\varepsilon)}{\delta_n(\varepsilon)}, \qquad n = 1, 2, \dots \tag{2.34}$$

Consider the case when a function f depends not only on ε, but also on variable x, which can be a scalar or vector function.

Definition 2.7. Sum of the form $\sum\limits_{n=0}^{\infty} a_n(x)\delta_n(\varepsilon)$, where $a_n(x)$ does not depend on ε, and $\delta_n(\varepsilon)$ is the asymptotic sequence, is be called as *an asymptotic decomposition of the function* $f(x;\varepsilon)$, if

$$f(x;\varepsilon) = \sum_{n=0}^{N} a_n(x)\delta_n(\varepsilon) + o(\delta_N(\varepsilon)) \quad \text{for} \quad \varepsilon \to 0 \tag{2.35}$$

or if, equivalently,

$$f(x;\varepsilon) = \sum_{n=0}^{N} a_n(x)\delta_n(\varepsilon) + O(\delta_{N+1}(\varepsilon)) \quad \text{for} \quad \varepsilon \to 0 \,. \tag{2.36}$$

Notice that the asymptotic decomposition of the function $\delta_n(\varepsilon)$ with respect to the asymptotic sequence may be presented in the following way:

$$f(x;\varepsilon) \sim \sum_{n=0}^{\infty} a_n\delta_n(\varepsilon) \quad \text{for} \quad \varepsilon \to 0 \,. \tag{2.37}$$

For a given function $f(x;\varepsilon)$ and a given asymptotic sequence $\delta_n(\varepsilon)$, the coefficients of decomposition (2.34) are defined in a unique way.

2.2.4 Nonuniform asymptotic decompositions

In many problems of applied mathematics, physics and engineering, asymptotic decompositions of functions searched may depend on spatial variables, time and obviously on a perturbation parameter ε. Such asymptotic decompositions are called *uniformly exact*, if an error yielded by them is uniformly small with respect to these variables.

In what follows, a more precise mathematical formulation of the earlier introduced description will be given. Consider both investigated function $f(x;\varepsilon)$, where x is a scalar or a vector variable independent on ε, and a chosen asymptotic sequence $\delta_n(\varepsilon)$, for which $\delta_n(\varepsilon) = o(\delta_{n-1}(\varepsilon))$ for $\varepsilon \to 0$.

Let us construct an asymptotic decomposition

$$f(x;\varepsilon) \sim \sum_{n=0}^{\infty} a_n(x)\delta_n(\varepsilon) \quad \text{for} \quad \varepsilon \to 0 \,. \tag{2.38}$$

Definition 2.8. An asymptotic decomposition (2.38) is called *uniformly exact*, if

$$f(x;\varepsilon) = \sum_{n=0}^{N-1} a_n(x)\delta_n(\varepsilon) + R_N(x;\varepsilon) \tag{2.39}$$

$$\text{for} \quad R_N(x;\varepsilon) = O(\delta_N(\varepsilon))$$

is a uniform decomposition for all considered values of x.
Otherwise, we say that that the decomposition is *nonuniformly exact* (*irregular, singular*), and the area of x variations for which the condition (2.39) is violated is called *the area of nonuniformity.*

In order to satisfy the nonuniformity condition (2.39), it is necessary that the term (for all n) $a_n(x)\delta_n(\varepsilon)$ is small in comparison to the previous term

$a_{n-1}(x)\delta_{n-1}(\varepsilon)$. Since, according to the definition $\delta_n(\varepsilon) = o(\delta_{n-1}(\varepsilon))$ for $\varepsilon \to 0$, in order to satisfy nonuniformity conditions it is sufficient that for all x the expression $a_n(x)$ is bounded or not more singular than $a_{n-1}(x)$. The mentioned requirement guarantees that each member of the asymptotic decomposition (2.38) improves the previous term of the decomposition.

Further, a simple example of uniformly exact asymptotic decomposition will be given for $\varepsilon \to 0$ (formulas 2.9 and 2.10 will be used):

$$\cos(x+\varepsilon) = \cos x \cos\varepsilon - \sin x \sin\varepsilon =$$

$$= \cos x \left(1 - \frac{\varepsilon^2}{2!} + \frac{\varepsilon^4}{4!} - \ldots\right) - \sin x \left(\varepsilon - \frac{\varepsilon^3}{3!} + \frac{\varepsilon^5}{5!} - \ldots\right) =$$

$$= \cos x - \varepsilon \sin x - \frac{\varepsilon^2}{2!}\cos x + \frac{\varepsilon^3}{3!}\sin x + \frac{\varepsilon^4}{4!}\cos x + O(\varepsilon^5)\ . \qquad (2.40)$$

Notice that decomposition (2.40) is uniformly exact, because all coefficients standing by powers of ε are bounded functions with values less than 1.

Consider now the decomposition of function $f(x;\varepsilon) = \sqrt{x+\varepsilon}$ for small ε. According to the formula (2.3), one gets

$$f(x;\varepsilon) = \sqrt{x+\varepsilon} = \sqrt{x(1+\varepsilon/x)} = \sqrt{x}(1+\varepsilon/x)^{1/2} =$$

$$= \sqrt{x}\left(1 + \frac{\varepsilon}{2x} - \frac{\varepsilon^2}{8x^2} + \frac{\varepsilon^3}{16x^3} - \ldots\right) \quad \text{for } \varepsilon \to 0\ . \qquad (2.41)$$

Observe that each term of the above decomposition (except for the first one) has the feature of singularity for $x = 0$ and is more singular than the previous one for $x \to 0$. Decomposition (2.37) is a nonuniform one for $\varepsilon/x = O(1)$, since all its terms for $\varepsilon/x = O(1)$ have same order of $O(\varepsilon)$. Notice that space dimensions can be different where the decomposition is uniform. For instance, in the case of decomposition

$$f(x;\varepsilon) = x^{\varepsilon} \equiv e^{\varepsilon \ln x} = 1 + \varepsilon \ln x + \frac{\varepsilon^2 \ln^2 x}{2} + \ldots$$

all its terms have same order $O(1)$ for $\varepsilon \ln x \sim O(1)$, i.e. $x = O(e^{-1/\varepsilon})$.

2.2.5 Operations on asymptotic decompositions

While applying asymptotic methods in solving various problems, it is assumed that asymptotic decompositions are substituted into equations and boundary conditions, and that then many basic operations like multiplication, division, rising to a power, differentiation or integration are carried out. However, using the mentioned operations uncritically may lead to erroneous results, i.e., to the occurrence of nonuniform asymptotic decompositions (see the earlier section). Let us discuss this problem in more detail [66].

Giving a sequence $\delta_n(\varepsilon)$ and giving asymptotic decompositions $f(x;\varepsilon)$, $q(x;\varepsilon)$ for $\varepsilon \to 0$ of the form

$$f(x;\varepsilon) \sim \sum_{n=0}^{\infty} a_n(x)\delta_n(\varepsilon) \quad \text{for} \quad \varepsilon \to 0 \ , \tag{2.42}$$

$$q(x;\varepsilon) \sim \sum_{n=0}^{\infty} b_n(x)\delta_n(\varepsilon) \quad \text{for} \quad \varepsilon \to 0 \ . \tag{2.43}$$

one may construct formula

$$\alpha f(x;\varepsilon) + \beta q(x;\varepsilon) \sim \sum_{n=0}^{\infty} [\alpha a_n(x) + \beta b_n(x)]\delta_n(\varepsilon) \ ; \quad \{\alpha, \beta\} = const \ .$$

If $f(x;\varepsilon)$, $a_n(x)$ can be integrated with respect to x, then

$$\int_{\alpha}^{x} f(x;\varepsilon)dx \sim \sum_{n=0}^{\infty} \delta_n(\varepsilon) \int_{\alpha}^{x} a_n(x)dx \quad \text{for} \quad \varepsilon \to 0 \ .$$

If $f(x;\varepsilon)$, $a_n(x)$ can be integrated with respect to ε, then

$$\int_{0}^{\varepsilon} f(x;\varepsilon)d\varepsilon \sim \sum_{n=0}^{\infty} a_n(x) \int_{0}^{\varepsilon} \delta_n(\varepsilon)d\varepsilon \quad \text{for} \quad \varepsilon \to 0 \ .$$

Multiplication of asymptotic sequences (decompositions) is, in general, not defined. It is realizable when an asymptotic sequence is also obtained as a result, i.e., it takes place for all asymptotic sequences $\delta_n(\varepsilon)$, for which a product $\delta_n\delta_m$ creates either an asymptotic sequence or an asymptotic decomposition.

It seems that a set of rising powers of the form ε^n belongs to the most popular and most frequently used. If for $\varepsilon \to 0$ one has

$$f(x;\varepsilon) \sim \sum_{n=0}^{\infty} a_n(x)\varepsilon^n = a_0 + a_1\varepsilon + a_2\varepsilon^2 + \ldots \tag{2.44}$$

$$q(x;\varepsilon) \sim \sum_{n=0}^{\infty} b_n(x)\varepsilon^n = b_0 + b_1\varepsilon + b_2\varepsilon^2 + \ldots \tag{2.45}$$

then for $\varepsilon \to 0$ one gets an asymptotic series of the form

$$f(x;\varepsilon)q(x;\varepsilon) \sim \sum_{n=0}^{\infty} c_n(x)\varepsilon^n = c_0 + c_1\varepsilon + c_2\varepsilon^2 + O(\varepsilon^3) \ ,$$

where

$$c_0 = a_0b_0 \ , \quad c_1 = a_0b_1 + a_1b_0 \ , \quad c_2 = a_0b_2 + a_1b_1 + a_2b_0 \ , \quad \ldots$$

or, more generally,

$$c_n(x) = \sum_{i=0}^{n} a_i(x) b_{n-i}(x) \ . \tag{2.46}$$

Dividing one asymptotic series by another one is not generally allowed and is carried out only formally. A denominator is moved to a numerator, applying transformation rules for powers. While carrying out a decomposition in accordance with the formula (2.3), two polynomials are multiplied, i.e., numerator and modified denominator. Generally, a division of asymptotic series is allowed if one gets a new asymptotic series as a result.

A division of power series is carried out in a rather simple way. If, for instance, for $\varepsilon \to 0$ decompositions (2.44), (2.45) hold, then for $\varepsilon \to 0$ a new asymptotic series is obtained of the form

$$\frac{f(x,\varepsilon)}{q(x,\varepsilon)} \sim \sum_{n=0}^{\infty} d_n(x)\varepsilon^n = d_0 + d_1\varepsilon + d_2\varepsilon^2 + O(\varepsilon^3) \ , \tag{2.47}$$

where:

$$d_0 = \frac{a_0}{b_0}, \quad d_1 = \frac{a_1}{b_0} - \frac{a_0 b_1}{b_0^2}, \quad d_2 = \frac{a_2}{b_0} + \frac{a_0 b_1^2}{b_0^3} - \frac{a_0 b_2}{b_0^2} - \frac{a_1 b_1}{b_0^2} \ .$$

In what follows, we prove the obtained result by applying the formula (2.3)

$$\frac{f(x,\varepsilon)}{q(x,\varepsilon)} = (a_0 + a_1\varepsilon + a_2\varepsilon^2 + \ldots)(b_0 + \varepsilon b_1 + \varepsilon^2 b_2 + \ldots)^{-1} =$$

$$= \frac{a_0}{b_0}\left(1 + \frac{a_1}{a_0}\varepsilon + \frac{a_2}{a_0}\varepsilon^2 + \ldots\right)\left(1 + \frac{b_1}{b_0}\varepsilon + \frac{b_2}{b_0}\varepsilon^2 + \ldots\right)^{-1} =$$

$$= \frac{a_0}{b_0}\left(1 + \frac{a_1}{a_0}\varepsilon + \frac{a_2}{a_0}\varepsilon^2 + \ldots\right)\left[1 - \frac{b_1}{b_0}\varepsilon + \varepsilon^2\left(\frac{b_1^2}{b_0^2} - \frac{b_2}{b_0}\right) + \ldots\right] =$$

$$= \frac{a_0}{b_0}\left[1 + \varepsilon\left(\frac{a_1}{a_0} - \frac{b_1}{b_0}\right) + \varepsilon^2\left(\frac{a_2}{a_0} + \frac{b_1^2}{b_0^2} - \frac{b_2}{b_0} - \frac{a_1 b_1}{a_0 b_0}\right) + \ldots\right] =$$

$$= \frac{a_0}{b_0} + \varepsilon\left(\frac{a_1}{b_0} - \frac{a_0 b_1}{b_0^2}\right) + \varepsilon^2\left(\frac{a_2}{b_0} + \frac{a_0 b_1^2}{b_0^3} - \frac{a_0 b_2}{b_0^2} - \frac{a_1 b_1}{b_0^2}\right) + O(\varepsilon^3) \ .$$

In general, raising powers is not allowed, since its formal introduction may yield a nonuniform asymptotic decomposition (series). As a simple example, one may consider a decomposition of the function $\sqrt{x+\varepsilon}$ for $\varepsilon \to 0$ (see formula 2.41).

In addition, a differentiation process with respect to variable x or to perturbation ε does not allow for the introduction of a precise definition. Therefore, one assumes a priori that analyzed functions and their derivatives are of the same order. However, this strict assumption may lead to the occurrence of nonuniform asymptotic series. It often happens that although perturbed functions are small in the whole area of x variations, their gradients defined through their derivatives may achieve large values in certain domains of defined subspaces of investigated functions (nonuniform subspaces of asymptotic decompositions). Therefore, it is recommended to carry out a strict analysis or to introduce corrections after a formal construction of a solution.

Possibility of the occurrence of nonuniform series, which is, rather unfortunately, observed in the majority of considered cases, positively influenced the wide development of *the theory of singular perturbations.* Some of them will be considered in the next three sections of the book.

Let us briefly outline some successive steps connected with transformations of asymptotic series using the example below, which will help to solve some tasks given at the end of this chapter.

Let us determine three first terms (different from zero) of asymptotic series of a rather complex function of the form

$$f(\varepsilon) = \frac{\ln\left[(1+2\varepsilon)\cos\sqrt{\varepsilon}\right]}{\sqrt[3]{1+\sin 2\varepsilon}} \quad \text{for } \varepsilon \to 0 .$$

According to the properties of logarithms and for (2.3), (2.9), (2.10) and (2.11) one gets

$$f(\varepsilon) = \left[\ln(1+2\varepsilon) + \ln\cos\sqrt{\varepsilon}\right](1+\sin 2\varepsilon)^{-1/3} =$$

$$= \left[2\varepsilon - \frac{(2\varepsilon)^2}{2} + \frac{(2\varepsilon)^3}{3} + \ln\left(1 - \frac{\varepsilon}{2} + \frac{\varepsilon^2}{4!} - \frac{\varepsilon^3}{6!}\right)\right](1+2\varepsilon)^{-1/3} =$$

$$= \left[2\varepsilon - 2\varepsilon^2 + \frac{8}{3}\varepsilon^3 + \left(-\frac{\varepsilon}{2} + \frac{\varepsilon^2}{24} - \frac{\varepsilon^3}{720}\right) - \frac{1}{2}\left(-\frac{\varepsilon}{2} + \frac{\varepsilon^2}{24}\right)^2 + \frac{1}{3}\left(-\frac{\varepsilon}{2}\right)^3\right]$$

$$\left(1 - \frac{2}{3}\varepsilon + \frac{8}{9}\varepsilon^2\right) = \left(\frac{3}{2}\varepsilon - \frac{25}{12}\varepsilon^2 + \frac{119}{45}\varepsilon^3\right)\left(1 - \frac{2}{3}\varepsilon + \frac{8}{9}\varepsilon^2\right) =$$

$$= \frac{3}{2}\varepsilon - \frac{37}{12}\varepsilon^2 + \frac{161}{30}\varepsilon^3 + O(\varepsilon^4) .$$

(One may apply symbolic computations, using for instance "Mathematica," to get the result very quickly.)

2.2.6 Asymptotic series. Comparison of asymptotic and convergent series. Advantages of application of asymptotic series and decompositions

An asymptotic series can be viewed as a particular case of an asymptotic decomposition. Allowing in (2.30) or (2.31) $N \to \infty$, one gets an infinite asymptotic series of the function $f(\varepsilon)$. This series may be *convergent* in a classical sense for a certain subspace of ε values, or it may be *divergent* for all

ε. A function $f(\varepsilon)$ can be represented by a *convergent series* in the following way:

$$f(\varepsilon) = \lim_{N\to\infty} \sum_{i=0}^{N} a_i q_i(\varepsilon) \quad \text{for a given fixed value of } \varepsilon \ .$$

In the case of *a divergent series* a function $f(\varepsilon)$ can be represented through an asymptotic decomposition with the increasing accuracy for small ε, so that in a limit one gets

$$f(\varepsilon) = \lim_{\varepsilon\to 0} \sum_{i=0}^{N} a_i q_i(\varepsilon) \quad \text{for a given fixed value of } N \ .$$

It is worth mentioning that the expectation of a formal mathematical convergence in practical applications belongs to rather unrealistic requirements. It happens that convergence in a strict mathematical sense depends on behavior of infinitely small terms, whereas in practice a researcher deals only with first terms of an asymptotic decomposition and usually believes that these terms tend quickly to a real solution of the investigated problem. However, very often representing a function through several terms of an asymptotic series only is more useful than the application of the whole series. The main advantage of asymptotic decompositions relies on the error estimation, which is of order of the first neglected term, and hence it tends to zero quickly with a decrease of ε. For a fixed value of ε an error can be decreased through the increase of the number of decomposition elements. If a series is divergent, then there is a critical instant when the further increase of terms causes an increase of error and the process should be stopped.

Consider, as an example, a representation of the function $f(\varepsilon)$ through *a series of small* ε, where

$$f(\varepsilon) = 1 - \operatorname{erf}\left(\frac{1}{\varepsilon}\right) = 1 - \frac{2}{\sqrt{\pi}} \int_0^{1/\varepsilon} e^{-t^2} dt \ .$$

In the above, $\operatorname{erf} x = \frac{2}{\sqrt{\pi}} \int_0^x e^{-t^2} dt = \frac{2}{\sqrt{\pi}} \left(x - \frac{x^3}{3} + \frac{1}{2!}\frac{x^5}{5} - \frac{1}{3!}\frac{x^7}{7} + \ldots \right)$ governs *errors' function* [109] widely used in the theory of probability.

This series has been obtained through the decomposition of the integrand exponential function in accordance with the formula (2.7) and then through successive integrations. One finally gets

$$f(\varepsilon) = 1 - \frac{2}{\sqrt{\pi}} \sum_{n=0}^{\infty} \frac{(-1)^n}{(2n+1)n!\varepsilon^{2n+1}} = A \ .$$

The obtained series A is not an asymptotic one. Series A is absolutely *convergent*, which can be shown through the estimation of its limit (for $n \to \infty$) for its two neighboring absolute element values (d'Alembert principle).

On the other hand, successive integration of the function (2.47) by parts can yield *an infinite asymptotic series* of the form

$$f(\varepsilon) = \frac{\varepsilon e^{-1/\varepsilon^2}}{\sqrt{\pi}} \left[1 + \sum_{n=1}^{\infty} (-1)^n \frac{1 \cdot 3 \ldots \cdot (2n-1)}{2^n} \varepsilon^{2n} \right] = B \,.$$

Series B is divergent for arbitrary ε. In spite of that, since for small ε first terms of series B decrease quickly, one may obtain more accurate representation of $f(\varepsilon)$, than that achieved through the application of series A. The reason is that the convergence of series A is very slow. For example, for $\varepsilon = 0.01$ one may take two terms of series B in order to determine $f(\varepsilon)$ with a relative error of order 10^{-8}. The same accuracy is achieved by series A only if one takes into consideration its 280 elements (see Figs. 2.1, 2.2).

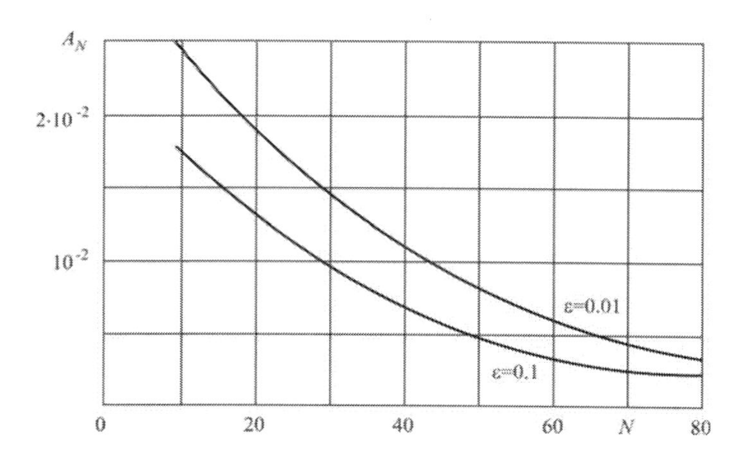

FIGURE 2.1: Dependence of the sum of elements of convergent series A on the number of its elements.

On the other hand, as it has been already mentioned, in the case of a divergent sequence B it is impossible to increase its accuracy through the increase of the number of its elements. Beginning from a certain number N, additional terms increase an estimation error of the function $f(\varepsilon)$. One may conclude from Fig. 2.2 that $N \approx 20$.

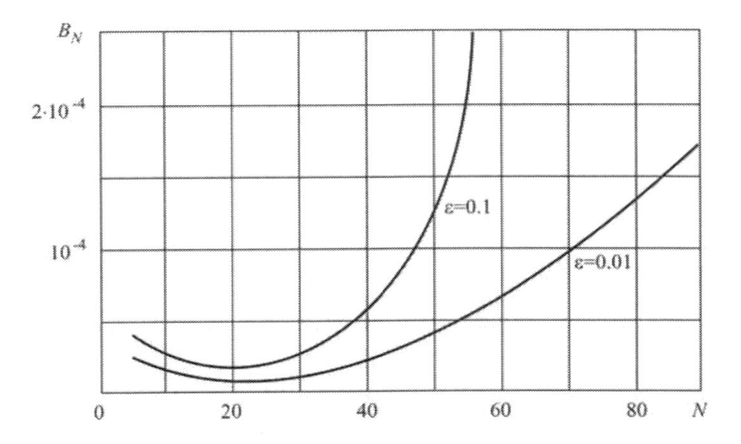

FIGURE 2.2: Sum of divergent series B.

Exercises

2.1. Determine with the accuracy of 0.001 and without mathematical tables the values of the following functions:

$$\cos 80^\circ, \quad \sin 36^\circ, \quad \tan 10^\circ. \quad \ln 1.1, \quad \ln 0.8, \quad \arctan 0.2, \quad e^{0.1}, \quad \sqrt{16.2}\ .$$

2.2. Compute approximate values with the accuracy of 0.001 of the integral $\int\limits_0^1 f(x)dx$, where $f(x)$ is the following :

$$e^{-x^2/2}, \quad e^{x^3/3}, \quad \frac{1}{x}\sin x, \quad \cos^3 x, \quad \arcsin x^2, \quad \tan x^3,$$

$$\frac{1}{2}\arctan\frac{x^2}{2}, \quad \ln\left(1+\frac{x^2}{3}\right)\ .$$

2.3. Determine first terms different from zero of an asymptotic decomposition into power series of a solution $y(x)$ of the following associated Cauchy problems:

$$y' = \cos x + y^2, \quad y(0) = 1; \quad y' = \sin x - y^2, \quad y(0) = 0\ ;$$

$$2y' = e^x - y, \quad y(0) = 1; \quad y' = x - \ln\left(1 - \frac{x}{2}\right), \quad y(0) = 1\ .$$

2.4. Using symbols of order "O capital," determine an order of magnitude of the below functions for $\varepsilon \to 0$:

$$25\varepsilon^{1/2}, \quad (2\pi+3)\varepsilon^2, \quad \ln(1-2\varepsilon), \quad \frac{1-\cos 2\varepsilon}{1+\sin 3\varepsilon},$$

$$1-\varepsilon^{5/2}-\cos(2\varepsilon),\quad \arcsin\left(\frac{\varepsilon}{(1+\varepsilon)^{3/2}}\right),\quad e^{\ln(1+2\varepsilon)}\ .$$

2.5. Introduce a decomposition of the following functions for $\varepsilon \to 0$:

$$2,\ 3\varepsilon,\ \frac{1}{\varepsilon},\ 3^{-1/\varepsilon},\ 2^{1/2},\ \frac{2}{\varepsilon^{5/2}},\ \sin\frac{\varepsilon}{\varepsilon^{5/2}},$$

$$\varepsilon^3\ln\varepsilon^{-2},\quad \arcsin(3\varepsilon^5)\ .$$

2.6. Determine two first terms different from zero of an asymptotic decomposition for $\varepsilon \to 0$, for the following functions:

$$\sqrt{1-\frac{\varepsilon}{3}+2\varepsilon^3},\quad \cos\sqrt[3]{1-\varepsilon t},\quad \sin\left(1-2\varepsilon+\varepsilon^3\right),\quad \ln\frac{1-3\varepsilon}{\sqrt[3]{1+2\varepsilon^2}},$$

$$e^{\ln(1-\varepsilon^3)},\quad 1+\varepsilon^3-\cos 2\varepsilon,\quad \arcsin\frac{2\varepsilon}{\sqrt[3]{1-\varepsilon}},\quad \ln\left(1+\ln(1+\sin\varepsilon)\right)\ .$$

2.7. Determine with the accuracy of second order the solutions to the following equations: $2x = 1+\varepsilon x^2$, $4-\varepsilon x^2 = x$, $2-\varepsilon x-4x^2 = 0$ and make their comparison for $\varepsilon = 0.1$ and $\varepsilon = 0.01$.

Chapter 3

Regular and singular perturbations

Experience earned during the analysis of various problems of applied mathematics, physics and engineering with the use of asymptotic approaches indicates that perturbation techniques can be divided into two main groups, i.e., those devoted to the analysis of *regular problems* and *singular* problems (*irregular*).

Regular problems are associated with the use of a classical method of small parameter, i.e., they allow for finding an asymptotic decomposition uniform in whole domain.

Singular problems (irregular) cannot be solved through a classical small parameter method only in some bounded domains of a sought function, and those subspaces are called *singular subspaces of asymptotic decompositions.*

Some introductory problems related to nonuniformity of an asymptotic decomposition have been already addressed in item 2.2.4. In this section, we aim at the brief introduction to illustrate and discuss methods to solve the problems associated with the occurrence of singular problems.

Nowadays, there are many various methods of the theory of singular perturbations allowing for the construction of uniformly applicable asymptotic decompositions in various branches of applied mathematics, physics and engineering. In what follows, some of these methods will be outlined. As it will further occur, some singular perturbations presented can be either completely independent, or some of them are matched through a different interpretation of the same idea. Recall that only few of the first terms of an asymptotic decomposition are usually taken for further considerations (even for complex problems, only two terms are taken). It is worth noticing that asymptotic series even if they are divergent give more advantages in comparison to uniform and absolutely convergent series (see section 2.2.6).

The organization of this section follows. In the first subsection, general definitions and descriptions of regular and singular perturbations are given. Then, some solutions to problems associated with vibrations of nonlinear one degree of freedom are constructed. Furthermore, an application of many methods of singular perturbations such as rescaled parameters, deformed variables, rescaling, full approximation, multiple scales variations of constants, and averaging are demonstrated and discussed. Special attention is paid to matching of asymptotic decompositions. In section 3.9, sources occurring in asymptotic decompositions are outlined. An overview of singular perturbations appearing in asymptotic decompositions is carried out in an object-lesson method without a rigorous mathematical treatment. In fact, the last one simply does not exist in many cases.

3.1 Introduction. Asymptotic approximation with respect to a parameter

Let us begin with rigorous definitions of regular and singular perturbations (see the end of chapter 1). For this purpose, we consider two equations defining a searched function $u(x)$ of the form:

$$\text{equation} \qquad E_0: \quad L_0[u] = f_0 \,, \tag{3.1}$$

$$\text{equation} \qquad E_\varepsilon: \quad L_0[u] + \varepsilon L_1[u] = f_0 + \varepsilon f_1 \,, \tag{3.2}$$

where: L_0, L_1 are known operators, $f_0(x)$, $f_1(x)$ are known functions, ε is a small (perturbation) parameter (we assume further that $\varepsilon > 0$). A searched function u and an independent variable x in general, can be vectors. The nonperturbed equation E_0 can be treated as the simplified model of a certain process, whereas the perturbed equation E_ε may play a role of an extended model of the mentioned process. Both terms $\varepsilon L_1[u]$ and εf_1 appearing in equation (3.2) are called *perturbations.*

Since equations E_0, E_ε are differential (ordinary or partial), one has to attach to the considered problem boundary and/or initial conditions (including a parameter ε).

Equations E_0, E_ε with defined boundary and initial conditions are referred to as *problems* E_0, E_ε, respectively. Assume that these problems are considered in a special space $D(x)$, tj. $u = u(x)$, $x \in D$. A solution to problem E_0 is denoted by $u_0(x)$, whereas a solution to the problem E_ε is denoted by $u_\varepsilon(x)$.

Fundamental question of the theory of singular perturbations follows: if the following difference $u_\varepsilon(x) - u_0(x) \to 0$ in a certain norm for $\varepsilon \to 0$. As a norm $||u(x)||$ of function $u(x)$ the Euclidean one will be taken, which for the vector $u(u_1(x), \ldots, u_k(x))$ reads $||u(x)|| = \sqrt{u_1^2(x) + \ldots + u_k^2(x)}$, and for a scalar function $u(x)$ it reads $||u(x)|| = |u(x)|$. Notice that the recalled norms depend on x ($x \in D$).

Definition 3.1. Problem E_ε is understood as *a regular perturbation*

$$\sup_D ||u_\varepsilon(x) - u_0(x)|| \to 0 \quad \text{for} \quad \varepsilon \to 0 \,. \tag{3.3}$$

Otherwise, problem E_ε is referred to as *a singular* one [170].

In formula (3.3) $\sup\limits_D f(x)$ denotes an upper branch of a function $f(x)$ in space D. In the case of regular perturbations a solution $u_0(x)$ of the problem

E_0 for small values of ε is close to the solution $u_\varepsilon(x)$ of problem E_ε in whole space D. In the case of singular perturbations of $u_0(x)$ for smallε, $u_0(x)$ is not situated close to $u_\varepsilon(x)$, in some part of the space D (this subspace is called *a subspace of nonuniformity of an asymptotic decomposition*).

The above considerations are illustrated by two simple examples.

Example 3.1. Find a solution to Cauchy problem of the following form

$$E_\varepsilon: \quad \frac{du}{dx} = -u + \varepsilon x\,, \quad x \in [0,1]\,, \quad u(0) = 1\,. \tag{3.4}$$

The linear first order differential equation (3.4) can be easily solved, and having satisfied the initial condition (3.4), one gets

$$u_\varepsilon(x) = (1+\varepsilon)e^{-x} + \varepsilon(x-1)\ . \tag{3.5}$$

The associated problem is obtained from (3.4), taking E_0 and it reads

$$E_0: \quad \frac{du}{dx} = -u\,, \quad x \in [0,1]\,, \quad u(0) = 1\,.$$

A solution to the above problem has the following form

$$u_0(x) = e^{-x}\ . \tag{3.6}$$

Substituting (3.5) and (3.6) into (3.3) yields

$$\sup_{[0,1]} \|u_\varepsilon(x) - u_0(x)\| = \varepsilon \max_{[0,1]} \left|e^{-x} + x - 1\right| \to 0 \quad \text{for } \varepsilon \to 0\ .$$

It means that according to the definition (3.3), the problem (3.4) is a regular one, and the perturbation εx is a regular one.

Example 3.2. Find a solution to the following Cauchy problem:

$$E_\varepsilon: \quad \varepsilon\frac{du}{dx} = -u + x\,, \quad x \in [0,1]\,, \quad u(0) = 1\,. \tag{3.7}$$

The linear differential equation (3.7) is solved in a similar way to that of (3.4). Satisfying the initial condition (3.7) gives

$$u_\varepsilon(x) = (1+\varepsilon)e^{-x/\varepsilon} + x - \varepsilon\ . \tag{3.8}$$

Note that the problem E_0 appeared in the following algebraic equation

$$E_0: \quad 0 = -u + x\,, \quad x \in [0,1]\,, \tag{3.9}$$

and therefore the initial condition can be omitted. Note also that

$$u_0(x) = x\ . \tag{3.10}$$

Substituting (3.5) and (3.10) into (3.3), for $\varepsilon \to 0$, one gets

$$\sup_{[0,1]} \|u_\varepsilon(x) - u_0(x)\| = \max_{[0,1]} \left|(1+\varepsilon)e^{-x/\varepsilon} - \varepsilon\right| = 1 \,.$$

According to the definition (3.3), problem (3.7) is singular, and, consequently, perturbation $\varepsilon \frac{du}{dx}$ is a singular one.

Drawings of solutions $u_0(x)$, $u_\varepsilon(x)$ for small $\varepsilon > 0$ are shown in Fig. 3.1.

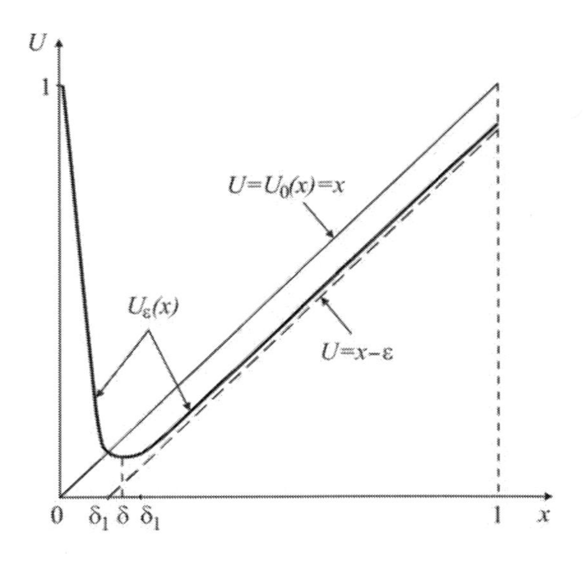

FIGURE 3.1: Graphs of solutions $u_0(x)$, $u_\varepsilon(x)$.

Figure 3.1 clearly exhibits a characteristic feature of singular perturbations. It is visible that in the area $D(x) = \{0 \le x \le 1\}$, where problems E_ε and E_0 (problems (3.7) and (3.9)) are defined, there is a subspace $[0, \delta)$, $\delta = \delta(\varepsilon)$, where the solution $u_0(x)$ v $u_\varepsilon(x)$, and a difference between them increases for $\varepsilon \to 0$.

In what follows, we are going to find a solution to the problem E_ε using a classical method of small parameter in the following asymptotic series:

$$u(x;\varepsilon) = \sum_{k=0}^{n} \varepsilon^k u_k(x) + 0(\varepsilon^{n+1}) = u_0(x) + \varepsilon u_1(x) + \ldots \tag{3.11}$$

Substituting (3.11) into (3.7), and equating to zero the coefficients standing by the same powers of ε, one gets: for ε^0: $u_0 = x$, for ε^1: $u_1 = -u_0' = -1$, for

ε^n: $u_n = -u'_{n-1} = 0$ $(n = 2, 3, \ldots)$. To sum up, in the series (3.11), according to the classical perturbation method, one obtains

$$u(x) = x - \varepsilon \,. \tag{3.12}$$

Notice that the asymptotic series (3.11) taking form of (3.12) is nonuniformly applied for $x = O(\varepsilon)$, because for $x = O(\varepsilon)$ two terms of expansion have the same order. For $0 \le x \le O(\varepsilon)$ this series is useless in further considerations.

Subspaces similar to those of $[0, \delta(\varepsilon)]$ from example 3.2 are usually referred to as *singular layers*. In order to construct nonuniformly exact solutions of problems with singular layers, or other singular problems, one has to apply special methods of the theory of singular perturbations. Many of them are illustrated in this chapter, and particularly in section 3.9 (in example 2, the method of combining the asymptotic series is shown).

It should be emphasized that during solving singular problems, the definitions of *uniform* and *nonuniform* asymptotic series play an important role (see 2.2.4).

Let us now introduce, using definition 3.1, a so-called asymptotic approximation along a parameter.

Let $u_\varepsilon(x)$ be a solution to problem E_ε, defined in space D, and let in a subspace $D_1 \subset D$ a certain function $u(x;\varepsilon)$ be defined.

Definition 3.2. A function $u(x;\varepsilon)$ is called *an asymptotic approximation with respect to parameter* ε of a solution $u_\varepsilon(x)$ in D_1, if

$$\sup_D \|u_\varepsilon(x) - u(x;\varepsilon)\| \to 0 \quad \text{for} \quad \varepsilon \to 0 \,.$$

If in addition $\sup_D \|u_\varepsilon(x) - u(x;\varepsilon)\| = O(\varepsilon^n)$, then it is said that $u(x;\varepsilon)$ is an asymptotic approximation for $u_\varepsilon(x)$ in D_1 with the accuracy of order of ε^n.

Let us explain the introduced definition using the examples given earlier. In Example 3.1, the function $u_0(x)$ occurring in (3.6) plays the role of an asymptotic approximation for solution $u_\varepsilon(x)$ in the whole interval $D(x) = \{0 \le x \le 1\}$ with the accuracy of ε. In Example 3.2, the function $u_0(x)$ occurring in (3.10) is the asymptotic approximation for $u_\varepsilon(x)$ with the accuracy of ε only in a subspace $D^0 = \{\delta(\varepsilon) < x < 1\}$, i.e., outside the boundary layer $D^i = \{0 \le x < \delta(\varepsilon)\}$.

Both construction and application of asymptotic methods of the theory of singular perturbations allows to construct uniformly exact asymptotic decomposition with the accuracy of required order of ε^n in the whole space D of problem E_ε.

3.2 Nonuniformities of a classical perturbation approach

Consider the following problem of *a Duffing's type*: Determine the function $u(t;\varepsilon)$, satisfying nonlinear differential equation and initial condition of the form

$$u'' + u + \varepsilon u^3 = 0\,, \quad 0 \le t < \infty\,, \tag{3.13}$$

$$u(0) = a\,, \quad u'(0) = 0\,. \tag{3.14}$$

The stated problem will be solved using a classical perturbation method. Function $u(t;\varepsilon)$ is represented by an asymptotic series with respect to parameter ε, characterizing the nonlinearity of spring of the form

$$u(t;\varepsilon) = u_0(t) + \varepsilon u_1(t) + \varepsilon^2 u_2(t) + O(\varepsilon^3)\,. \tag{3.15}$$

Differentiating (3.15) with respect to time t, one gets

$$u' = u_0' + \varepsilon u_1' + \varepsilon^2 u_2' + O(\varepsilon^3), \quad u'' = u_0'' + \varepsilon u_1'' + \varepsilon^2 u_2'' + O(\varepsilon^3)\,, \tag{3.16}$$

where $'$ denotes differentiation with respect to time t.

Limiting our considerations to the terms of order $O(\varepsilon^3)$ inclusively, one gets

$$u^3 = \left(u_0 + \varepsilon u_1 + \varepsilon^2 u_2 + O(\varepsilon^3)\right)^3 = u_0^3\left(1 + \varepsilon\frac{u_1}{u_0} + \varepsilon^2\frac{u_2}{u_0} + O(\varepsilon^3)\right)^3 =$$

$$= u_0^3\left[1 + 3\left(\varepsilon\frac{u_1}{u_0} + \varepsilon^2\frac{u_2}{u_0} + O(\varepsilon^3)\right) + 3\left(\varepsilon\frac{u_1}{u_0} + \varepsilon^2\frac{u_2}{u_0} + O(\varepsilon^3)\right)^2 + O(\varepsilon^3)\right] =$$

$$= u_0^3\left[1 + 3\varepsilon\frac{u_1}{u_0} + 3\varepsilon^2\left(\frac{u_2}{u_0} + \frac{u_1^2}{u_0}\right) + O(\varepsilon^3)\right]\,.$$

Observe that the above series is obtained through the application of formula (2.3). Substituting (3.15) and (3.16), as well as the lastly obtained series into (3.13), (3.14) and through the comparison of coefficients standing by same powers of ε.

In the case of terms without ε the following zero order problem is obtained

$$u_0'' + u_0 = 0, \quad u_0(0) = a, \quad u_0'(0) = 0\,. \tag{3.17}$$

In the case of terms with first power of ε the first order approximation may be found while solving the following problem (for correction $u_1 t$):

$$u_1'' + u_1 = -u_0^3, \quad u_1(0) = 0, \quad u_1'(0) = 0\,. \tag{3.18}$$

The terms associated with ε^2 yield the second order approximation (for correction $u_2 t$) defined by the equation:

$$u_2'' + u_2 = -3u_0^2 u_1\,, \quad u_2(0) = 0\,, \quad u_2'(0) = 0\,. \tag{3.19}$$

Analogically, one may obtain an equation yielding $u_3(t)$, etc. Notice that all obtained problems (3.17)–(3.19) are linear, although beginning from the second equation they are nonhomogeneous. The complexity of nonhomogeneity increases with the increase of computation steps and depends on the solutions obtained in previous steps.

In our further considerations, only the first and second order approximations will be computed.

The differential equation (3.17) is a particular case of differential equation (1.3), and its solution has the following general form

$$u_0 = u_0(t; c_1, c_2) = c_1 \cos t + c_2 \sin t \ . \tag{3.20}$$

Let us transform the solution (3.20) to a more suitable form

$$u_0 = u_0(t; a, \beta) = a\cos(t + \beta), \quad \text{where } a = \sqrt{c_1^2 + c_2^2}, \ \ \beta = -\arctan\frac{c_2}{c_1} \ .$$

The initial conditions (3.20) allow to define constants: $c_1 = a$, $c_2 = 0$, and the solution to *a zero order approximation of Duffing equation* reads

$$u_0(t) = a\cos t \ . \tag{3.21}$$

The solution (3.21) governs harmonic oscillations with constant amplitude a and frequency $\omega_0 = 1$,which does not correspond to real oscillation process governed by equation (3.13). In order to increase the accuracy of solution (3.21), *the first order approximation* problem yielding correction $u_1(t)$ will be analyzed.

Substituting (3.21) into (3.18), we obtain

$$u_1'' + u_1 = -a^3\cos^3 t \ .$$

In order to find a particular solution of this linear nonhomogeneous equation, one has to expand $\cos^3 t$ using trigonometric transformations of the form

$$\cos 3\alpha = \cos(\alpha + 2\alpha) = \cos\alpha\cos 2\alpha - \sin\alpha\sin 2\alpha =$$

$$= \cos\alpha(2\cos^2\alpha - 1) - 2\sin^2\alpha\cos\alpha =$$

$$= 2\cos^3\alpha - \cos\alpha - 2(1-\cos^2\alpha)\cos\alpha = 4\cos^3\alpha - 3\cos\alpha \ .$$

Recall that

$$\cos^3\alpha = \frac{1}{4}(3\cos\alpha + \cos 3\alpha) \ .$$

As a result, instead of (3.18) one gets

$$u_1'' + u_1 = -a^3\cos^3 t \equiv -\frac{a^3}{4}(3\cos t + \cos 3t), \quad u_1(0) = u_1'(0) = 0 \ . \tag{3.22}$$

As it is well known, a solution to the linear differential homogeneous equation (3.22) consists of a sum of a general solution u_{10} of homogeneous equation and a certain particular solution u_{1*} of nonhomogeneous equation of the form

$$u_1 = u_{10} + u_{1*} \ . \tag{3.23}$$

A general solution reads

$$u_{10}(t; c_3, c_4) = c_3 \cos t + c_4 \sin t \ . \tag{3.24}$$

A particular solution u_{1*} will be sought. Using the method of nondefined coefficients and knowing the right-hand side of equation (3.22), one gets [60]:

$$u_{1*} = At \sin t + B \cos 3t \, ; \quad \{A, B\} = const \ . \tag{3.25}$$

Differentiating (3.25) two times and substituting the result into (3.22), one gets first $A = -3a^3/8$, $B = a^3/32$, and then, according to (3.23)–(3.25), the following is obtained

$$u_1(t; c_3, c_4) = c_3 \cos t + c_4 \sin t - \frac{3}{8} a^3 t \sin t + \frac{a^3}{32} \cos 3t \ .$$

Initial conditions (3.22) yield $c_3 = -a^3/32$, $c_4 = 0$.

Finally, the solution to (3.22) reads

$$u_1(t) = \frac{a^3}{32}(\cos 3t - \cos t) - \frac{3a^3}{8} t \sin t \ . \tag{3.26}$$

Substituting the determined u_0 and u_1 into (3.15) the following asymptotic decomposition of the searched function with the accuracy of $O(\epsilon)$ is found

$$u(t; \varepsilon) = a \cos t + \varepsilon \left[\frac{a^3}{32}(\cos 3t - \cos t) - \frac{3a^3}{8} t \sin t \right] + O(\varepsilon^2) \ . \tag{3.27}$$

A solution to *the second approximation* (3.19) is sought in an analogous way. However, due to its complexity it will not be given explicitly. It is worth noticing that beside typical trigonometric functions, its terms are proportional to the combinations of $t \sin t$, $t \sin 3t$, $t^2 \cos t$.

It occurs that the solution (3.27), obtained through a classical method of small parameter t, is not uniformly suitable in the whole range of independent value definition, i.e., it is nonhomogeneous with respect to t. The first term of the asymptotic decomposition (3.27) is of order 1, and according to the classical perturbation method, the second term of series (3.27) should be of order $O(1)$, i.e., the introduced correction should be essentially smaller than the first determined term. Since the expression in square brackets (3.27) is proportional to t, it may unboundedly increase with $t \to \infty$, i.e., $\varepsilon u_1 / u_0 \to \infty$ for $t \to \infty$ (ε is a fixed value), which contradicts the idea of perturbation parameter. From (3.27), one may conclude that for $\varepsilon t \sim O(1)$, or equivalently for $t \sim O(1/\varepsilon)$, the second term of decomposition (3.27) is of order of the first term, which contradicts the assumption of perturbation method.

In other words, decomposition (3.27) is uniformly useful only for such values of time t, for which $\varepsilon t < O(1)$, i.e., for $t < O(1/\varepsilon)$.

One may also show that during successive computations of terms of asymptotic decomposition (3.15), they consist also of the terms proportional to $t^n \cos t$, $t^n \sin t$, $n = 2, \ldots$, which increase unboundedly with $t \to \infty$.

These terms are called *secular* (in French *siécle* denotes a century). This name appeared for the first time during the investigation of various problems in astronomy. In the mentioned problems, the terms of order εt appear. However, in astronomy, ε is usually very small and the product εt begins to play an essential role in computations after a long time of centuries order.

It should be emphasized that the occurrence of secular terms is characteristic for nonlinear problems of vibrations. For this reason, in order to have an asymptotic decomposition uniformly suitable with respect to time t, secular terms are not allowed. Notice that according to the theory of finding particular solutions of second order differential equations with constant coefficients [119], the reason for the occurrence of a secular term in $u_1(t)$ the existence of a term proportional to $\cos t$ (see the right-hand side of the equation 3.22). Secular terms may appear also in this case if nonhomogeneous part includes terms proportional to $\sin t$.

However, the stated problem can be solved using the method of "elongated" parameters.

3.3 Method of "elongated" parameters

Method of "elongated" parameters is sometimes called *Lindstedt-Poincaré method*, due to creators.

Consider the following differential equation

$$u'' + \omega_0^2 = \varepsilon f(u, u') \ , \quad \varepsilon << 1 \ , \tag{3.28}$$

governing the nonlinear vibrations of a certain nonlinear system. In the above, $u(t; \varepsilon)$ denotes a deviation from initial position defined through the initial condition $t = 0$, $0 \leq t < \infty$, ω_0 is the frequency of vibrations of a linear system, ε is a small parameter characterizing the nonlinearity of a system, f is an arbitrary nonlinear function.

Duffing's equation (3.13) is a particular case of equation (3.28), i.e., for the frequency of the linear system equal to $\omega_0 = 1$, and for $f(u, u') = -u^3$.

A solution to equation (3.28) using small parameter includes secular terms in the asymptotic series (3.15) in $u(t; \varepsilon)$ (see, for instance, 3.27).

It is well known that nonlinearity changes the frequency of the considered system, beginning with the value of ω_0 corresponding to a linear system up to the value of $\omega(\varepsilon)$ slightly differing from ω_0. In order to determine this dependence, in 1882, Lindstedt introduced the frequency $\omega(\varepsilon)$ directly to the equation (3.28) through a transformation of independent variable

$$s = \omega(\varepsilon)t \ , \tag{3.29}$$

where $\omega(\varepsilon)$ is a certain constant introduced only through the value of small parameter ε. In the next step, the amplitude and frequency of oscillations are developed into series with respect to small parameter ε of the following form

$$u(t;\varepsilon) = u_0(s) + \varepsilon u_1(s) + \varepsilon^2 u_2(s) + \dots \ , \tag{3.30}$$

$$\omega = \omega_0 + \varepsilon\omega_1 + \varepsilon^2\omega_2 + \dots \ , \tag{3.31}$$

$$s = \omega t = (\omega_0 + \varepsilon\omega_1 + \varepsilon^2\omega_2 + \dots)t \ . \tag{3.32}$$

Notice that decomposition (3.32), where $\omega_1, \omega_2, \dots$ are arbitrary numbers, may be interpreted as a certain "elongation" of parameter ω.

Arbitrary numbers (*elongation coefficients*) $\omega_1, \omega_2, \dots$ may be chosen through some established conditions. The choice of coefficients $\omega_1, \omega_2, \dots$ recommended by Lindstedt relies on omitting secular terms in series (3.32) i.e., it guaranties a uniform suitability with respect to t. Poincaré proved that the series (3.30) proposed by Lindstedt belongs to asymptotic series.

It should be emphasized that various forms of the same method are applied in different branches of physics and engineering (see examples provided in monographs [118, 119]). The main idea of this method relies on both appropriate choice of a parameter associated with a magnitude of perturbation and on the development of dependent variables with respect to the mentioned parameter (see for instance 3.30 and 3.31). Coefficients $\omega_1, \omega_2, \dots$ (elongation coefficients of parameter ω or another parameter) are chosen to secure the uniformly suitable asymptotic series (3.30). The mentioned approach is called the method of *"elongated" parameters*.

Let us illustrate a technique of elongated parameters method using an example of Duffing's problem (3.13) and (3.14).

Let us first introduce a new independent variable s of the form

$$s = \omega t \ , \tag{3.33}$$

where ω is a certain constant value depending only on the magnitude of perturbation ε. Let us introduce the solution $u(t;\varepsilon) \equiv F(s;\varepsilon)$ and parameter ω in the form of an asymptotic series

$$u(t;\varepsilon) \equiv F(s;\varepsilon) = F_0(s) + \varepsilon F_1(s) + \varepsilon^2 F_2(s) + \dots \ , \tag{3.34}$$

$$\omega = 1 + \varepsilon\omega_1 + \varepsilon^2\omega_2 + \dots \tag{3.35}$$

In the case of a Duffing's oscillator, the frequency of linear oscillations of the system (first term of series 3.35) reads $\omega_0 = 1$.

Let us now use an independent variable s instead of t, and let us apply the following differentiation rule

$$\frac{du}{dt} = \frac{dF}{dt} = \frac{dF}{ds} \cdot \frac{ds}{dt} = \omega F' , \quad u'' = \omega^2 F'' ,$$

where primes denote differentiation with respect to variable s. Duffing's problem (3.13) and (3.14) takes the following form

$$\omega^2 F'' + F + \varepsilon F^3 = 0 , \qquad F(0) = a , \quad F'(0) = 0 . \tag{3.36}$$

Observe that the frequency $\omega(\varepsilon)$ is directly introduced to a differential equation.

Substituting (3.34) and (3.35) into (3.36) yields

$$F_0'' + 2\varepsilon\omega_1 F_0'' + \varepsilon F_1'' + \ldots + F_0 + \varepsilon F_1 + \ldots + \varepsilon F_0^3 + \ldots = 0 .$$

Comparing the terms standing by ε^0 and ε^1, the following zero and first order approximations are obtained

$$F_0'' + F_0 = 0 , \quad F_0(0) = a , \quad F_0'(0) = 0 , \tag{3.37}$$

$$F_1'' + F_1 = -F_0^3 - 2\omega_1 F_0'' , \quad F_1(0) = F_1'(0) = 0 . \tag{3.38}$$

A solution to the problem (3.37) coincides with the solution (3.21) to the problem (3.18), obtained with the use of a small parameter method, since in zero order approximation, the "elongation" with respect to parameter does not appear. Let

$$F_0(s) = a \cos s . \tag{3.39}$$

Substituting (3.39) into (3.38) and using some trigonometric identities, one gets

$$F_1'' + F_1 = -a^3 \cos^3 s + 2a\omega_1 \cos s \equiv 2a\omega_1 \cos s - \frac{a^3}{4}(\cos 3s + 3\cos s)$$

or

$$F_1'' + F_1 = a \cos s \left(2\omega_1 - \frac{3}{4}a^2\right) - \frac{a^3}{4} \cos 3s . \tag{3.40}$$

Since during the solution of equation (3.40) a secular term is generated by the term proportional to $\cos s$, standing on the right hand side of the equation (3.40), it should be removed.

As a result, the following condition yielding unknown elongation coefficient ω_1, is obtained

$$2\omega_1 - \frac{3}{4}a^2 = 0 \quad \Rightarrow \quad \omega_1 = \frac{3}{8}a^2 . \tag{3.41}$$

(Trivial case is excluded from our considerations.)

Satisfying the condition (3.41), equation (3.40) yields the following equation

$$F_1'' + F_1 = -\frac{a^3}{4}\cos 3s, \quad F_1(0) = F_1'(0) = 0 \ . \tag{3.42}$$

In the next step one gets

$$F_1 = F_{10} + F_{1*}, \quad F_{10} = -c_1 \cos s + c_2 \sin s \ . \tag{3.43}$$

However, in this case a particular solution to nonhomogeneous differential equation (3.42) is sought in the form

$$F_{1*} = B\cos 3s \ .$$

One gets $B = a^3/32$ and the fulfillment of the initial conditions (3.42) yields

$$F_1(s) = \frac{a^3}{32}(\cos 3s - \cos s) \ . \tag{3.44}$$

Substituting (3.39), (3.44) and (3.41) into (3.34) and (3.35), the following *uniformly suitable solution to Duffing's problem* with respect to t and with the accuracy of $O(\varepsilon)$ is obtained

$$u(t;\varepsilon) = a\cos s + \varepsilon\frac{a^3}{32}(\cos 3s - \cos s) + O(\varepsilon^2) \ , \tag{3.45}$$

$$s = t\left(1 + \frac{3a^2}{32}\varepsilon + O(\varepsilon^2)\right) \ . \tag{3.46}$$

Formula (3.46) yields the frequency dependence on ε and this dependence can be improved through computations of successive approximations to a real value.

Notice that during the construction of an asymptotic decomposition (3.45) (the same holds for 3.27), while solving the nonhomogeneous equation (3.42) with respect to F_1, the solution F_{10} of the associated homogeneous equation (term $-\varepsilon a^3 \cos s/32$ in 3.45) with arbitrary constants c_1, c_2 (independent on ε) has been taken into account in a general solution. One may, however, proceed in a different manner. Namely, in all formulas of order ε, ε^2, one may omit initial conditions except for the solution u_0, and treat the associated arbitrary conditions as dependent on ε. Therefore, a solution to (3.45) takes the form

$$u(t;\varepsilon) = a\cos(s+\beta) + \varepsilon\frac{a^3}{32}(\cos 3s + 3\beta) + O(\varepsilon^2) \ . \tag{3.47}$$

In the remaining part of the book the last approach is used.

3.4 Method of deformed variables

The main idea of deformed variables, often called *PLG method* (it comes from abbreviation of names of H. Poincaré, M. Lighthill and Y. Guo), is introduced in references [103, 104, 119, 120, 121, 122, 161]. In our considerations, we consider only the case of a function of two independent variables x, y and small parameter ε. Let the function $u(x, y; \varepsilon)$ satisfy a certain partial differential equation and the given boundary conditions. According to the classical small parameter approach, a sought solution is introduced in the form of the asymptotic decomposition

$$u(x, y; \varepsilon) = u_0(x, y) + \varepsilon u_1(x, y) + \varepsilon^2 u_2(x, y) + O(\varepsilon^3); \quad \varepsilon \ll 1 \tag{3.48}$$

and let us assume that this series is uniformly suitable with respect to one of the variables, for instance x. In this case, both the function u and the variable ε are developed into a perturbation series through the introduction of a new independent variable X of the form

$$u(x, y; \varepsilon) \equiv U(X, y; \varepsilon) = U_0(X, y) + \varepsilon\, U_1(X, y) + \varepsilon^2\, U_2(X, y) + O(\varepsilon^3), \tag{3.49}$$

$$x = X + \varepsilon \mu_1(X) + \varepsilon^2 \mu_2(X) + O(\varepsilon^3). \tag{3.50}$$

Functions $\mu_1(X), \mu_2(X), \ldots$ occurring in (3.50) are referred to as *deformation functions*. Recall that in the method of elongated parameters, in transformation (3.32), $\omega_1, \omega_2, \ldots$ play the role of unknown constants. From this point of view, the method of deformed variables may be treated as a generalization of the method of "elongated" parameters. If a uniformly suitable solution is obtained through the method of "elongated" parameters, then it always may be obtained through the method of deformed variables, but not the other way round.

Let us apply (3.49), (3.50) and let us compute partial derivatives using the following rules: $u_x = U_x \frac{dX}{dx} = U_x / \left(\frac{dx}{dX}\right)$, $u_y = U_y$, and so on. As a result, instead of the initial problem, one gets a problem yielding $U(X, y; \varepsilon)$. We focus on the construction of uniformly suitable asymptotic series of the form of (3.14). Notice that $U_0(X, y) = u_0(x, y)$. It means that a deformation does not achieve zero order approximation and the solution U_0 coincides with u_0. A problem yielding the function $U_1(X, y)$ will include an arbitrary deformation $\mu_1(X)$. Before the determination of U_1, one takes μ_1 to bound the ratio of U_1/U_0. In general case, we also require that the ratios of U_n/U_{n-1}, μ_n/μ_{n-1}, $n = 1, 2, \ldots$ should be bounded, i.e., the successive approximation should be less singular than the previous one.

Let us apply the method of deformed variables to solve Duffing's problem (3.13), (3.14). Both dependent and independent variables are sought in the form

$$u(t; \varepsilon) \equiv F(S; \varepsilon) = F_0(S) + \varepsilon F_1(S) + O(\varepsilon^2), \tag{3.51}$$

$$t = S + \varepsilon\mu_1(S) + O(\varepsilon^2) \ . \tag{3.52}$$

According to the rule of differentiation of complex and inversed functions, one gets

$$u' = \frac{du}{dt} = \frac{du}{dS}\frac{dS}{dt} = \frac{dF}{dS}\bigg/\frac{dt}{dS} = \frac{F_0' + \varepsilon F_1' + O(\varepsilon^2)}{1 + \varepsilon\mu_1' + O(\varepsilon^2)} =$$

$$F_0' + \varepsilon(F_1' - \mu_1' F_0') + O(\varepsilon^2) \ ,$$

$$u'' = F_0'' + \varepsilon(F_1'' - \mu_1'' F_0 - 2\mu_1' F_0'') + O(\varepsilon^2) \ . \tag{3.53}$$

Substituting (3.51) and (3.53) into (3.13) and (3.14), and equating to zero the coefficients standing by the same powers of ε, the following zero order (terms standing by ε^0) and first order (terms standing by ε^1) approximations are obtained

$$F_0'' + F_0 = 0 \ ; \quad F_0(0) = a \ , \quad F_0'(0) = 0 \ , \tag{3.54}$$

$$F_1'' + F_1 = \mu_1'' F_0 - 2\mu_1' F_0'' - F_0^3 \ ; \quad F_1(0) = F_1'(0) \approx 0 \ . \tag{3.55}$$

Notice that the first order approximation is not deformed

$$F_0 = a\cos S \ . \tag{3.56}$$

Substituting (3.56) into (3.55), the following equation yielding F_1 is obtained

$$F_1'' + F_1 = a\cos S\left(\mu_1'' + 2\mu_1' + \frac{3}{4}a^2\right) + \frac{a^3}{4}\cos 3S \ . \tag{3.57}$$

Since in the right-hand side of equation (3.57), a term proportional to $\cos S$, appears, i.e., a secular term occurs in F_1 (see 3.29), therefore on function $\mu_1(S)$ the following requirement is added

$$\mu_1'' + 2\mu_1' + \frac{3}{4}a^2 = 0 \ . \tag{3.58}$$

The simplest choice of $\mu_1(S) = AS$, where $A = const$, yields $\mu_1(S) = -3a^2S/8$. Owing to this choice of μ_1, the solution (3.44) serving for the determination of F_1 is obtained. Substituting the found values of F_0, F_1, μ_1 into (3.51) and (3.52), a solution coinciding with either (3.45), (3.46) or (3.47) constructed with the method of "elongated" parameters is obtained. It is worth noticing that certain ambiguity of the choice of deformation is a characteristic feature of the method, and may be actually treated as its advantage.

Nowadays, the method of deformed values is widely used during the analysis of nonlinear phenomena in various branches of fluid mechanics, aerodynamics, impacting waves, theory of elasticity, physics, i.e., during the solution of various complex problems of nonlinear vibrations governed by either ordinary or partial differential equations [118, 119, 120, 121, 122, 123].

In chapter 4 of this book, we give some more examples of this method. In what follows, we reconsider the model example given by Lighthill of the following form [166, 167]:

$$(x + \varepsilon y)\frac{dy}{dx} + q(x)y = r(x) \ .$$

We focus on finding the solution $y(x;\varepsilon)$ to the following initial problem

$$(x + \varepsilon y)y' + y = 1\,; \quad y(1) = 2\ . \tag{3.59}$$

The choice of the illustrated example is two folded. Namely, one may compare results yielded by different singular perturbation methods, and their errors with respect to an exact solution. Notice that the equation (3.59) has *a complex singularity* located along straight line $x = -\varepsilon y$ (Fig. 3.2). This singularity, as it has been mentioned in previous section, causes the occurrence of nonuniformity while applying a classical perturbation technique.

The equation (3.59) has the form of $y' = f\left(\frac{ax+by+c}{\alpha x+\beta y+\gamma}\right)$, where $a = \gamma = 0$, $b = -1$, $c = \alpha = 1$, $\beta = \varepsilon$. The method of its solution is given, for instance, in reference [85, p. 39].

In our case, $\Delta = a\beta - b\alpha = 1 \neq 0$, and hence the following variables transformation is applied: $au + bv = ax + by + c$, $\alpha u + \beta v = = \alpha x + \beta y + \gamma$. Cramer's formulas yield: $x = u + (b\gamma - c\beta)/\Delta$, $y = \nu + (c\alpha - a\gamma)/\Delta$.

As a result, the homogeneous first order differential equation is obtained $\frac{dv}{du} = f\left(\frac{a+bv/u}{\alpha+\beta v/u}\right)$, which can be solved through a standard transformation $\frac{\nu}{u} = z(u)$. Finally, satisfying the initial condition (3.59) yields the following exact solution of the form [166]

$$y = \sqrt{\left(\frac{x}{\varepsilon}\right)^2 + 2\frac{1+x}{\varepsilon} + 4} - \frac{x}{\varepsilon}\ . \tag{3.60}$$

Let us apply the classical perturbation technique to find a solution in the following form $y(x;\varepsilon) = \sum\limits_{n=0}^{N} \varepsilon^n y_n(x) + O(\varepsilon^{N+1})$. Substituting this series into (3.59) and equating to zero the terms standing by the same powers of and comparing terms standing by same powers of ε, the following approximation equations are obtained

$$xy_0' + y_0 = 1\,, \quad y_0(1) = 2\,;$$

$$xy_1' + y_1 = -y_0y_0'\,, \quad y_1(1) = 0\,;$$

$$xy_2' + y_2 = -y_1y_0' - y_0y_1'\,, \quad y_2(1) = 0\,.$$

After solving successively three linear first order differential equations and satisfying initial conditions problem (3.59) has the following solution

$$y(x;\varepsilon) = \frac{1+x}{x} - \varepsilon\frac{(1-x)(1+3x)}{2x^3} + \varepsilon^2\frac{(1+x)(1-x)(1+3x)}{2x^5} + O(\varepsilon^3)\ . \tag{3.61}$$

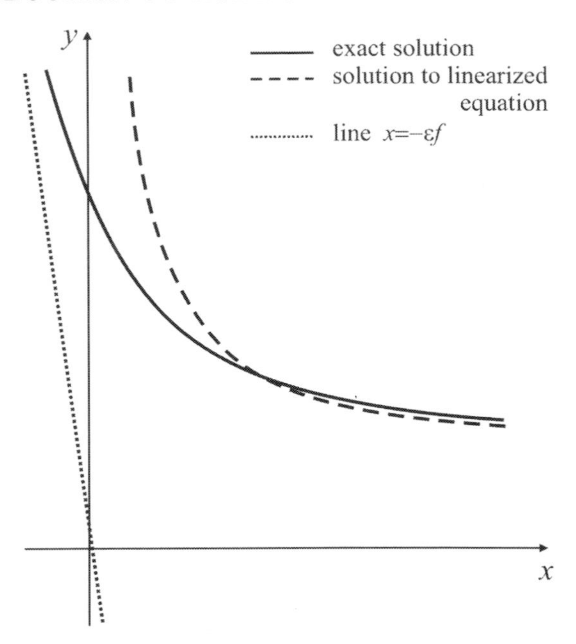

FIGURE 3.2: Integral curves of equation (3.59).

Although series (3.61) is convergent, its radius of convergence tends to zero for $x \to 0$. The series (3.61) is uniformly suitable for small values of x. For $x = O(\varepsilon^{1/2})$, all its terms have the same order of magnitude of $O(\varepsilon^{-1/2})$, whereas for $x < O(\varepsilon^{1/2})$, its next terms exceed the previous ones.

Let us apply now the method of deformed variables. We introduce *a deformed variable* S and we have developed both independent x and dependent y variables into series of the form

$$y(x;\varepsilon) \equiv F(S;\varepsilon) = F_0(S) + F_1(S) + O(\varepsilon^2) \ , \tag{3.62}$$

$$x = S + \varepsilon\mu_1(S) + \varepsilon^2\mu_2(S) + O(\varepsilon^3) \ . \tag{3.63}$$

In what follows, our construction is limited to the uniformly suitable solution for F_0.

We are going to substitute (3.62) and (3.63) into (3.59), and to apply the method of deformed variables. Notice that

$$y'(x) = F'(S)\frac{dS}{dx} = F'(S)\frac{1}{dx/dS} = \frac{F'(S)}{1+\varepsilon\mu_1' + \ldots} = F'(S)(1-\varepsilon\mu_1' + \ldots) \ .$$

Equation (3.59) yields

$$(S+\varepsilon\mu_1+\ldots+\varepsilon F_0+\ldots)(F_0'+\varepsilon F_1'+\ldots)(1-\varepsilon\mu_1'+\ldots)+F_0+\varepsilon F_1+\ldots = 1$$

Equating to zero the terms standing by ε^0 and ε, one gets

$$SF_0' + F_0 = 1 \ ,$$

$$SF_1' + F_1 = F_0'S\mu_1' - \mu_1 F_0' - F_0 F_0' \equiv \mu_1'(1 - F_0) - \mu_1 F_0' - F_0 F_0' \equiv$$

$$\equiv \left(\mu_1(1 - F_0) - \frac{1}{2}F_0^2 + c_1\right)' \ .$$

In order to determine F_0, equations $(SF_0)' = 1$, $F_0(1) = 2$ are used, and their solution coincides with solution y_0 obtained through (3.61)

$$F_0(S) = \frac{1+S}{S} \ . \tag{3.64}$$

A solution yielding F_1 with an accent of (3.64) reads

$$(SF_1)' = -\left(\frac{\mu_1}{S} + \frac{1 + 2S + s^2}{2S^2} + c_1\right)' \ ,$$

where: $F_1 = \frac{c}{S} - \frac{1}{S^2}\left(\mu_1 + \frac{1+2S}{2S}\right)$.

In order to keep for $S \to 0$ the order of singularity of F_1 lower than that of F_0, it is required that a singularity of second term in F_1 (for $S \to 0$) does not achieve $O\left(\frac{1}{S}\right)$.

This requirement can be satisfied through a proper choice of so far arbitrary *deformation coefficient* $\mu_1(S)$. One may realize this deformation in different ways.

The most natural way gives

$$\mu_1 = -\frac{1+2S}{2S} \ , \tag{3.65}$$

and it yields the uniformly suitable first order approximation of the form

$$f(x;\varepsilon) = \frac{1+S}{S} + O(\varepsilon) \ , \tag{3.66}$$

$$x = S - \varepsilon\frac{1+2S}{2S} + O(\varepsilon^2) \ . \tag{3.67}$$

Solving the obtained squared equation with respect to S and choosing root $S = x$ for $\varepsilon = 0$, one gets

$$S = \frac{x + \varepsilon + \sqrt{(x+\varepsilon)^2 + 2\varepsilon}}{2} \ . \tag{3.68}$$

Next, by multiplying numerator and denominator of the equation $x + \varepsilon - \sqrt{(x+\varepsilon)^2 + 2\varepsilon}$, the following first order uniformly suitable approximation is obtained

$$f(x;\varepsilon) = 1 + \frac{1}{S} = \sqrt{\left(\frac{x}{\varepsilon}\right)^2 + \frac{2}{\varepsilon}(1+x) + 1} - \frac{x}{\varepsilon} \ . \tag{3.69}$$

Although the obtained solution does not fully coincide with an exact one, both of them coincide well with magnitude order of $O(\varepsilon)$.

Process of solution corrections may be further extended, and the problem related to the ambiguity of deformation choice will be discussed in the next item.

It occurs that the illustrated method of deformed variables is tedious in applications during the determination of higher order approximations, as well as during the introduction of more independent variables.

Let $u(x, y; \varepsilon)$ satisfy a certain partial differential equation and let the series (3.48) be uniformly suitable with respect to both x, and y. According to the method of deformed coordinates, a solution is sought in the form

$$u(x, y; \varepsilon) \equiv U(X, Y; \varepsilon) = U_0(X, Y) + \varepsilon U_1(X, Y) + \varepsilon^2 U_2(X, Y) + O(\varepsilon^3) \ , \quad (3.70)$$

$$x = X + \varepsilon \mu_1(X, Y) + \varepsilon^2 \mu_2(X, Y) + O(\varepsilon^3) \ , \quad (3.71)$$

$$y = Y + \varepsilon \nu_1(X, Y) + \varepsilon^2 \nu_2(X, Y) + O(\varepsilon^3) \ . \quad (3.72)$$

In this case, μ_i, ν_i; $i = 1, 2, \ldots$ are arbitrary functions of X, Y. Let us determine partial derivatives U_X, U_Y in accordance with the rule of differentiation of a complex function (now and later, a subscript denotes a partial derivative)

$$U_X = u_x x_X + u_y y_Y \ , \quad U_Y = u_x x_Y + u_y y_Y \ . \quad (3.73)$$

Two equations of (3.73) are linear algebraic equations with two unknowns u_x, u_y, where U_X, U_Y are defined through (3.70), whereas x_X, x_Y, y_X, y_Y are determined from the equations (3.71), (3.72). Applying Cramer's rule, one gets

$$\begin{gathered} u_x = U_{0X} + \varepsilon(U_{1X} - U_{0Y} y_{1X} - U_{0X} x_{1X}) + \\ + \varepsilon^2 \left[U_{2X} - x_{1X} U_{1X} - y_{1X} U_{1Y} + U_{0X}(x_{1X}^2 + x_{1Y} y_{1X} - x_{2X}) + \right. \\ \left. + U_{0Y}(x_{1X} y_{1X} + y_{1X} y_{1Y} - y_{2X}) \right] + O(\varepsilon^3) \ . \end{gathered}$$

The derivative u_y, is similar, whereas the derivative u_{xx} consists of 27 terms. In what follows, the construction of solution in the form of a partial differential equation in new deformed variables U, X, Y is complicated, and hence a modification of this method has been proposed.

3.5 Method of scaling and full approximation

Method of scaling is simpler than the method of deformed coordinates. The advantages of its use are clearly outlined when one deals with only one independent variable. This method enables also to achieve good results in the case of problems requiring the deformation of two independent variables, i.e., when an application of classical method of deformed coordinates usually fails (see section 3.3.4). Method proposed in reference [132], especially after the publication of the review paper [62], evoked the interest of many researchers. The main idea of this method is the introduction of deformations of independent variables directly into the series obtained with the use of classical method of small parameter.

In reference [62], the scaling method is illustrated by various examples consisting of a few variables. However, in order to introduce, both in a simple and illustrative way, the discussed method, two cases with one and two independent coordinates are further discussed. The first example is studied in more detail, whereas in the second one only final results is given.

3.5.1 Deformation of one independent variable

Let a certain unknown function $f(x, x_i; \varepsilon)$, where x, x_i $(i = 1, \ldots, n)$ are independent variables, and ε is a small parameter, satisfy a certain nonlinear equation(s) and certain boundary and/or initial conditions. In a general case, the analytical solution of this problem is not possible. In this case, the classical method of small parameter is applied, but a yielded solution is usually uniformly unsuitable with respect to one or few independent variables.

Assume that after application of a classical perturbation technique, the following asymptotic solution is constructed

$$f(x; \varepsilon) = f_0(x) + \varepsilon f_1(x) + \varepsilon^2 f_2(x) + O(\varepsilon^3) , \tag{3.74}$$

which is nonuniformly applicable with respect to x (variables for the sake of simplicity are omitted).

Let us apply multiple scaling. Let us introduce *the deformation* of variable X and function $F(X; \varepsilon)$ of the form

$$x = X + \varepsilon\mu_1(X) + \varepsilon^2\mu_2(X) + O(\varepsilon^3) , \tag{3.75}$$

$$f(x; \varepsilon) \equiv F(X; \varepsilon) = F_0(X) + \varepsilon F_1(X) + \varepsilon^2 F_2(X) + O(\varepsilon^3) , \tag{3.76}$$

where μ_1, μ_2, ... are arbitrary functions, i.e., *deformation coefficients*.

The originality of the discussed method lies mainly in the direct introduction of deformation (3.75) and (3.76) into series (3.74). The found solution is transformed into a power series with respect to ε using Taylor formulas and then it is corrected by deformation coefficients μ_1, μ_2, ... in order to achieve a uniformly suitable series. In other words, deformation coefficients are chosen to eliminate singular terms of asymptotic expansion (3.76).

Substituting (3.75) into (3.74), one gets

$$f(x; \varepsilon) = f_0\left(X + \varepsilon\mu_1 + \varepsilon^2\mu_2 + O(\varepsilon^3)\right) + \varepsilon f_1\left(X + \varepsilon\mu_1 + O(\varepsilon^2)\right) +$$

$$+\varepsilon^2 f_2\left(X + O(\varepsilon)\right) + O(\varepsilon^3) . \tag{3.77}$$

Applying Taylor formulas in (3.77), one gets

$$f(x;\varepsilon) = f_0(X) + \varepsilon\mu_1 f_0'(X) + \varepsilon^2\mu_2 f_0'(X) + \frac{\varepsilon^2\mu_1^2}{2} f_0''(X) + \ldots +$$

$$+\varepsilon f_1(X) + \varepsilon^2\mu_1 f_1'(X) + \ldots + \varepsilon^2 f_2(X) + O(\varepsilon^3) =$$

$$f_0(X) + \varepsilon\left[\mu_1 f_0'(X) + f_1(X)\right] +$$

$$+\varepsilon^2\left[\mu_2 f_0'(X) + \frac{\mu_1^2}{2} f_0'' + \mu_1 f_1' + f_2(X)\right] + O(\varepsilon^3) . \tag{3.78}$$

Comparing (3.76) with (3.78), one obtains

$$F_0(X) = f_0(X) , \tag{3.79}$$

$$F_1(X) = f_1(X) + \mu_1 f_0'(X) , \tag{3.80}$$

$$F_2(X) = f_2(X) + \mu_1 f_1'(X) + \mu_2 f_0'(X) + \frac{\mu_1^2}{2} f_0''(X) . \tag{3.81}$$

The procedure described so far can be extended to yield the quantities of order $O(\varepsilon^3)$, and so on, only if corresponding terms of the asymptotic series (3.74) are known.

According to the formula (3.79), a deformation is not exhibited in the first approximation (F_0 coincides with f_0 with the assumed accuracy). Next term of the series (3.80) is constructed via the condition of limitations of F_1/F_0. It may be achieved by an appropriate choice of deformation coefficient μ_1. Next, for a chosen μ_1 and then μ_2 one may achieve the bounded ratios of F_2/F_1, μ_2/μ_1. As a result, a uniformly suitable solution with the required accuracy order for the sought function $f(x;\varepsilon)$ is achieved.

As an example, Duffing's equation will be analyzed. The solution of this equation, applying a small parameter method with the accuracy of $O(\varepsilon)$ has the form of (3.27)

$$u(t;\varepsilon) = u_0(t) + \varepsilon u_1(t) + O(\varepsilon^2) =$$

$$= a\cos(t) + \varepsilon a^3\left[\frac{1}{32}(\cos 3t - \cos t) - \frac{3}{8} t\sin t\right] + O(\varepsilon^2) , \tag{3.82}$$

and it is uniformly suitable for $\varepsilon t = O(1)$.

Let us introduce a deformed variable S defined by the formula

$$t = S + \varepsilon\mu_1(S) + O(\varepsilon^2)$$

and let us seek for the solution of the form

$$u(t;\varepsilon) = F_0(S) + \varepsilon F_1(S) + O(\varepsilon^2) .$$

According to (3.79), one gets

$$F_0(S) = u_0(S) = a\cos S .$$

Due to (3.80), one obtains

$$F_1(S) = u(S) + \mu_1 u_0'(S) = \frac{a^3}{32}(\cos 3S - \cos S) - \frac{3a^3}{8} S \sin S - \mu_1 a \sin S \ .$$

In order to delete a secular term occurring in F_1 and proportional to S, the following requirement should be satisfied

$$\mu_1 = -\frac{3a^2}{8} S \ . \tag{3.83}$$

As a result, one gets

$$F_1(S) = \frac{a^3}{32}(\cos 3S - \cos S), \tag{3.84}$$

and the uniformly suitable solution with the accuracy of $O(\varepsilon)$ reads

$$u(t;\varepsilon) = a \cos S + \varepsilon \frac{a^3}{32}(\cos 3S - \cos S) + O(\varepsilon^2) \ , \tag{3.85}$$

where

$$S = t\left(1 + \varepsilon \frac{3a^2}{8} + O(\varepsilon^2)\right) \ . \tag{3.86}$$

The obtained formula coincides with (3.45)–(3.47). In other words, all of the three lustrated methods, i.e., the method of elongated parameters, deformed variables and multiple scaling yield the same result.

3.5.2 Deformation of two independent variables

Assume that for a certain unknown function $f(x,y;\varepsilon)$ the following asymptotic series has been constructed

$$f(x,y;\varepsilon) = f_0(x,y) + \varepsilon f_1(x,y) + \varepsilon^2 f_2(x,y) + O(\varepsilon^3) \ , \tag{3.87}$$

which is nonuniformly suitable in a certain subspace of solution definition with respect to x and y.

Let us introduce the deformed independent variables X, Y as well as dependent variable $F(X,Y;\varepsilon)$ due to formulas

$$x = X + \varepsilon\mu_1(X,Y) + \varepsilon^2\mu_2(X,Y) + O(\varepsilon^3) \ , \tag{3.88}$$

$$y = Y + \varepsilon\nu_1(X,Y) + \varepsilon^2\nu_2(X,Y) + O(\varepsilon^3) \ , \tag{3.89}$$

$$f(x,y;\varepsilon) \equiv F(X,Y;\varepsilon) = F_0(X,Y) + \varepsilon F_1(X,Y) + \varepsilon^2 F_2(X,Y) + O(\varepsilon^3) \ , \tag{3.90}$$

where μ_1, μ_2, ν_1, ν_2 are arbitrary functions (deformation coefficients).

Let us substitute (3.88) and (3.89) into (3.87), and let us transform (3.87) using Taylor's series. Results have the following forms

$$F_0(X,Y) = f_0(x,y) \ , \tag{3.91}$$

$$F_1(x,y) = f_1(x,y) + \mu_1 f'_{0X} + \nu_1 f'_{0Y} \ , \tag{3.92}$$

$$F_2(X,Y) = f_2(X,Y) + \mu_1 f'_{1X} + \nu_1 f'_{1Y} + \frac{1}{2}\mu_1^2 f''_{0XX} +$$

$$+\mu_1\nu_1 f''_{0XY} + \frac{1}{2}\nu_1^2 f''_{0YY} + \mu_2 f'_{0X} + \nu_2 f'_{0Y} \ . \tag{3.93}$$

Notice that a deformation is not exhibited by the first approximation.

Notice that deformation coefficients μ_1, ν_1, μ_2, ν_2 in (3.92), (3.93) are chosen according to Lighthill principle: ratios F_i/F_{i-1}, μ_i/μ_{i-1}, ν_i/ν_{i-1} ($i = 1, 2, \dots$) should be bounded in the whole domain of solution, which allows to remove the singularities in asymptotic series (3.90) and to transform it into the uniformly suitable series with the required accuracy.

The advantages of multiple scaling are visible during analysis of initial problem (3.59). Its solution defined through the perturbation method (3.61) is nonuniformly suitable for $x \sim O(\varepsilon^{1/2})$.

Let us introduce the deformation (3.62), (3.63). Let us substitute the introduced deformation to (3.61) and let us apply the formulas (3.79)–(3.81). The formula (3.79) yields

$$F_0(\varepsilon) = f_0(S) = (1+S)/S \ . \tag{3.94}$$

According to (3.80) and (3.94), one gets

$$F_1(S) = f_1(S) + \mu_1 f'_0(S) = \frac{(S-1)(1+3S)}{2S^3} - \frac{\mu_1}{S^2} = \frac{3S^2 - 2S - 2\mu_1 S - 1}{2S^3}. \tag{3.95}$$

In order to guarantee that for F_1, $S \to 0$ will have singularity smaller than F_0, i.e., not larger than $O(1/S)$, the following equation should be satisfied $2S + 2\mu_1 S + 1 = 0$,

$$\mu_1 = -\frac{1+2S}{2S} \ . \tag{3.96}$$

The choice coincides with (3.65) and yields directly results of (3.66)–(3.69).

A solution improvement is relatively simple owing to the application of multiple scaling. According to (3.81) and with the account of (3.94)–(3.96), one gets

$$F_2(S) = f_2(S) + \mu_1 f'(S) + \mu_2 f'_0(S) + \frac{\mu_1^2}{2} f''_0(S) = -\frac{3 + 4\mu_2 S}{6S^3}.$$

In order to keep the order of F_2 for $S \to 0$ smaller than $O(1/S)$, one may take $\mu_2 = \frac{AS^2-3}{4S}$, where A is the arbitrary constant. Finally, the uniformly suitable solution with the accuracy of $O(\varepsilon^2)$ has the following form

$$f(x;\varepsilon) = \frac{1+S}{S} + \varepsilon\frac{3}{2}S + O(\varepsilon^2), \quad x = S - \varepsilon\frac{1+2S}{2S} + \varepsilon^2\frac{AS^2-3}{4S} + O(\varepsilon^3) \ .$$

3.5.3 Method of full approximation

Let us turn back to the deformation in the first approximation, since there is a choice of deformation. We require that $F_1(S) = 0$. As a result, one gets

$$\mu_1 = \frac{3S^2 - 2S - 1}{2S}, \tag{3.97}$$

and the solution

$$x = S + \varepsilon\frac{3S^2 - 2S - 1}{2S} + O(\varepsilon^2)\ , \tag{3.98}$$

$$f(x;\varepsilon) = \frac{1+S}{S} + O(\varepsilon^2)\ , \tag{3.99}$$

which after excluding S from (3.98) and (3.99) (as it was in the case of solution (3.66), (3.67)) yields the exact solution (3.60)!). It means also that it is not worthy to determine more terms of series (3.98), (3.99). To conclude, one may obtain exact solution to problem (3.59) already in the first approximation (by a proper deformation choice).

In references [120, 126, 154] a full approximation method is developed, and its main idea is the optimal choice of asymptotic transformation (deformation) of dependent and independent variables. We focus mainly on a transformation of a nonlinear boundary value problem into the linear one, which allows to solve the problem.

After getting the linear solution, an inversed transformation yields the sought nonlinear problem. One may easily observe that in the method of full approximation (and in the method of deformed coordinates), the deformation of variables is carried out. Essential singularity of full approximation approach means achieving a uniformly suitable (and even exact in simple problems of (3.59) type) solution of nonlinear boundary value problem already in the first approximation.

Finally, let us briefly summarize the fundamental elements of full approximation method [120, 154]:

- The formulation of a boundary value problem in physical space;
- The choice of the deformation of variables and the investigation of full variables deformation existence (a proof of the real possibility of the effective construction of an optimal asymptotic algorithm yielding already in first approximation a nonlinear uniformly suitable solution);
- The formulation of the problem in the space of deformed variables (space of approximation);
- The solution to the problem in space of deformed variables;
- Inversed transition with respect to physical space and getting of sought solutions.

Some special cases related to the application of the full approximation method and various examples of its applications are discussed and illustrated in references [120, 126, 154].

3.6 Multiple scale methods

3.6.1 Introduction

The main idea of *multiple scale method* is the introduction of different time scales in order to construct uniformly suitable asymptotic decompositions to solve one of the chosen problems. A variety of multiple scales method is associated with various possibilities of the introduction of these scales [89]. Let us consider two examples. The first of them includes the analysis of a linear oscillator with damping governed by the equation (1.8) $u'' + 2\mu u' + u = 0$, where $u = u(t;\mu)$ is the amplitude of oscillations, μ is a nondimensional damping coefficient describing the ratio of resistance medium force and inertial forces (section 1.2).The second example deals with Duffing's problem. Assume that μ is small, i.e., oscillations of a constant damping are considered. Let us introduce $\varepsilon = \mu$, let us introduce the initial equation in the form

$$u'' + 2\varepsilon u' + u = 0 \ , \quad \varepsilon \ll 1 \ . \tag{3.100}$$

In what follows, we are going to compare the obtained solution with the exact one and to illustrate various variants of the multiple scale method.

Let us determine a general solution of the linear second order differential equation with constant coefficients (3.100). The corresponding characteristic equation $k^2 + 2\varepsilon k + 1 = 0$ possesses two complex conjugate roots $k_{1,2} = -\varepsilon \pm i\sqrt{1-\varepsilon^2}$ and therefore a general solution has the following form [127]

$$u(t;\varepsilon) = e^{-\varepsilon t}\left[c_1 \cos(\sqrt{1-\varepsilon^2}t) + c_2 \sin(\sqrt{1-\varepsilon^2}t)\right] \ ; \quad \{c_1, c_2\} = const \ .$$

Let us transform it to the form that is more suitable

$$u(t;\varepsilon) = a \cdot e^{-\varepsilon t} \cos\left(\sqrt{1-\varepsilon^2 t} + \beta\right) \ , \tag{3.101}$$

where a, β are arbitrary constants ($a = \sqrt{c_1^2 + c_2^2}$, $\beta = -\arctan(c_2/c_1)$).

Determine the solution to the equation (3.100) using a classical perturbation method

$$u(t;\varepsilon) = u_0(t) + \varepsilon u_1(t) + \varepsilon^2 u_2(t) + O(\varepsilon^3) \ . \tag{3.102}$$

Substituting (3.102) into (3.100) and comparing the terms standing by the same powers, one gets

$$u_0'' + u_0 = 0 \ , \tag{3.103}$$

$$u_1'' + u_1 = -2u_0' \ , \tag{3.104}$$

$$u_2'' + u_2 = -2u_1' \ . \tag{3.105}$$

Solving successively the zero order equation (3.103), the first (3.104) and second order equations (3.104) and (3.105), one may apply the method of constant variations or the method of undetermined coefficients [119]. Substituting results to (3.102), one gets

$$u(t;\varepsilon) = a\cos(t+\beta) - \varepsilon at\cos(t+\beta) +$$

$$+\frac{1}{2}\varepsilon^2 a\left[t^2\cos(t+\beta) + t\sin(t+\beta)\right] + O(\varepsilon^3) \ . \tag{3.106}$$

Notice that the solution (3.106) obtained by the application of multiple scale method involves (similarly to Duffing's type equation (3.27), secular terms and is useless for large time values, i.e., for $t \geq O\ (1/\varepsilon)$. It should be emphasized that according to (3.106) the solution u depends not only on t and ε separately, but also on their combinations εt, $\varepsilon^2 t$. Observe that it holds for the case of the exact solution (3.101), since after the expansion, for $\varepsilon \to 0$ for fixed t series (3.106) is obtained. For $\varepsilon \to 0$ and fixed t, according to formulas (2.7), (2.3) and (2.4), one gets

$$e^{-\varepsilon t} = 1 - \varepsilon t + \frac{1}{2!}(\varepsilon t)^2 + \frac{(\varepsilon t)^3}{3!} + \dots \ , \tag{3.107}$$

$$\cos\left(\sqrt{1-\varepsilon^2}\, t + \beta\right) = \cos\left(t - \frac{1}{2}\varepsilon^2 t - \frac{1}{8}\varepsilon^4 t + \beta + \dots\right) =$$

$$= \cos\left(t - \frac{1}{2}\varepsilon^{\,2} t + \beta\right) + \frac{1}{8}\varepsilon^{\,4} t \sin\left(t - \frac{1}{2}\varepsilon^{\,2} t + \beta\right) + \dots \tag{3.108}$$

Notice that $e^{\varepsilon t}$ can be approximated through the finite number of terms in the decomposition (3.107) only when $t < O\ (1/\varepsilon)$. In order to construct a decomposition valid for $t < O\ (1/\varepsilon)$, the product εt should be considered as one quantity $T_1 = \varepsilon t = O(1)$. Then, an arbitrary finite decomposition $e^{-\varepsilon t}$ holds for time interval $1/\varepsilon$ and it has the form

$$\exp(-\varepsilon t) = \exp(-T_1) \ .$$

Uselessness of asymptotic decomposition (3.106) is yielded through the analysis of formulas (3.107) and (3.108). Although the infinite series (3.107) and (3.108) are convergent for arguments of exponential by function and cosinus (formulas 2.7 and 2.10), a uniform approximation to the function $e^{-\varepsilon t}$, $\cos\left(\sqrt{1-\varepsilon^2}t + \beta\right)$ using a finite number of terms of series (3.107), (3.108) (using truncated series 3.107, 3.108) cannot be achieved. It is clearly demonstrated in Figs. 3.3, 3.4, where the digits standing by curves denote the numbers of series (3.107), (3.108) used in computations.

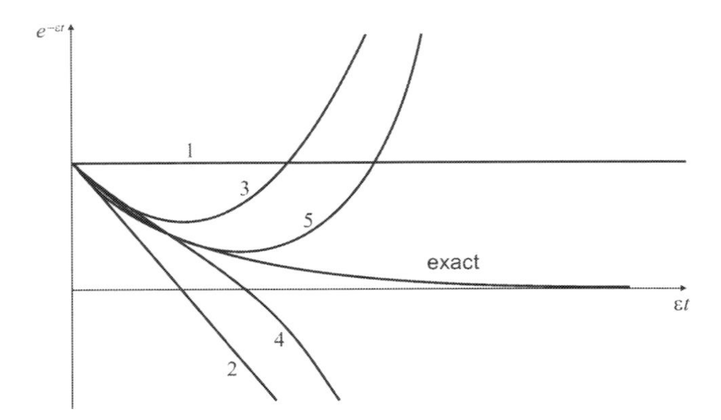

FIGURE 3.3: Exact solution and its approximations through different numbers of series (3.107) terms.

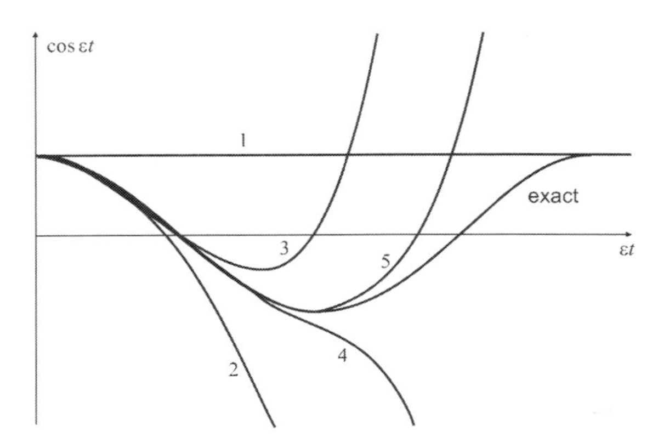

FIGURE 3.4: Exact solution and its approximations through different numbers of series (3.108) terms.

Substituting (3.107) and (3.108) into (3.101), one gets the decomposition (3.106). In order to get the "truncated" decomposition (3.107) holding for time ε^{-1}, the product εt should be treated as one quantity $T_1 = \varepsilon t = O(1)$. In order to get the "truncated" decomposition (3.108) holding for time of order ε^{-2}, the product $\varepsilon^2 t$ should be treated as one quantity $T_2 = \varepsilon^2 t = O(1)$, because the second (correction) term in (3.108) of order $O(\varepsilon^4 t)$ is of order $O(\varepsilon^2)$ or smaller in comparison to the first one with the order 1 up to the time estimation of order $O(\varepsilon^{-2})$.

In the considerations carried out here, it is assumed that $u(t, \varepsilon)$ does not depend explicitly on t, εt, $\varepsilon^2 t, \ldots$ and ε. Therefore, in order to get a "truncated" decomposition valid for time interval of order $O(\varepsilon^{-n})$, where n is natural number, it is more rational to assume that u depends on $n + 1$ various scales of time $T_0, T_1, \ldots, T_n$, where

$$T_n = \varepsilon^n t\,; \quad n = 0, 1, 2, \ldots \tag{3.109}$$

and search for its solution of the form

$$u(t;\varepsilon) = \hat{u}(T_0, T_1, \ldots, T_n; \varepsilon) = \sum_{k=0}^{n-1} \varepsilon^k u_k(T_0, T_1, \ldots, T_n) + O(\varepsilon T_n)\ . \tag{3.110}$$

The last term in a decomposition (3.110) indicates that the obtained result holds for time interval of order $O(\varepsilon^{-n})$. In order to increase this interval and to keep the series uniformity, one may use the successive time scales, i.e., $T_{n+1}, \ldots$.

Timescale $T_1 = \varepsilon t$ corresponds to the slower time interval than scale $T_0 = t$, whereas scale $T_2 = \varepsilon^2 t$ corresponds to slower time variations than scale T_1, and so on. Formula (3.110) indicates that the stated problem, governed by ordinary differential equation equation (6.5.1), is transformed to a problem governed by a partial differential equation. Below, the steps of successive differentiations are shown

$$\frac{d}{dt} = \frac{\partial}{\partial T_0} \cdot \frac{dT_0}{dt} + \frac{\partial}{\partial T_1} \cdot \frac{dT_1}{dt} + \ldots = \frac{\partial}{\partial T_0} + \varepsilon \frac{\partial}{\partial T_1} + \varepsilon^2 \frac{\partial}{\partial T_2} + \ldots \tag{3.111}$$

$$\frac{d^2}{dt^2} = \frac{d}{dt}\left(\frac{d}{dt}\right) = \frac{\partial^2}{\partial T_0^2} + 2\varepsilon \frac{\partial^2}{\partial T_0 \partial T_1} + \varepsilon^2 \left(2\frac{\partial^2}{\partial T_0 \partial T_2} + \frac{\partial^2}{\partial T_1^2}\right) + \ldots \tag{3.112}$$

By substituting (3.110)–(3.112) into (3.100) and comparing coefficients standing by the same series, partial differential equations are obtained, which serve for getting $u_0, u_1, \ldots, u_n$.

The solution to these equations will include arbitrary functions dependent on timescales $T_1, \ldots, T_n$. In order to define these functions, one may require to satisfy the condition, as it has been already made in the case of deformed coordinates (3.31), that singularities of higher approximations do not achieve singularities of first approximation. In other words, one requires that

$$\frac{u_i}{u_{i-1}} < \infty\ , \tag{3.113}$$

where $i = 1, 2, \ldots, n$ correspond to timescales $T_0, T_1, \ldots, T_n$.

The condition (3.113) is equivalent to that of removing secular terms from decomposition (3.110).

Let us finish the considerations of this section by the following remark. Although the application of multiple scale method complicates the problem (a transition from ordinary to partial differential equations), this is compensated by the occurrence of functions depending on various time scales $T_1, T_2, \ldots$, and their proper choice may modify asymptotic decomposition (3.110) to a form uniformly suitable for time intervals of our interest.

3.6.2 Derivative decomposition along one and two variables

Since in the earlier mentioned version of multiple scale method both the sought function (3.110) and its derivative (see (3.111)) are cast into asymptotic decomposition, then this variant of asymptotic approach is called *a derivative decomposition.*

This method may be even more generalized if during the introduction of different timescales, an asymptotic series $\delta_k(\varepsilon)$, is applied, and not power series with respect to ε, i.e., we assume that

$$T_k = \delta_k(\varepsilon)t\,, \quad \text{where for} \quad \varepsilon \to 0\,, \quad \delta_{k+1}(\varepsilon) = o(\delta_k(\varepsilon))\,. \tag{3.114}$$

Then, *a derivative operator* has the form

$$\frac{d}{dt} = \sum_{k=0}^{m} \delta_k(\varepsilon)\frac{\partial}{\partial T_k}\,. \tag{3.115}$$

Equations (3.114) and (3.115) can be even more generalized by the following assumption

$$T_k = \delta_k(\varepsilon)g_k\left[\mu_k(\varepsilon)t\right]\,, \tag{3.116}$$

$$\frac{d}{dt} = \sum_{k=0}^{n} \delta_k(\varepsilon)\mu_k(\varepsilon)g_k'\left[\mu_k(\varepsilon)t\right]\frac{\partial}{\partial T_k}\,, \tag{3.117}$$

where $\mu_k(\varepsilon)$ is another asymptotic series. Notice that the formula (3.116) allows for the introduction of linear and nonlinear timescales [118].

The form of exact solution (3.101) suggests another variant of multiple scale method. Namely, in this solution, time occurs in the form of two combinations

$$\xi = \varepsilon t, \quad \eta = \sqrt{1-\varepsilon^2}t = \left(1 - \frac{1}{2}\varepsilon^2 - \frac{1}{8}\varepsilon^4 + \ldots\right)t\,.$$

This observation motivated Cole and Kevorkian [90, 89], to introduce two timescales to trace large time intervals of the form

$$u(t;\varepsilon) = \hat{u}(\xi,\eta;\varepsilon) = \sum_{k=0}^{n-1} \varepsilon^k u_k(\xi,\eta) + O(\varepsilon^n)\,, \tag{3.118}$$

where:

$$\xi = \varepsilon t\,, \quad \eta = (1 + \varepsilon^2\varpi_2 + \varepsilon^3\varpi_3 + \ldots + \varepsilon^n\varpi_n)t\,. \tag{3.119}$$

In the above, ϖ_n are the earlier unknown constant values, which allow by a proper choice to obtain a suitable decomposition (3.118). In the considered case the timescale ξ is slower than that of η, co which allows for the introduction of analogy between this and averaging methods (chapter 3.8). There is a series of works illustrating the equivalence of results obtained with the use of these two methods [118].

Formulas (3.118) and (3.119) yield

$$\frac{d}{dt} = \varepsilon \frac{\partial}{\partial \xi} + \left(1 + \varepsilon^2 \varpi_2 + \varepsilon^3 \varpi_3 + \ldots + \varepsilon^n \varpi_n\right) \frac{\partial}{\partial \eta} , \tag{3.120}$$

and therefore often the described version of two scales is referred to as *a derivative decomposition with respect to two variables* [118]. Also this procedure, analogically to earlier described procedure of the derivative decomposition with respect to one variable (see (3.100)–(3.112)), can be generalized.

The solution is sought in the following form

$$\xi = \mu(\varepsilon) t , \quad \eta = \sum_{k=0}^{n} \delta_k(\varepsilon) g_k \left[\mu(\varepsilon) t\right] , \tag{3.121}$$

$$\frac{d}{dt} = \mu(\varepsilon) \frac{\partial}{\partial \xi} + \left(\sum_{k=0}^{n} \delta_k(\varepsilon) \mu(\varepsilon) g_k' \left[\mu(\varepsilon) t\right] \right) \frac{\partial}{\partial \eta} . \tag{3.122}$$

Notice that the formulas (3.121), (3.122) clearly indicate different variants of the multiple scales method. Let us repeat the remark included in the reference [118]: "Multiple scale method has been so popular that it has been discovered again every six months. It is applicable in wide areas of physics, engineering and applied mathematics."

The huge amount of problems solved with this method is demonstrated in the reference [118].

3.6.3 Application to the problems of vibrations

Let us begin with the application of multiple scales method to analyze the oscillations of a mechanical one-degree-of-freedom system governed by equation (3.100). Note that in the case of linear differential equation of the form (3.100) one may apply different timescales (3.109), without the decomposition of a sought function into series (3.84). Substituting (3.111) and (3.112) into (3.100), one gets

$$\left[\frac{\partial^2}{\partial T_0^2} + 2\varepsilon \frac{\partial^2}{\partial T_0 \partial T_1} + \varepsilon^2 \left(\frac{\partial^2}{\partial T_1^2} + 2 \frac{\partial^2}{\partial T_0 \partial T_2} \right) + \ldots \right] u + u =$$

$$= -2\varepsilon \left(\frac{\partial}{\partial T_0} + \varepsilon \frac{\partial}{\partial T_1} + \varepsilon^2 \frac{\partial}{\partial T_2} \right) u .$$

Comparing the coefficients standing by same powers of ε, one gets

$$\frac{\partial^2 u}{\partial T_0^2} + u = 0 , \tag{3.123}$$

$$2 \frac{\partial^2 u}{\partial T_0 \partial T_1} = -2 \frac{\partial u}{\partial T_0} , \tag{3.124}$$

$$\frac{\partial^2 u}{\partial T_1^2} + 2\frac{\partial^2 u}{\partial T_0 \partial T_2} = -2\frac{\partial u}{\partial T_1} \ . \tag{3.125}$$

A solution to equations (3.123)–(3.125) may be found in a complex form. For this purpose, the following identities are used

$$\cos\alpha = \frac{1}{2}(e^{i\alpha} + e^{-i\alpha}), \quad \sin\alpha = \frac{1}{2}(e^{i\alpha} - e^{-i\alpha}) . \tag{3.126}$$

Let us verify (3.126). Substituting (2.7) $x = i\alpha$, where $i = \sqrt{-1}$, one gets

$$e^{i\alpha} = 1 + \frac{i\alpha}{1!} + \frac{(i\alpha)^2}{2!} + \frac{(i\alpha)^3}{3!} + \frac{(i\alpha)^4}{4!} + \frac{(i\alpha)^5}{5!} + \ldots \tag{3.127}$$

Knowing that $i^2 = -1$, the terms of (3.127) are successively integrated to yield

$$e^{i\alpha} = \left(1 - \frac{\alpha^2}{2!} + \frac{\alpha^4}{4!} + \ldots\right) + i\left(\alpha - \frac{\alpha^3}{3!} + \frac{\alpha^5}{5!} - \ldots\right) .$$

According to the formula

$$e^{i\alpha} = \cos\alpha + i\sin\alpha \tag{3.128}$$

we have

$$e^{-i\alpha} = \cos\alpha - i\sin\alpha \ . \tag{3.129}$$

Adding and subtracting formulas (3.128) and (3.129), called *Euler's formulas*, one gets (3.126).

Let us present the general solution $u = a(T_1, T_2)\cos[T_0 + \beta(T_1, T_2)]$ of the differential equation (3.123) in a complex form by applying the first of dependencies (3.126) and using the series transformation rule (for the sake of simplicity, the arguments standing by variables α, β are omitted).

Observe that

$$u = a\cos(T_0 + \beta\ ) = \frac{a}{2}\left[e^{i(T_0+\beta)} + e^{-i(T_0+\beta)}\right] = \frac{a}{2}e^{i\beta}e^{iT_0} + \frac{a}{2}e^{-i\beta}e^{-iT_0},$$

i.e.,

$$u = Ae^{iT_0} + \bar{A}e^{-iT_0} \ . \tag{3.130}$$

Denoting by $A, \bar{A}$ two conjugated quantities

$$A(T_1, T_2) = \frac{a}{2}e^{i\beta}, \quad \bar{A}(T_1, T_2) = \frac{a}{2}e^{-i\beta} \ . \tag{3.131}$$

Substituting (3.130) into (3.124), one gets

$$\left(\frac{\partial A}{\partial T_1} + A\right)e^{iT_0} + \left(\frac{\partial \bar{A}}{\partial T_1} + \bar{A}\right)e^{-iT_0} = 0 \ . \tag{3.132}$$

Since the equation (3.132) holds for an arbitrary T_0, and since the exponents are always positive, then the coefficients standing by the exponents in the equation (3.132) should be equal to zero, i.e.:

$$\frac{\partial A}{\partial T_1} + A = 0\,, \qquad \frac{\partial \bar{A}}{\partial T_1} + \bar{A} = 0\;. \tag{3.133}$$

Since the second dependence of (3.133) is complex conjugated with the first one, and if the first equation of (3.133)

$$\frac{\partial A}{\partial T_1} + A = 0 \tag{3.134}$$

is satisfied, then the second equation of (3.133) is also automatically satisfied.
A solution to (3.134) reads

$$A = a(T_2)e^{-T_1}\;. \tag{3.135}$$

Substituting (3.130) and accounting for (3.135) into (3.125), one gets

$$\left(\frac{\partial^2 A}{\partial T_1^2} + 2i\frac{\partial A}{\partial T_2} + 2\frac{\partial A}{\partial T_1}\right)e^{iT_0} + \left(\frac{\partial^2 \bar{A}}{\partial T_1^2} - 2i\frac{\partial \bar{A}}{\partial T_2} + 2\frac{\partial \bar{A}}{\partial T_1}\right)e^{-iT_0} = 0\,,$$

and hence

$$\frac{\partial^2 A}{\partial T_1^2} + 2i\frac{\partial A}{\partial T_2} + 2\frac{\partial A}{\partial T_1} = 0\;. \tag{3.136}$$

Substituting (3.136) into (3.125), one gets

$$2i\frac{\partial a}{\partial T_2} - a = 0\;. \tag{3.137}$$

The solution to equation (3.137) reads

$$a(T_2) = a_0 e^{-\frac{iT_2}{2}}\,, \tag{3.138}$$

where a_0 is an arbitrary constant.
Substituting $a(T_2)$ into (3.135), yields $A = = a_0 e^{-T-(iT_2/2)}$, and hence equation (3.130) takes the form

$$u = a_0 e^{-T_1} e^{i\left(T_0 - \frac{T_2}{2}\right)} + \bar{a}_0 e^{-T_1 - i\left(T_0 - \frac{T_2}{2}\right)}\;. \tag{3.139}$$

Let us take a_0 in the form $a_0 = \frac{1}{2}ae^{i\beta}$, and hence $\bar{a}_0 = \frac{1}{2}ae^{-i\beta}$. According to the first formula of (3.126) and (3.139), one gets

$$u = \frac{1}{2}ae^{-T_1}\left[e^{i\left(T_0 - \frac{T_2}{2} + \beta\right)} + e^{-i\left(T_0 - \frac{T_2}{2} + \beta\right)}\right] = ae^{-T_1}\cos\left(T_0 - \frac{T_2}{2} + \beta\right)\;.$$

Remembering that $T_0 = t$, $T_1 = \varepsilon t$, $T_2 = \varepsilon^2 t$, the earlier obtained expression can be transformed to the following form

$$u(t;\varepsilon) = ae^{-\varepsilon t} \cos\left(t - \frac{\varepsilon^2 t}{2} + \beta\right) ,$$

which is in agreement with the exact solution of (3.6.2) with the accuracy of $O(\varepsilon^2)$ (a, β are arbitrary constants).

Let us finally reconsider Duffing's equation (3.120). We are going to find its solution of the following form

$$u(t;\varepsilon) = u_0(T_0, T_1) + \varepsilon u_1(T_0, T_1) + O(\varepsilon T_1)\,; \quad T_0 = t\,, \quad T_1 = \varepsilon t\,, \tag{3.140}$$

computing only the first approximation. Substituting (3.140), (3.111) and (3.112) into (3.100) one gets

$$\frac{\partial^2 u_0}{\partial T_0^2} + \varepsilon \frac{\partial^2 u_1}{\partial T_0^2} + 2\varepsilon \frac{\partial^2 u_0}{\partial T_0 \partial T_1} + u_0 + \varepsilon u_1 + \varepsilon u_0^3 + O(\varepsilon^2) = 0\,.$$

Equating to zero the coefficients standing by ε^0 and ε^1, one gets

$$\frac{\partial^2 u_0}{\partial T_0^2} + u_0 = 0\,, \tag{3.141}$$

$$\frac{\partial^2 u_1}{\partial T_0^2} + u_1 = -2\frac{\partial^2 u_0}{\partial T_0 \partial T_1} - u_0^3\,. \tag{3.142}$$

A general solution to the equation (3.141) has the form

$$u_0 = a(T_1) \cos\left[T_0 + \beta(T_1)\right]\,. \tag{3.143}$$

Notice that in our case, the quantities a and β are not constant, but they are the functions of slow scale T_1, since the function $u_0 = u_0(T_0, T_1)$, and its second derivative appearing in (3.141) are expressed by a variable T_0. Dependencies $a(T_1)$, $\beta(T_1)$ are known and they are determined in the process of solving the equation (3.142) by elimination of the secular terms occurring in u_1.

Substituting (3.143) into (3.142) and remembering that

$$\cos^3 \alpha = \frac{1}{4}\left(3\cos\alpha + \cos 3\alpha\right),$$

one gets

$$\frac{\partial^2 u_1}{\partial T_0^2} + u_1 = -2\frac{\partial^2}{\partial T_0 \partial T_1}\left[a\cos(T_0 + \beta)\right] - \frac{3}{4}a^3 \cos(T_0 + \beta) - \frac{1}{4}a^3 \cos(3T_0 + 3\beta)\,. \tag{3.144}$$

Let us transform the first term standing on the right-hand side of (3.144). Applying the rule of a complex function differentiation and remembering that $a = a(T_1)$, $\beta = \beta(T_1)$, one gets

$$-2\frac{\partial^2}{\partial T_0 \partial T_1}\left[a\cos(T_0+\beta)\right] = -2\frac{\partial}{\partial T_1}\left\{\frac{\partial}{\partial T_0}\left[a\cos(T_0+\beta)\right]\right\} =$$

$$= 2\frac{\partial a}{\partial T_1}\sin(T_0+\beta) + 2a\frac{\partial \beta}{\partial T_1}\cos(T_0+\beta)\ .$$

The equation (3.144) is transformed into the form

$$\frac{\partial^2 u_1}{\partial T_0^2} + u_1 = 2\frac{\partial a}{\partial T_1}\sin(T_0+\beta) + \left(2a\frac{\partial \beta}{\partial T_1} - \frac{3}{4}a^3\right)\cos(T_0+\beta) -$$

$$-\frac{1}{4}a^3\cos(3T_0+3\beta)\ . \tag{3.145}$$

As it has been already mentioned, in order to avoid secular terms in u_1, the coefficients standing by $\sin(T_0+\beta)$ and $\cos(T_0+\beta)$ in (3.145) should be compared. This operation and forms of $a = a(T_1)$, $\beta = \beta(T_1)$ yield

$$\frac{da}{dT_1} = 0\ , \tag{3.146}$$

$$2a\frac{d\beta}{dT_1} - \frac{3}{4}a^3 = 0\ . \tag{3.147}$$

A particular solution to a nonhomogeneous differential equation (3.145) is sought with the method of undetermined coefficients and it yields the following solution

$$u_1 = \frac{1}{32}a^3\cos(3T_0+3\beta)\ . \tag{3.148}$$

Equation (3.146) yields

$$a = a_0 = const\ , \tag{3.149}$$

whereas equation (3.147) for $a_0 \neq 0$ gives $\frac{d\beta}{dT_1} = \frac{3}{8}a_0^2$, and finally

$$\beta = \frac{3}{8}a_0^2 T_1 + \beta_0\ , \tag{3.150}$$

where β_0 is an arbitrary constant.

Substituting (3.143), (3.148)–(3.150) into (3.140), and taking $T_0 = t$, $T_1 = \varepsilon t$ into consideration, one obtains

$$u = a_0\cos\left(t + \frac{3}{8}\varepsilon\, t a_0^2 + \beta_0\right) + \frac{1}{32}\varepsilon\, a_0^2\cos\left(3t + 3\beta_0 + \frac{9}{8}\varepsilon\, t a_0^2\right) + O(\varepsilon^2 t)\ , \tag{3.151}$$

which agrees fully with decompositions (3.47), (3.85), obtained by the method of Lindstedt-Poincaré and the rescaling method.

The analysis of equation (3.151) yields a conclusion that for $t = O(\varepsilon^{-2})$ the introduced error is of order $O(1)$, i.e., the same as the first term of decomposition (3.151). It means that for $t \geq O(\varepsilon^{-2})$, the decomposition (3.151) is unsuitable. If $t = O(\varepsilon^{-1})$, then the error is of order $O(\varepsilon)$, i.e., of the second decomposition (3.151) term order. To conclude, the decomposition (3.151), uniformly suitable for $t \leq O(\varepsilon^{-1})$, should contain only first term

$$u(t;\varepsilon) = a_0 \cos\left(t + \frac{3}{8}\varepsilon t a_0^2 + \beta\right) + O(\varepsilon) . \tag{3.152}$$

In other words, in order to construct a uniformly suitable decomposition of the first order without solving the equation with respect to u_1, the secular terms occurring in decomposition (3.140) should be removed, and a dependence between u_0 and timescale T_1 should be found.

Proceeding analogically while constructing a uniformly suitable solution of first order through either the method of "elongated" parameters (subchapter 3.3) or the method of rescaling (see (3.5)), one has to remove the secular terms from the equation yielding u_1, and, next, determine the correction ϖ_1, characterizing the dependence of the frequency of oscillations on small parameter ε.

3.7 Variations of arbitrary constants

Method of variations of arbitrary constants has been initially applied to solve linear nonhomogeneous differential equations, assuming the solutions associated with homogeneous conditions to be known.

In what follows, the following nonhomogeneous, second order differential equation will be considered

$$y'' + p(x)y' + q(x)y = R(x) , \tag{3.153}$$

where p, q, R are known continuous functions in certain interval of x variations.

Let $y_1(x)$, $y_2(x)$ be fundamental solutions to homogeneous equation of the form

$$y'' + p(x)y' + q(x)y = 0 . \tag{3.154}$$

It means that y_1, y_2 are linearly independent solutions to equation (3.154). General solution of equation (3.154) reads

$$y_0 = A_1 y_1(x) + A_2 y_2(x) , \tag{3.155}$$

where A_1, A_2 are arbitrary constants.

A particular solution to equation (3.153) is sought in the form

$$y_* = A_1(x)y_1(x) + A_2(x)y_2(x) , \qquad (3.156)$$

i.e., we assume that previous constants A_1, A_2 will now be functions of x and they should be determined.

Differentiating (3.156) with respect to x, one gets

$$y'_* = A_1y'_1 + A'_1y_1 + A_2y'_2 + A'_2y_2 . \qquad (3.157)$$

Since there are three unknown functions A_1, A_2, y_* and only two equations (3.153), (3.156), then the additional condition applied to A_1, A_2, y_* and independent on (3.153) and (3.156) is necessary.

The following requirement is stated

$$A'_1y_1 + A'_2y_2 = 0 . \qquad (3.158)$$

Then, from (3.157), one gets

$$y'_* = A_1y'_1 + A_2y'_2 . \qquad (3.159)$$

Differentiating (3.159), one gets

$$y''_* = A_1y''_1 + A_2y''_2 + A'_1y'_1 + A'_2y'_2 . \qquad (3.160)$$

By substituting (3.160), (3.157), (3.156) into (3.153), and according to the fact that y_1, y_2 are solutions to the equation (3.154), one gets

$$A'_1y'_1 + A'_2y'_2 = R . \qquad (3.161)$$

As a result, two algebraic equations (3.158), (3.161) serving for determination of two unknown functions A'_1, A'_2 are obtained.

Solving this system of equations with Cramer's rule, one gets

$$A'_1 = -\frac{R(x)y_2(x)}{W(x)}, \quad A'_2 = -\frac{R(x)y_1(x)}{W(x)}, \qquad (3.162)$$

where:

$$W(x) = \begin{vmatrix} y_1 & y_2 \\ y'_1 & y'_2 \end{vmatrix} = y_1(x)y'_2(x) - y'_1(x)y_2(x) . \qquad (3.163)$$

Recall that function $W(x)$ defined by (3.163) is called *Wronski function* or *Wronskian*.

Integrating (3.162) and substituting the found A_1, A_2 into (3.156), the following particular solution is found

$$y_*(x) = y_2(x)\int_{x_0}^{x} \frac{y_1(t)R(t)}{W(t)}dt - y_1(x)\int_{x_0}^{x} \frac{y_2(t)R(t)}{W(t)}dt . \qquad (3.164)$$

Finally, a general solution of equation (3.153) reads

$$y(x) = c_1 y_1(x) + c_2 y_2(x) + y_*(x) \ , \tag{3.165}$$

where c_1, c_2 are arbitrary constants.

The described procedure has been generalized and applied in the series of works during the solutions of various ordinary and partial differential equations.

Consider now the application of a variation of arbitrary constants to solve the Duffing problem (3.13). For $\varepsilon = 0$ (zero order approximation), a solution reads

$$u(t) = a\cos(t+\beta) \ , \tag{3.166}$$

where a, β are certain arbitrary constants.

Differentiating (3.166) with respect to t, one gets

$$u'(t) = -a\sin(t+\beta) \ . \tag{3.167}$$

As it has been already mentioned, the key feature of the described method is the assumption that a solution to equation (3.13) for $\varepsilon \neq 0$ has the form (3.166), where $a = a(t)$, $\beta = \beta(t)$. Therefore, in our case, three functions $u(t)$, $a(t)$, $\beta(t)$, should be determined, but we have only two equations (3.13), (3.166).

Therefore, an additional requirement matching the mentioned three functions is required. As this required third equation, we take equation (3.13) and (3.166). It means that u, u' have the same form as in linear case for $\varepsilon = 0$. It finally leads to obtaining two first order differential equations yielding two independent functions of the form $a(t)$, $\beta(t)$.

Differentiating (3.166) with respect to t, in accordance with the complex functions differentiation rules of $a = a(t)$, $\beta = \beta(t)$, one gets

$$u' = -a\sin(t+\beta) + a'\cos(t+\beta) - a\beta'\sin(t+\beta) \ . \tag{3.168}$$

Comparing (3.167) and (3.168) yields

$$a'\cos(t+\beta) - a\beta'\sin(t+\beta) = 0 \ . \tag{3.169}$$

Differentiating (3.167) with respect to time, one gets

$$u'' = -a'\sin(t+\beta) - a\cos(t+\beta) - a\,\beta'\cos(t+\beta) \ . \tag{3.170}$$

Substituting (3.170) and (3.166) into Duffing equation (3.13), one gets

$$a'\sin(t+\beta) + a\beta'\cos(t+\beta) = \varepsilon a^3\cos^3(t+\beta) \ . \tag{3.171}$$

Obtained differential equations (3.169) and (3.171) are of the first order with respect to a', β'. Multiplying (3.169) by $\cos(t+\beta)$, one gets

$$a'\cos^2(t+\beta) - a\beta'\sin(t+\beta)\cos(t+\beta) = 0 \ .$$

Then we multiply (3.171) by $\sin(t+\beta)$ to get

$$a'\sin^2(t+\beta) + a\beta'\sin(t+\beta)\cos(t+\beta) = \varepsilon a^3 \sin(t+\beta)\cos^3(t+\beta)\ .$$

Summing these equations yields

$$a' = \varepsilon\, a^3 \sin(t+\beta)\cos^3(t+\beta)\ . \tag{3.172}$$

Substituting (3.172) into (3.169) for $a \neq 0$, one finds

$$\beta' = \varepsilon\, a^2 \cos^4(t+\beta)\ . \tag{3.173}$$

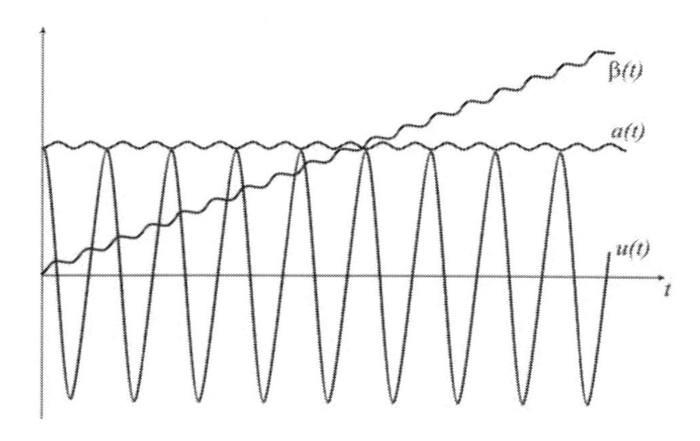

FIGURE 3.5: Variations in time of functions $u(t)$, $a(t)$, $\beta(t)$.

Finally, the method of constants variation reduces the problem of integration of second order differential equation (3.13) to that of the integration of two nonlinear differential first order equations (3.172) and (3.173) for functions $a(t)$, $\beta(t)$ with initial conditions (3.14). One may observe that, instead of one nonlinear equation (3.13) we have two of them (3.172) and (3.173) now. However, the following advantages are achieved.

1. Now, one may apply any of the asymptotic approaches to find a solution analytically.

2. During the numerical integration of equations (3.172) and (3.173), one may apply a larger integration step than in the case of equation (3.13) serving for determination of $u(t)$, since main terms of function $a(t)$, $\beta(t)$ change in time more slowly than $u(t)$.

Since $|\sin(t+\beta)| \leq 1$, $|\cos(t+\beta)| \leq 1$, then for the bounded value of a, one gets $a'(t) = O(\varepsilon)$, $\beta'(t) = O(\varepsilon)$. This situation is also illustrated in Fig. 3.5, taken from reference [119].

3.8 Averaging methods

Rapid development of various variants of *averaging method* has been motivated by searching for a universal method suitable for the analysis of both linear and nonlinear vibrations occurring in many branches of industry, civil and mechanical engineering, ships and vehicles construction, airplanes and space technology. The requirement of such qualitative and quantitative analysis appears also during the investigation of beam vibrations or rod vibrations externally vibrations externally driven in a periodic or stochastic manner by the forces of variable amplitudes and frequencies.

Asymptotic methods of averaging also play an important role during the investigation of equations of the following form

$$\begin{cases} \overline{a}' = \varepsilon A(\overline{a}, \overline{b}; \varepsilon) \\ \overline{b}' = B_0(\overline{a}, \overline{b}) + \varepsilon B(\overline{a}, \overline{b}; \varepsilon) \ , \end{cases} \tag{3.174}$$

where $\bar{a}$, $\bar{b}$ are certain vectors, ε is a small parameter, and A, B, B_0 are changeable functions. Similar systems of equations often appear during the analysis of various problems in physics and mechanics. In particular, they are typical in a wide field of problems of vibrations. Their particularities are manifested through *slow* (components of vector $\bar{a}$) and *fast* (components of vector $\bar{b}$) changes. During solving equations (3.153) the right-hand sides are averaged.

3.8.1 Methods of Van der Pol and Krylov-Bogolubov-Mitropolskiy (KBM)

Dutch engineer Van der Pol discovered the economical method of solving nonlinear problems of one-degree-of-freedom. Namely, he proposed to reduce the general problem to that of a particular system (3.174), and next, the system (3.174) has been simplified by averaging its right-hand sides with respect to the "fast variable." Although Van der Pol method is clear and suitable for computations, it is based on intuition rather than on a rigorous statement.

N.M. Krylov and N.N. Bogulobov [34, 91] proposed more general approach to analyze equations (3.174). Its main characteristic feature is finding a suitable change of variables allowing for the separation of "fast" and "slow" variables, and then constructing an asymptotic series with the first term the same as that of the Van der Pol method solution. The mentioned approach is developed mainly in the works of Yu. A. Mitropolskiy and others [109, 116, 117].

Let us briefly illustrate *Van der Pol method*. In what follows, vibrations $u = u(t; \varpi; \varepsilon)$ governed by the second order differential equation of the form

$$u'' + \varpi^2 u = \varepsilon f(u, u') \ , \tag{3.175}$$

are analyzed, where ϖ is real constant, ε is a small parameter, and f denotes a certain bounded function.

It is obvious that (3.175) describes quasi-linear vibrations. For $\varepsilon = 0$, from (3.175), one gets the equation

$$u'' + \varpi^2 u = 0 \,, \tag{3.176}$$

which governs linear and harmonic vibrations of the form

$$u = a\cos\varphi \,, \quad \varphi = \varpi(t + t_0) \tag{3.177}$$

with the amplitude a and phase φ, where ϖ is the frequency of vibrations, and a and t_0 are arbitrary constants.

It is assumed that for small ε, a solution to equations (3.175) will describe a certain vibration process of the form (3.177). One may also expect that the amplitude a of this process will change slowly for small values of ε. The phase of the vibration process also undergoes changes in time. The described vibration process is characterized in full by the instant values of amplitude a and phase φ. Therefore, they may serve as variables governing vibration process assuming slow changes.

Let us briefly discuss (3.175) new variables $a(t)$, $\varphi(t)$. In order to define them, one has to add one more equation to the relation (3.177).

Applying *the method of constants variation* we require that a and φ should be matched through relation

$$u' = -\varpi a \sin\varphi \,, \tag{3.178}$$

which holds always for the constant amplitude.

In what follows, we introduce differential equations satisfied by functions $a(t)$, $\varphi(t)$. Differentiating of (3.177) in accordance with the complex functions differentiation rule and comparing of the obtained result with (3.178), one gets

$$a' \cos\varphi - a\varphi' \sin\varphi + \varpi\, a \sin\varphi = 0 \,. \tag{3.179}$$

Obtained condition (3.179) is a compatibility condition of (3.177) and (3.178).

Differentiating (3.178) one more time in agreement with the complex functions differentiation and substituting (3.177) into (3.175), the following one more equation is obtained

$$-a'\varpi \sin\varphi - \varpi a\varphi' \cos\varphi + a\varpi^2 \cos\varphi = \varepsilon f(a\cos\varphi; -\varpi a \sin\varphi) \,. \tag{3.180}$$

System of equations (3.179), (3.180) is a linear one with two unknowns a', φ'. Let us multiply (3.179) by $\sin\varphi$, whereas (3.180) by $\frac{1}{\varpi}\cos\varphi$. Combinations of these two equations yield φ', and a':

$$\begin{cases} a' = -\dfrac{\varepsilon}{\varpi} f(a\cos\varphi, -a\varpi\sin\varphi)\sin\varphi \equiv \dfrac{\varepsilon}{\varpi} f_1(a,\varphi) \,, \\ \varphi' = \varpi - \dfrac{\varepsilon\cos\varphi}{\varpi a} f(a\cos\varphi, -a\varpi\sin\varphi) \equiv \varpi - \dfrac{\varepsilon}{\varpi a} f_2(a,\varphi) \,. \end{cases} \tag{3.181}$$

The system of two equations (3.181) is equivalent to equation (3.175). It is a particular case of the system (3.174). In the last case, amplitude a is the "slow" variable, whereas phase φ is the "fast" variable. Notice that parameters of the system (3.181) are periodic functions of φ regardless of the form of function $f(a, \varphi)$.

The first equation of $f(a, \varphi)$ shows that variable a changes slowly, because its derivative $a' = O(\varepsilon)$. During this time interval when phase φ will change to the value of 2π, both amplitude and vibration character will change slightly. Therefore, the right hand sides of system (3.181) can be substituted by their averaged value within the period. In other words, instead of system (3.181), the following one will be considered

$$\begin{cases} a' = -\dfrac{\varepsilon}{\varpi}\overline{f}_1(a) \\ \varphi' = \varpi - \dfrac{\varepsilon}{\varpi a}\overline{f}_2(a) \ , \end{cases} \tag{3.182}$$

where:

$$\begin{cases} \overline{f}_1(a) = \dfrac{1}{2\pi}\displaystyle\int_0^{2\pi} f(a\cos\varphi, -a\varpi\sin\varphi)\sin\varphi d\varphi, \\ \overline{f}_2(a) = \dfrac{1}{2\pi}\displaystyle\int_0^{2\pi} f(a\cos\varphi, -a\varpi\sin\varphi)\cos\varphi d\varphi \ . \end{cases} \tag{3.183}$$

Equations (3.182) are called *truncated equations* or *Van der Pol equations.* They are simpler than initial equations (3.181), since the first of them can be integrated independently on the second one. In the system (3.182), slow and fast motion components are separated, and therefore averaging methods are also called *asymptotic methods of motion separation*. By integrating the first equation of (3.182), time history of amplitude $a(t)$ is found, which is sufficient in many practical problems. Phase is also defined by integration. However, in practice, the velocity of its variations is more important. It is directly defined by the second equation of the system (3.182).

To briefly conclude the considerations of this section, let us emphasize that Van der Pol method yields a solution of equation (3.175), and it consists of the following steps:

1. transition of variable u to variables a, φ;

2. obtaining the exact solutions (3.181) with the method of constants variation;

3. the exchange of exact equations with the truncated ones (3.182) and their integration.

3.8.2 Duffing's problem and the averaging procedure

Let us apply one of averaging methods to solve Duffing's problem. We have already described a transition from the variable u to the variables $a(t)$, $\beta(t)$ and then the use of constants variations yielded exact equations (3.172), (3.173).

Let us now construct solutions. In order to integrate analytically the equations (3.172), (3.173), the first number of powers of trigonometric functions occurring there will be decreased:

$$\sin\varphi\cos^3\varphi = \frac{1}{2}\sin\varphi\cos\varphi(1+\cos 2\varphi) =$$

$$= \frac{1}{4}\sin 2\varphi + \frac{1}{4}\sin 2\varphi\cos 2\varphi = \frac{1}{4}\sin 2\varphi + \frac{1}{8}\sin 4\varphi \ , \tag{3.184}$$

$$\cos^4\varphi = \frac{(1+\cos 2\varphi)^2}{4} = \frac{1}{4}\left(1 + 2\cos 2\varphi + \frac{1+\cos 4\varphi}{2}\right) =$$

$$= \frac{3}{8} + \frac{1}{2}\cos 2\varphi + \frac{1}{8}\cos 4\varphi \ . \tag{3.185}$$

Applying the introduced trigonometric identities (3.184), (3.185) and the change of variables $\varphi = t + \beta$, (3.172), (3.173) are transformed to the following form

$$a' = \frac{1}{8}\varepsilon a^3(2\sin 2\varphi + \sin 4\varphi) \ , \tag{3.186}$$

$$\beta' = \frac{1}{8}\varepsilon\, a^2(3 + 4\cos 2\varphi + \cos 4\varphi) \ . \tag{3.187}$$

Since the geometric functions are bounded, then $a' = O(\varepsilon)$, $\beta' = O(\varepsilon)$, i.e., for small ε functions $a(t)$, $\beta(t)$ slowly change in time. It means that for $t \in [0;\pi]$ they can be treated approximately as constant. Integrating in interval $[0;\pi]$ the equations (3.186) and (3.187), one gets

$$\frac{1}{\pi}\int_0^\pi a'\, dt = \frac{\varepsilon}{8\pi}\int_0^\pi a^3(2\sin 2\varphi + \sin 4\varphi)dt \ , \tag{3.188}$$

$$\frac{1}{\pi}\int_0^\pi \beta'\, dt = \frac{\varepsilon}{8\pi}\int_0^\pi a^2(3 + 4\cos 2\varphi + \cos 4\varphi)dt \ . \tag{3.189}$$

Since a, β in the interval $[0;\pi]$ can be treated approximately as constant values, then in (3.188) and (3.189) the quantities a, a', β' are excluded from the integral, i.e., the equations (3.188), (3.189) are averaged in interval $[0;\pi]$, to yield

$$a' = \frac{\varepsilon a^3}{8\pi}\int_0^\pi (2\sin 2\varphi + \sin 4\varphi)dt \ , \tag{3.190}$$

$$\beta' = \frac{3}{8}\varepsilon a^2 + \frac{\varepsilon a^2}{8\pi}\int_0^{\pi}(4\cos 2\varphi + \cos 4\varphi)dt\ . \tag{3.191}$$

Owing to the earlier introduced change $\varphi = t + \beta$ and to the condition that β is constant for $t \in [0;\pi]$, one gets $dt = d\varphi$; $\varphi \in [\beta; \pi + \beta]$. Then, the equations (3.190) and (3.191) take the form

$$a' = \frac{\varepsilon a^3}{8\pi}\int_{\beta}^{\pi+\beta}(2\sin 2\varphi + \sin 4\varphi)d\varphi = -\frac{\varepsilon a^3}{8\pi}\left(\cos 2\varphi + \frac{1}{4}\cos 4\varphi\right)\Bigg|_{\beta}^{\pi+\beta} = 0\ , \tag{3.192}$$

$$\beta' = \frac{3}{8}\varepsilon a^2 + \frac{\varepsilon a^2}{8\pi}\int_{\beta}^{\pi+\beta}(4\cos 2\varphi + \cos 4\varphi)d\varphi =$$

$$= \frac{3}{8}\varepsilon a^2 + \frac{\varepsilon a^2}{8\pi}\left(2\sin 2\varphi + \frac{1}{4}\sin 4\varphi\right)\Bigg|_{\beta}^{\pi+\beta} = \frac{3}{8}\varepsilon a^2\ . \tag{3.193}$$

The equation (3.192) yields

$$a = a_0 = const\ , \tag{3.194}$$

whereas the integration of (3.193) gives

$$\beta = \frac{3}{8}\varepsilon a_0^2 t + \beta_0\ , \tag{3.195}$$

where a_0, b_0 are arbitrary constants.

Substituting the values of a and β to (3.166), the first order approximation is obtained in the following form

$$u(t;\varepsilon) = a_0 \cos\left(t + \beta_0 + \frac{3}{8}\varepsilon a_0^2 t\right)\ , \tag{3.196}$$

which fully coincides with that obtained through Lindstedt-Poincaré method (3.47), method of rescaling and multiple scale method.

Note that the results (3.192), (3.193) can be obtained without averaging procedure through the analysis of dependencies (3.186), (3.187) and conclusions while applying the method of constants variations (Fig. 3.5). In the first approximation, the change of quantities a', β' is determined by the slowly changing terms appearing in the right-hand sides of (3.186) and (3.187). The slowly changing part of equations (3.187) is equal to zero, i.e., $a' = 0$, and $\frac{3}{8}\varepsilon a^2$, $\beta' = \frac{3}{8}\varepsilon a^2$. The fast changing terms appearing in the right-hand sides of (3.186), (3.187) are proportional to sinus and cosinus and cause small oscillations around straight lines $a = a_0$, $\beta = \frac{3}{8}\varepsilon a_0^2 t$ (Fig. 3.5).

Observe that the idea of separating "fast" and "slow" variables in averaging procedures motivated also the introduction of "fast" and "slow" scales in the multiple scales method. Although two of the mentioned ideas are realized differently, in many cases they give equivalent results [118, 120].

3.9 Matching asymptotic decompositions

In the majority of considered problems, the influence of perturbation has been small. However, its permanent action may lead to the accumulation of nonlinear effects, and hence, the asymptotic decompositions constructed with the use of the classical perturbation approach are often not useful in large timescales or for large distances. There is a wide spectrum of problems in physics, aero- and hydrodynamics, and in theory of elasticity, among others, where additional essential nonlinear phenomena are exhibited in small (with respect to the area) subspaces of a considered space. In order to study this kind of problem, one of the most important methods of the theory of singular perturbations has been developed, i.e., *the method of matching asymptotic decompositions*. The ideas related to this method, as well as many researchers of many countries have developed the associated mathematical tools.

L. Prandtl belongs to the pioneers of this scientific branch with his study of *a boundary layer*. Namely, he analyzed a viscous flow around a body, and a flow field has been divided by him into two subspaces, i.e., that lying in the vicinity of the body, where flow viscosity (*boundary layer*) plays an essential role, and the rest of the space, where flow viscosity can be neglected. The initially stated problem has been converted into two simpler problems, governed by simpler differential equations. The solutions to each of the separated problems have been constructed in a relatively simple manner, but a new problem of their matching by keeping a continuous transition between two subspaces that occurred.

The method of matched asymptotic decompositions generalizes the problem initially formulated by Prandtl. Mathematical background of this method has been developed by K. Friedrichs [71, 72]; P. Lagerstrom [94, 95]; M. Van-Dyke [166, 167, 168], J. Langer [100]; A. Nayfeh [119, 120, 121, 122, 123]; M.I. Vishik and L.A. Lusternik [173, 174]; A.B. Vasil'eva [169]; Burgers [56]; Keller [88]; Verhulst [171, 172] and others [166, 170, 63, 118, 119].

3.9.1 Fundamental notions and terminology

The aim of this section is to explain in a possibly simple manner and without rigorous mathematical tools the ideas and algorithms of the application of matched asymptotic decompositions and the terminology used by this theory.

Firstly, the example given in section 3.1 is revisited. Find a solution $u = u(x;\varepsilon)$ to the problem

$$\varepsilon\frac{du}{dx} + u = x\,, \quad u(o) = 1\,; \quad x \in [0;1]\,, \tag{3.197}$$

where ε is small parameter, $\varepsilon \ll 1$.

In section 3.3.1 it has been emphasized that when solving the stated problem with a classical perturbation method, in its defined space $D(x) = \{0 \leq x \leq 1\}$ two subspaces occur $D^o(x) = \{\delta_1(\varepsilon) \leq x \leq 1\}$ i $D^i(x) = \{0 \leq x \leq \delta_2(\varepsilon)\}$, where $\delta_1, \delta_2 > 0$, $\delta_1 < \delta_2$; $\delta_1, \delta_2 \to 0$ for $\varepsilon \to 0$ (an object-lesson manner role of δ_1, δ_2 is shown in Fig. 3.1. The subspace $D^i(x)$, called further *a boundary layer*), obtained through classical tools of perturbation technique (3.11), (3.12) is nonuniform, and therefore it requires a special approach during further analysis.

Let us introduce certain fundamental notions and definitions.

External variables – dimensionless dependent and independent variables introduced by the initial characteristics of variables characterizing a problem. In the considered example, x, $u(x;\varepsilon)$ are external dependent variables.

External decomposition – asymptotic decomposition of dependent variable (the sought function u) for $\varepsilon \to 0$ for the fixed values of independent external variables with respect to a certain external series of scaling functions $\mu_k(\varepsilon)$ of the form

$$u(x;\varepsilon) = u^o(x;\varepsilon) = \sum_{k=0}^{n} \mu_k(\varepsilon)\, u_k(x) + O(\mu_{n+1}(\varepsilon)) \ . \tag{3.198}$$

In our example, the external decomposition reads (it will be shown further in section 6.31 that it consists of two terms)

$$U(x;\varepsilon) = u^o(x;\varepsilon) = u_0(x) + \varepsilon u_1(x) \ , \tag{3.199}$$

and a superscript "o" comes from the word "outer". In this case, one finds

$$u^o(x;\varepsilon) = x + \varepsilon \ . \tag{3.200}$$

External space is a space, where external decomposition (3.198) holds. In our case, this subspace reads $D^o = \{\delta_1(\varepsilon) \leq x \leq 1\}$.

External limit is a limit of the external asymptotic decomposition (3.198), for $\varepsilon \to 0$ and for fixed external independent variables, i.e.,

$$\lim_{\varepsilon \to 0} u^o(x;\varepsilon) = \lim_{\varepsilon \to 0}(x + \varepsilon) = x \ . \tag{3.201}$$

Internal space is a space, where the application of external decomposition is improper (i.e., it is nonuniformly suitable), and where internal asymptotic decomposition holds. In the considered case $D^i(x) = \{0 \leq x \leq \delta_2(\varepsilon)\}$ (superscript "i" comes from the word "inner").

Internal variables are dependent dimensionless variables and the independent ones elongated (compressed) through certain functions in such a way as to keep an order of 1 in an internal space.

Internal variables are presented in the following way

$$x = \alpha(\varepsilon)X \ , \quad u(x;\varepsilon) = \beta(\varepsilon)U(X;\varepsilon) \ , \tag{3.202}$$

where the elongated (compressed) coefficients $\alpha(\varepsilon)$ and $\beta(\varepsilon)$ are chosen through the analysis of external decomposition (3.198) or through the application of the rule of minimal singularity realization. The choice of internal variables is further discussed in section 9.6. In the considered example $\alpha(\varepsilon) = \beta(\varepsilon) = \varepsilon$, since for $x = O(\varepsilon)$ the external decomposition (3.200) is nonuniform (both terms in (3.200) are of same order) and function $u^o(x;\varepsilon) = O(\varepsilon)$.

Internal equation and internal boundary value problem is governed by equations yielding $U(X;\varepsilon)$ and valid in D^i and these equations associated with boundary conditions are satisfied through $U(X;\varepsilon)$ in D^i, respectively. They are obtained according to the exact formulation of the problem with the use of variables transformation (3.202).

Internal decomposition is an asymptotic decomposition of dependent internal variable $U(X;\varepsilon)$ for $\varepsilon \to 0$ and for values of internal variables with respect to a certain "internal" asymptotic series of scaling comparison functions $\nu_l(\varepsilon)$

$$U(X;\varepsilon) = \sum_{l=0}^{m} \nu_l(\varepsilon) U_l(X) + O\left(\varepsilon^{m+1}\right) .$$

Then one gets

$$u(x;\varepsilon) = u^i(X;\varepsilon) = \beta(\varepsilon) U(X;\varepsilon) = \beta(\varepsilon) \left(\sum_{l=0}^{m} \nu_l(\varepsilon) U_l(X) + O(\nu_{m+1}(\varepsilon)) \right) . \tag{3.203}$$

In the considered example

$$u(x;\varepsilon) = u^i(X;\varepsilon) = \varepsilon \sum_{l=0}^{m} \varepsilon^l U_l(X) = \varepsilon U_0(X) , \tag{3.204}$$

since, as it will be proved further, $U_l(X) = 0$ for $l = 1, 2, \ldots$

Internal limit is a limit of an internal asymptotic decomposition (3.203), for $\varepsilon \to 0$, for fixed values of internal independent variables.

Overlapping space of decompositions is a space where both external and internal decompositions are suitable for application and they approximate in a satisfactory manner a sought solution to the problem defined through interval of (δ_1, δ_2).

Complex decomposition is a decomposition transformed into external (internal) decomposition for $\varepsilon \to 0$ or fixed external (internal) variables.

External and internal equations, boundary conditions of k-th order ($k \in Z$) are the equations, and the equations associated with the boundary conditions obtained from exact equations and boundary conditions after substituting external decompositions (3.198) into them, and after equating to zero the coefficients standing by $\mu_k(\varepsilon)$.

External solution of k-th approximation is a solution to the k-th order external boundary value problem.

Internal equation and external boundary conditions of l***-th order approximation*** ($l \in Z$) – the equation and equations associated with boundary conditions after the substitution of internal decompositions (3.203) and after equating to zero the coefficients standing by $\nu_l(\varepsilon)$.

Internal solution of l***-th order approximation*** is a solution of internal boundary value problem of l-th approximation.

Notice that in accordance with the method of matched asymptotic decompositions, a complex initial system is divided into two simpler problems in external space D^0 and in internal space D^i. After the construction of k-th approximation of external solution and of l-th approximation of internal solution (values of k, l are possibly maximal or sufficient for analysis requirement), the solutions are matched in a subspace of overlapping of spaces D^0 and D^i applying one of the methods of matching (matching rules will be further described in section 3.9.3). In the next step, if it is required, one may construct one complex asymptotic decomposition uniformly suitable in the whole space of the existence of solution (ways and example of construction of complex asymptotic decompositions are given in section 3.9.4).

Let us revisit the boundary value problem (3.197) associated with Cauchy problem. We are going to solve it, using matched asymptotic methods outlining the earlier introduced definitions and notion.

The outer *external expansion* is sought (3.198) in the form

$$u(x;\varepsilon) = u^0(x;\varepsilon) = \sum_{k=0}^{n} \varepsilon^k u_k(x) + O(\varepsilon^{n+1}) = u_0(x) + \varepsilon u_1(x) + \varepsilon^2 u_2(x) + O(\varepsilon^3) \; . \tag{3.205}$$

External series of scaling functions $\mu_k(\varepsilon) = \varepsilon^k$ is chosen due to the character of differential equation (3.197).

Substituting (3.204) into (3.197), one obtains the series of external boundary problems of k-th order (in this case, the initial condition (3.197) does not play any role) and external k-th order approximations read

$$\begin{aligned} k = 0 : &\quad u_0(x) = x \; , \\ k = 1 : &\quad u_1(x) = -u_0'(x) = -1 \; , \\ k = 2, 3, \dots : &\quad u_k(x) = -u_{k-1}(x) = 0 \; . \end{aligned}$$

The above considerations yield that *external decomposition* consists of two terms

$$u^0(x;\varepsilon) = x - \varepsilon \; . \tag{3.206}$$

Since for $x = O(\varepsilon)$ the external decomposition is nonuniform, then it can be applied only in *the external space* $D^o = \{\delta_1(\varepsilon) \le x \le 1\}$.

In *internal space* $D^i = \{0 \le x \le \delta_2(\varepsilon)\}$, *internal variables* X, $U(X;\varepsilon)$ are introduced through the dependencies

$$x = \varepsilon X \; , \quad u(x;\varepsilon) = u^i(x;\varepsilon) = \varepsilon U(X;\varepsilon) \; . \tag{3.207}$$

The form of the elongation functions $\alpha(\varepsilon) = \varepsilon$, $\beta(\varepsilon) = \varepsilon$ in (3.207) is motivated by the analysis of solution (3.206), since for $x = O(\varepsilon)$ we have $u^0(x;\varepsilon) = O(\varepsilon)$.

Substituting (3.207) into (3.197) and accounting for $\frac{du}{dx} = \frac{d(\varepsilon U)}{dX}\frac{dX}{dx} = \frac{dU}{dX}$, the following *internal boundary value problem* is obtained

$$\frac{dU}{dX} + U = X\,, \quad U(0) = \frac{1}{\varepsilon}\,. \tag{3.208}$$

Substituting *internal decomposition*

$$U(X;\varepsilon) = \sum_{l=0}^{m} \varepsilon U_l(X) + O(\varepsilon^{m+1}) \tag{3.209}$$

into (3.208), the internal boundary value problem of i-th order is obtained

$$\frac{dU_0}{dX} + U_0 = X\,, \quad U_0(0) = \frac{1}{\varepsilon}\,, \quad l = 0\,; \tag{3.210}$$

$$\frac{dU_1}{dX} + U_1 = 0\,, \quad U_l(0) = 0\,, \quad l = 1, 2, \ldots \tag{3.211}$$

A general solution of the first order linear equation (3.210) can be easily determined: $U_0(X, c) = ce^{-X} + X - 1$, where c is an arbitrary constant. After satisfying the initial conditions (3.210), a solution to the internal boundary value problem of zero order reads

$$U_0(x) = (1 + \frac{1}{\varepsilon})e^{-X} + X - 1\,. \tag{3.212}$$

After integration of a differential equation (3.211), one gets $U_l = c_l e^{-X}$, where c_l are arbitrary constants. The initial condition (3.210) yields c_l, $l = 1, 2, \ldots$

Finally, *the internal solution* holds in D^i, and according to (3.207), (3.209), (3.204) and (3.212), one gets

$$u(x;\varepsilon) = u^i(X;\varepsilon) = \varepsilon U_0(X) = (1+\varepsilon)e^{-X} + \varepsilon X - \varepsilon\,. \tag{3.213}$$

A natural question arises: How to match external (3.206) and internal (3.213) solutions?

Let us represent solutions u^0 and u^i in the same coordinates. For instance, in external coordinates we have

$$u^0 = x - \varepsilon\,, \quad u^i = (1+\varepsilon)e^{-x/\varepsilon} + x - \varepsilon\,.$$

Comparing u^i and $u_\varepsilon(x)$, one may observe that the internal solution coincides with the exact one(!) (This is not true, however).

Although the limits of these solutions for fixed value of x and for $\varepsilon \to 0$ coincide and they are equal to x, and the solutions coincide for $\varepsilon \to 0$ and $x > O(\varepsilon)$, i.e., in the external space D^0, they differ strongly in the internal space D^i.

The analyzed example does not illustrate matching in general case. Namely, in the considered system u^0 and u^i are defined uniquely and do not depend on each other, and the introduction of internal variables allowed to get exact solution to the problem! In problems that are more complicated (considered further), an external decomposition is not defined uniquely, i.e., it is composed of a set of decompositions. The process of their matching is focused on finding arbitrary constants, which means that one has to make a choice from a set of internal decompositions coinciding with the external one.

One may also meet the opposite situation, when an internal decomposition does not depend on external one, or even more complicated problem, when two of decompositions are mutually dependent.

3.9.2 Example with a boundary layer

One of the reasons for the occurrence of asymptotic decomposition singularity is the existence of a small parameter standing by the highest derivative of the considered differential equation. Applying classical method of small parameter, this derivative is "lost" in the first order approximating equation. The decrease of the order of differential equation does not allow to satisfy all boundary conditions. In the vicinity, where boundary conditions are not satisfied, an irregular subspace appears (a so-called *boundary layer*).

Consider an object-lesson example, consisting of the characteristic features of more complex problems. Find a solution of the following ordinary differential equation

$$\varepsilon y'' + y' + y = 0\,, \quad 0 \le x \le 1\,, \tag{3.214}$$

with the following boundary conditions

$$y(0) = a\,, \quad y(1) = b\,. \tag{3.215}$$

The equation (3.214) is the linear second order differential equation with constant coefficients, and the solution to two-point boundary value problem (3.214), (3.215) reads

$$y = \frac{(ae^{K_2} - b)e^{k_1 x} + (b - ae^{k_1})e^{k_2 x}}{e^{k_2} - e^{k_1}}\,; \quad k_{1,2} = \frac{-1 \pm \sqrt{1-4\varepsilon}}{2\varepsilon}\,. \tag{3.216}$$

Let us apply the classical method of perturbation, assuming that

$$y(x;\varepsilon) = y^0(x;\varepsilon) = \sum_{n=0}^{\infty} \varepsilon^n y_n(x)\,. \tag{3.217}$$

Substituting (3.217) into (3.214) and (3.215), and equating to zero the coefficients standing by the same powers of ε, the following series of simple problems occurs

$$y_0' + y_0 = 0\,, \quad y_0(1) = b\,;$$

$$y'_n + y_n = -y'_{n-1}\,, \quad y_n(1) = 0\,; \quad n \geq 1\,.$$

Solving these problems with respect to y_0 and y_1, equation (3.217) yields

$$y^0(x;\varepsilon) = y_0(x) + \varepsilon y_1(x) + o(\varepsilon^2) = be^{1-x} + \varepsilon be^{1-x}(1-x) + O(\varepsilon^2)\,. \quad (3.218)$$

Comparison of the obtained results with the exact solution (3.216) yields the conclusion that (3.218) approximates well the real solution everywhere except in the vicinity of the point $x = 0$ (Fig. 3.6), where the boundary condition $y(0) = a$ is not satisfied.

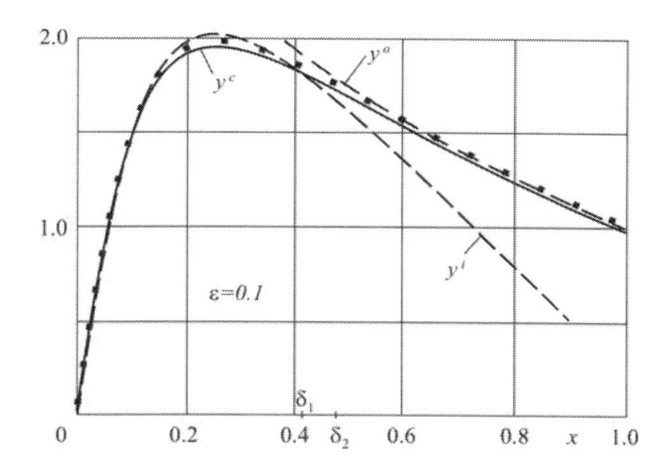

FIGURE 3.6: External y^o, internal y^i and composite y^c decompositions.

Let us apply the earlier introduced terminology, i.e., recall that x, y are external variables, (3.218) is an external asymptotic decomposition, $D^i = [0, \delta_2(\varepsilon)]$ is an internal subspace (boundary layer), $D^0 = [\delta_1(\varepsilon), 1]$ is an external subspace (δ_1, δ_2 are illustrated in Fig. 3.6).

In order to construct an internal decomposition in D^i, let us introduce new (dependent $X = x/\sigma(\varepsilon)$ and independent $Y(X;\varepsilon) = y/\nu(\varepsilon)$) variables. In the beginning, elongated functions $\sigma(\varepsilon)$, $\nu(\varepsilon)$ are not known and they will be determined in a computational process. Since the following relations hold

$$y' = \frac{dy}{dx} = \frac{dy}{dX}\frac{dX}{dx} = \frac{\nu}{\sigma}\frac{dY}{dX}\,; \quad y'' = \frac{\nu}{\sigma^2}\frac{d^2y}{dX^2}\,,$$

equation (3.214) can be transformed to the following one

$$\frac{\varepsilon}{\sigma^2(\varepsilon)}\frac{d^2y}{dx^2} + \frac{1}{\sigma(\varepsilon)}\frac{dY}{dX} + Y = 0\,. \quad (3.219)$$

The form of this equation allows to take $\nu(\varepsilon) = 1$.

Let us choose $\sigma(\varepsilon)$ according to *the rule of smallest singularity*. We are going to choose the internal first order approximation from (3.219) in order to keep the most essential elements omitted in external solution, at the same time preserving possibly small singularity. For $o(\varepsilon) \leq \sigma(\varepsilon) \leq O(\varepsilon)$ the highest derivative in (3.219) is preserved. Furthermore, if we take $\sigma(\varepsilon) = \varepsilon$, the first derivative will be also preserved.

Internal equation reads

$$\frac{d^2Y}{dX^2} + \frac{dY}{dX} + \varepsilon Y = 0 \; . \tag{3.220}$$

Internal variables are introduced in the following way

$$x = \varepsilon X \, , \quad y(x;\varepsilon) = Y(X;\varepsilon) = Y_0(X) + \varepsilon Y_1(X) + \ldots \tag{3.221}$$

Substituting (3.221) into (3.220) and into the internal boundary condition $y(0) = a$, the following series of simple initial problems is achieved

$$Y_0'' + Y_0' = 0 \, , \quad Y_0(0) = a \, ;$$

$$Y_n'' + Y_m' = -Y_{n-1} \, , \quad Y_n(0) = 0 \, ; \quad n \geq 1 \; .$$

Observe that internal equations are of second order, hence one may expect that also the second boundary condition of (3.215) can be satisfied $y(1) = b$ for $x = 1$.

However, this not allowed. The internal solution is unsuitable there and it cannot be used.

Solving the problems of Y_0 and Y_1, the following (3.221 taken into account) internal asymptotic decomposition in internal subspace D^i is obtained

$$y^i(X,\varepsilon) = Y_0(X) + \varepsilon Y_1(X) + O(\varepsilon^2) = a - A_0(1 - e^{-X}) +$$

$$\varepsilon \left\{ A_1(1 - e^{-X}) - \left[a - A_0(1 + e^{-X}) \right] X \right\} + O(\varepsilon^2) \; . \tag{3.222}$$

Notice that (3.222) is a set of infinite decompositions depending on constants A_0, A_1. The values of A_0, A_1 should be determined by matching the decompositions (3.222) and (3.218), which yields the internal decomposition coinciding with external one.

3.9.3 Fundamental rules and order of matching

As it has already been mentioned, a matching procedure allows to omit singularities appearing in either internal or external solutions, and to obtain one combined solution. For example, in the example considered above, matching should allow to determine the constants A_0, A_1, tj. i.e., it should yield one solution from a set of infinitely internal ones (3.222) coinciding with the external solution (3.218).

Notice that matching two solutions can be realized, when internal and external solutions have the same overlapping space (*space of general applicability*). How should a process of matching solutions be realized? Let us first give the matching rules outlined in the reference [166], and let us illustrate their applications using decompositions (3.218) and (3.222).

The most general rule is: *"The internal decomposition of an external decomposition is equal to the external decomposition of an internal decomposition."* This rule has a different form in different practical applications.

L. Prandtl has already formulated one of simplest rules: *"the internal limit of external limit is equal to the external limit of internal limit."*

Let us apply this rule. Equation (3.218) yields external boundary (x is fixed)

$$\lim_{\varepsilon \to 0} y^o(x;\varepsilon) = be^{1-x} = \overline{y^o}(x) \ .$$

Let us find the internal limit of external limit accounting for $x = \varepsilon X$ (X is fixed)

$$\lim_{\varepsilon \to 0} \overline{y^o}(\varepsilon X) = b \lim_{\varepsilon \to 0} e^{1-\varepsilon X} = be \ . \tag{3.223}$$

Next, let us find from (3.222) the internal limit (X is fixed)

$$\lim_{\varepsilon \to 0} y^i(X;\varepsilon) = a - A_0(1 - e^{-X}) = \overline{y^i}(X)$$

and let us find the internal limit of this limit (x is fixed)

$$\lim_{\varepsilon \to 0} y^{-i}\left(\frac{x}{\varepsilon}\right) = a - A_0 + A_0 \lim_{\varepsilon \to 0} e^{-\frac{x}{\varepsilon}} = a - A_0 \ . \tag{3.224}$$

According to the rule of limiting matching, let us compare (3.223) and (3.224), to get

$$A_0 = a - be \ . \tag{3.225}$$

These considerations show that this rule can be applied (but not always [166]) to match approximations to the first internal and external solutions.

In order to match higher order approximations, M. Van-Dyke proposed the following rule allowing for matching internal and external asymptotic decompositions keeping their arbitrary number of terms. It is as follows: *"the m-th term internal decomposition of n-th term external decomposition is equal to n-th term external decomposition of m-th term internal decomposition,"* where $m, n \in N$.

The outlined rule can be presented in the form

$$\left(y_n^o\right)_m^i = \left(y_m^i\right)_n^o \ . \tag{3.226}$$

Let us apply this rule to decompositions (3.218), where $n = 2$, and to (3.222), where $m = 2$.

In order to determine $(y_2^o)^i$ let us present (3.218) in the form of internal variables. They are expanded for $\varepsilon \to 0$, where X s fixed by keeping only two higher order terms

$$\left(y_2^i\right)_2^o = be^{1-\varepsilon X} + \varepsilon be^{1-\varepsilon X}(1-\varepsilon X) = be(e^{-\varepsilon X}+\varepsilon) = be(1-\varepsilon X+\varepsilon) \ . \quad (3.227)$$

In order to determine $(y_2^o)_2^i$, let us present (3.222) in the form of external variables, and let us expand the obtained formula into series for $\varepsilon \to 0$ and for fixed x keeping only two terms of higher order

$$\left(y_2^i\right)_2^o = a - A_0\left(1-e^{-\frac{x}{\varepsilon}}\right) + \varepsilon\left\{A_1(1-\varepsilon^{-\frac{x}{\varepsilon}}) - \left[a - A_0\left(1+e^{-\frac{x}{\varepsilon}}\right)\right]\frac{x}{\varepsilon}\right\} =$$

$$= a - A_0 + \frac{A_0}{e^{\frac{x}{\varepsilon}}} + \varepsilon A_1 - \frac{\varepsilon A_1}{e^{\frac{x}{\varepsilon}}} - ax + A_0 x + \frac{A_0 x}{e^{\frac{x}{\varepsilon}}} =$$

$$= (a - A_0)(1-x) + \varepsilon A_1 + (EST) \ , \quad (3.228)$$

where EST denote exponentially small terms, which are neglected.

In order to compare (3.227) and (3.228) one has to reduce the problem to the same variables. Transforming (3.228) to external variables, one gets

$$(y_2^o)_2^i = be(1-x) + \varepsilon be \ . \quad (3.229)$$

Comparing (3.229) and (3.228) we have

$$A_0 = a - be\,, \quad A_1 = be\,, \quad (3.230)$$

and finally, the following internal solution is defined

$$y^i = be + (a-be)e^{-X} + \varepsilon\left\{be\left(1-e^{-X}\right) - \left[be - (a-be)e^{-X}\right]X\right\} + O(\varepsilon^2) \ . \quad (3.231)$$

Notice that for $m = n = 1$ equation (3.226) yields the same result (3.225), similar to that of limiting matching. Formulas (3.229) and (3.228) yield also that for $m = 1$ and $n = 2$ determination of A_1 is not possible.

In the considered case, all external approximations do not depend on internal ones and can be determined independently, but the internal approximation in each step for $m = n \geq 3$ depends on them. In general case, the described process is not true, and matching internal and external decompositions is realized step by step through the following chain: $m = n = 1$, $m = 2$, $n = 1$; $m = n = 2$, and so on. Both decompositions are mutually interacting, causing the improvement of accuracy of asymptotic functions series matching internal and external decompositions.

In the considered example, the series of functions matching external and internal decompositions of the form $\{\varepsilon^n\}|_0^\infty$ coincides. However, this is not true in majority of the considered cases (see [106]). If it is not known how to number terms of decomposition for a function y, one may use the more general matching rule of the following form: *"The internal decomposition of* *order $\Delta(\varepsilon)$ of external decomposition of order $\delta(\varepsilon)$ is equal to the external decomposition of order $\delta(\varepsilon)$ of the internal one of order $\Delta(\varepsilon)$,"* where $\delta(\varepsilon)$, and $\Delta(\varepsilon)$ are two arbitrary functions of the same order. This rule may be also formulated as:

$$\left(y^0_{\delta(\varepsilon)}\right)^i_{\Delta(\varepsilon)} = \left(y^i_{\Delta(\varepsilon)}\right)^0_{\delta(\varepsilon)} . \tag{3.232}$$

Notice that this principle takes the form

$$\left(y^0_\varepsilon\right)^i_\varepsilon = \left(y^i_\varepsilon\right)^0_\varepsilon .$$

3.9.4 Construction of matched asymptotic expansion

The illustrated problem of development of a singular perturbation in the form of two matched decompositions, i.e., external y^o and internal y^i, may lead to the following fundamental question: Where and how is it possible to transit from one to the other decomposition?

A rough recipe for such transition is indicated by the point of intersection of curves representing external and internal solutions ("matching" solutions). However, in this case, the so-called *singular points* may appear. Furthermore, it may happen that curves y^o and y^i do not overlap.

However, matched decompositions possess *the areas of overlapping*, i.e., areas where both internal and external decompositions are suitable. The reason for overlapping decompositions is explained by a theorem given by S. Kaplun [86]. Namely, if an asymptotic decomposition is uniformly suitable in a certain interval of variations of x, then it is also suitable in some wider interval depending on a small parameter ε (the formulation and proof of the theorem can be found in [63]).

Occurrence of the area of overlapping of the decompositions y^o and y^i allows to construct one decomposition y^c instead of two, which is uniformly suitable in the whole area of definition of function y. This decomposition is called *composite.*

The construction of composite decomposition can be realized in various ways. Although the obtained results will be different, because there are many composite solutions, all of them will be equivalent with respect to an assumed order of smallness. In practice, two essentially different ways are realized: *additive* and *multiplicative.*

Additive complex decomposition y^c_a is a sum of external and internal decompositions with neglecting their common part, i.e., the last one is not taken two times

$$y^c_a = y^o + y^i - (y^o)^i = y^o + y^i - (y^i)^o . \tag{3.233}$$

According to

$$\left((y^o)^i\right)^o = (y^o)^i = (y^i)^o = \left((y^o)^i\right)^i ,$$

from (3.233), one gets

$$(y_a^c)^o = y^o + (y^i)^o - (y^o)^i = y^o \ , \tag{3.234}$$

$$(y_a^c)^i = (y^o)^i + y^i - (y^o)^i = y^i \ . \tag{3.235}$$

Dependencies (3.234) and (3.235) mean that the additive composite decomposition transits to the external (internal) decomposition in external (internal) area.

Multiplicative composite decomposition y_m^c is equal to the ratio of a product of external and internal decompositions, and their general part

$$y_m^c = \frac{y^o y^i}{(y^o)^i} = \frac{y^o y^i}{(y^i)^o} \ . \tag{3.236}$$

The obtained relation can not be used in the case of $(y^o)^i = (y^i)^o = 0$.

The accuracy of composite solution is smaller than that of external and internal solutions in proper spaces. For instance, for external (3.218) and internal (3.231) decompositions, the additive composite decomposition has the form

$$y^c = b\,[1 + \varepsilon(1-x)]\,e^{1-x} + [(a-be)(1+x) - \varepsilon be]\,e^{\frac{-x}{\varepsilon}} + O(\varepsilon^2) \ . \tag{3.237}$$

In Fig. 3.6 solutions to the problem (3.214), (3.215) for $a = 0$, $b = 1$, $\varepsilon = 0.1$, where y^o is the external decomposition (3.218), y^i is the internal decomposition (3.231), and y^c jest is the additive decomposition (3.237) are shown. Notice that a composite solution y^c overlaps practically with the exact solution (3.216).

3.9.5 Example with a singularity

We are going to study the initial problem (3.59)

$$(x + \varepsilon y)y' + y = 1 \, , \quad y(1) = 2 \ , \tag{3.238}$$

which has been considered in section 3.4, 3.5, using the matched asymptotic decompositions method. Recall that in section 3.4 the following external asymptotic decomposition (3.61) is obtained:

$$y^o(x;\varepsilon) = y_0(x) + \varepsilon y_1(x) + O(\varepsilon^2) =$$

$$= \frac{1+x}{x} - \frac{\varepsilon(1-x)(1+3x)}{2x^3} + O(\varepsilon^2) \ , \tag{3.239}$$

which has a singularity for $x = 0$ and is useless for $x = O(\varepsilon^{1/2})$.

In order to confirm the statement, one may determine x, for which the terms of decomposition (3.239) have the same magnitude order. For this purpose, the terms of (3.239) in the vicinity of $x = 0$ are compared. As a result, one gets $1/x \sim \varepsilon/x^3$. It means that the external decomposition is irregular for

$$x = O(\varepsilon^{1/2}) . \tag{3.240}$$

According to (3.239), one gets

$$y^o(x;\varepsilon) = O(\varepsilon^{-1/2}) . \tag{3.241}$$

Using (3.240) and (3.241), let is introduce internal variables X, F of order 1 in internal space in the vicinity of $x = 0$, assuming that

$$X = x/\varepsilon^{1/2} , \quad F(X;\varepsilon) = \varepsilon^{1/2} y(x;\varepsilon) . \tag{3.242}$$

Substituting (3.242) into (3.238) and according to formula

$$y' = \frac{dy}{dx} = \frac{1}{\varepsilon^{1/2}} \frac{dF}{dX} \frac{dX}{dx} = \frac{1}{\varepsilon} \frac{dF}{dX} ,$$

one gets the following internal equation

$$(X + F)\frac{dF}{dX} + F = \varepsilon^{1/2} . \tag{3.243}$$

The initial condition (3.238) holds in external space, and hence it may be omitted. An arbitrary constant appearing during the integration of (3.243) can be used further in order to match external and internal solutions.

The form of equation (3.243) allows to use the following internal decomposition

$$F(X;\varepsilon) = F_0(X) + \varepsilon^{1/2} F_1(X) + O(\varepsilon) . \tag{3.244}$$

Substituting (3.244) into (3.243)and comparing the coefficients standing by same powers of ε, the following series of internal equations is obtained

$$(X + F_0)F_0' + F_0 = 0 ,$$

$$(X + F_0)F_1' + F_1 F_0' + F_1 = 1 , \quad \text{and so on.} \tag{3.245}$$

In order to compare the obtained results with those given in sections 3.4, 3.5, it is sufficient to determine only F_0, which due to (3.245) has the following form

$$F_0(X) = \sqrt{X^2 + 2c_1} - X . \tag{3.246}$$

In order to match y^o with (3.239) and F_0 with (3.246) the rule of limiting matching will be applied. The external limit is found from (3.239) (x is fixed) of the form

$$\lim_{\varepsilon \to 0} y^o(x;\varepsilon) = \frac{1+x}{x} = \overline{y^o}(x) .$$

In what follows, we determine *the limit of the external limit*, taking into account that $x = \varepsilon^{1/2}X$, and next we return to the external variable (X is fixed), to get

$$\lim_{\varepsilon\to 0} \overline{y^o}(\varepsilon^{1/2}X) = \lim_{\varepsilon\to 0} \frac{1+\varepsilon^{1/2}X}{\varepsilon^{1/2}X} = \frac{1}{\varepsilon^{1/2}X} = \frac{1}{x} . \tag{3.247}$$

The internal limit, according to (3.242), (3.244) and (3.246), reads

$$y^i(X;\varepsilon) = \varepsilon^{-1/2}(\sqrt{X^2+2c_1} - X) .$$

Let us determine now *the external limit of internal limit* (x is fixed):

$$\lim_{\varepsilon\to 0}\left(\varepsilon^{-1/2}\left(\sqrt{\frac{x^2}{\varepsilon}+2c_1} - \frac{x}{\varepsilon^{1/2}}\right)\right) = \lim_{\varepsilon\to 0}\left(\sqrt{\frac{x^2}{\varepsilon^2}\left(1+\frac{2c_1\varepsilon}{x^2}\right)} - \frac{x}{\varepsilon}\right) =$$

$$\frac{x}{\varepsilon}\left(1+\frac{c_1\varepsilon}{x^2}+\dots\right) - \frac{x}{\varepsilon} = \frac{c_1}{x} . \tag{3.248}$$

Comparing (3.247) and (3.248), one gets $c_1 = 1$ and the following internal solution is obtained

$$y^i(x;\varepsilon) = \sqrt{(x/\varepsilon)^2 + 2/\varepsilon} - x/\varepsilon ,$$

which coincides with the accuracy of $O(\varepsilon)$ with both the external solution (3.60) and the solution (3.69), obtained with the use of the methods of deformed coordinates and rescaling.

At the end of this section, let us emphasize that methods of singular perturbations (3.238), may be ordered in the following sequence with respect of their simplicity and efficiency in applications: rescaling, matched asymptotic expansions, deformed coordinates method.

3.9.6 On the choice of internal variables

Assume that the asymptotic decomposition of a sought function $u(x;\varepsilon)$ has the following form

$$u(x;\varepsilon) = \sum_{i=0}^{N} \mu_i(\varepsilon)u_i(x) + O\left(\mu_{N+1}(\varepsilon)\right), \tag{3.249}$$

where $\mu_i(\varepsilon)$ is a known series of asymptotic scaling functions. As it has been already mentioned, this series does not need to be the simplest one, i.e., of the form ε^n, $n = 0,1,2,\dots$, and it is defined in the process of finding a strictly defined problem. Substituting (3.249) to exact equations and attached boundary conditions for $u(x;\varepsilon)$, the following successive problems serving for the determination of $u_0(x)$, $u_1(x)\dots$ are found. According to the earlier assumptions, one may always find a solution u_0. However, finding u_1, is impossible in majority of cases (section 3.4, 3.5 and the reference [106]). Therefore, computations are often limited to the determination of asymptotics $u_1^*(x) = \lim_{x \to x_*} u(x)$, where x_* is a certain value taken from a definition space of the function $u(x;\varepsilon)$, for which the solution (3.249) is not uniformly suitable.

Generally, one deals with two following fundamental cases:

First of them is focused on the study of solution $u_0(x)$. Namely, it is shown that u_0 and (or) its derivatives have singularities in the vicinity of a certain $x = x_*$, i.e., in this area, u_0 and (or) $u_0', u_0'', \ldots$ increase unlimitedly (see chapters 3 and 4, and [106]). In the mentioned area, the fundamental assumption of a linear theory is not satisfied. Namely, the function u_0 and its derivatives are not limited in the whole range of its definiteness, and decomposition (3.249) becomes the irregular one.

Second case shows by the analysis of asymptotic decomposition (3.249) or its asymptotic that in the vicinity of certain $x = x_*$ first two terms of decomposition (3.249) have same order

$$\mu_0(\varepsilon)u_0^*(x) \sim \mu_1(\varepsilon)u_1^*(x), \tag{3.250}$$

where:

$$u_0^*(x) = \lim_{x \to x_*} u_0(x)\,, \quad u_1^*(x) = \lim_{x \to x_*} u_1(x)\,,$$

(see simple example included in section 3.1, and a more complex one given in chapter 4). Condition (3.250), according to the definition (2.26) means, that the decomposition (3.249) is not regular in the vicinity of $x = x_*$.

Observe that in both cases, a problem associated with defining a magnitude decomposition nonuniformity (internal space), as well as with the choice of independent internal variable X and dependent variable $U(X;\varepsilon)$ occurs.

In general case, internal variables can be formulated as follows

$$x = \alpha(\varepsilon)X\,, \quad u(x;\varepsilon) = \beta(\varepsilon)U(X;\varepsilon)\,, \tag{3.251}$$

where expansion or compression coefficients $\alpha(\varepsilon)$, $\beta(\varepsilon)$ should be determined. There are two fundamental ways of determining these coefficients.

The most economical way of the analysis is associated with the formula (3.250). Assuming $\mu_0, \mu_1, \mu_0^*, \mu_1^*$ to be known, from (3.250) one may define $x = x_*(\varepsilon)$ and after its substitution into (3.249) and (3.250) one may define the order of $u(x;\varepsilon)$ in the vicinity of $x = x_*$ (on external D^o and internal D^i limits of subspaces of D). Both obtained results, as well as the condition stating that internal variables X, U should be of order 1 in an internal subspace, define at once functions $\alpha(\varepsilon)$ and $\beta(\varepsilon)$. Examples are given in chapter 4 and in reference [106].

If one of the described ways is not applicable, than the other one is recommended to define $\alpha(\varepsilon)$ and $\beta(\varepsilon)$. It focuses on the rule of smallness singularity of internal equation and has been illustrated while getting from (3.219) the internal equation (3.220).

3.10 On the sources of nonuniformities

According to the so far given examples and problems in this and previous chapters, references [75, 76, 118, 119, 166] and other, it is rather obvious that singular perturbations are typical in theory and applications. If to solve the mentioned problems classical perturbation techniques are used, then the obtained asymptotic decompositions will be nonuniformly suitable, i.e., there are subspaces where these decompositions are unsuitable.

In one-dimensional cases governed by ordinary differential equations such singularities occur in the vicinity of points. In two-dimensional cases governed by partial differential equations, the areas of nonuniformities on a plain lie in the vicinity of curves ([106], § 6.6). In space, the areas of irregularity are localized in the areas of a surface, a curve or a point.

It is important to illustrate and discuss the sources (reasons) of the occurrence of such nonuniformities.

One of them is the mathematical reason, since not all manipulations of asymptotic decompositions have a rigorous mathematical substantiation (chapter 2, section 2.5).

Other reasons are associated with the nature of the analyzed problems occurring during their mathematical formulation. Below, five more important are described.

1. ***Infinite space of a sought function definition.***

   As examples, various problems of nonlinear vibrations may serve (chapter, section 2.3, chapter 4, section 2, [18, 16, 118, 119]). A solution to the mentioned problems with the use of a classical perturbation approach contains terms proportional to time, and they are useless for large time intervals.

   Similar phenomena of localization of small nonlinear perturbations with increase of time and/or distance (accumulation effects) appear in problems related to nonlinear waves with finite amplitude movement, impact waves in fluids, elastic nonlinear waves and other (chapter 4, section 4.2, chapter 5, section 2,3, [106]).

2. ***Small parameter standing by the highest derivative of analyzed differential equation.***

   2.1. *Small parameter standing by the highest derivative of an ordinary differential equation.*

   In this case, the application of classical perturbation approach yields at once the decrease of an order of an analyzed differential equation and to linking *the boundary layers* (3.1), (3.214).

   2.2. *Small parameter standing by one of the highest derivatives of a partial differential equation.*

Consider briefly an equation important in application, governing the steady state of a gas of the form

$$(1 - M^2)\Phi_{X_X} + \Phi_{Y_Y} = f(X, T, \Phi, \Phi_X, \Phi_Y) , \qquad (3.252)$$

where $\Phi(X, Y)$ is the velocity potential, M is Mach number, $|1 - M| \ll 1$, and f is a nonlinear function.

The introduction of the small parameter $\varepsilon = |1 - M| \ll 1$ and construction of a solution Φ in the form of asymptotic decomposition

$$\Phi(X, Y; \varepsilon) = \varepsilon\Phi_1(X, Y) + \varepsilon^2\Phi_2(X, Y) + O(\varepsilon^3) , \qquad (3.253)$$

yields vanishing of the first approximation of the higher derivative Φ_{1XX} and decomposition (3.253) is unsuitable for application.

3. ***Change of a partial differential equation type.***

We consider the equation (3.252) for $|1 - M| \ll 1$. If $M < 1$ (subsonic flow), the equation (3.252) is elliptic. If $M > 1$ (supersonic flow), then the equation (3.10) is hyperbolic. It is easy to conclude, that after a transition through the point $M = 1$, when the change of differential equation type (3.252) occurs, one may expect the occurrence of nonuniformly suitable asymptotic decomposition (3.253) (recall that solutions of elliptic and hyperbolic types equations differ in essential way).

4. ***The occurrence of singularities in a direct asymptotic decomposition.***

It often happens that during the application of a classical method of small parameter, some singularities appear already in zero order approximation, which are not associated with an exact solution at all. Furthermore, in higher order approximations the mentioned singularities are even more exhibited. We consider example (3.59) where a solution is obtained through a direct use of a classical perturbation parameter (3.61). On the other hand, an exact solution (3.60) of the problem (3.59) does not have any singularities.

5. ***Singularities of initial and boundary conditions.***

Many examples of the occurrence of nonuniformities in asymptotic decomposition associated with initial and/or boundary conditions are considered in references [106, 118, 119, 166].

3.11 On the influence of initial conditions

In reference [138], the following problem is studied. Consider the following Duffing's equation

$$\ddot{x} + \omega_0^2 x + \varepsilon x^3 = 0 \tag{3.254}$$

with the following attached boundary conditions

$$x(0) = A\,, \quad \dot{x}(0) = 0\,. \tag{3.255}$$

The application of Lindstedt-Poincaré method [138, 119] allows for the approximation of the initial problem (3.254), (3.255) through the power series with respect to a perturbation parameter ε. However, one may use another way of initial conditions (3.255) application. If for $t = 0$ one takes (3.255), then the following *amplitude frequency characteristic* [1] is obtained

$$\omega = \omega_0 + \frac{3A^2}{8\omega_0}\varepsilon - \frac{21A^4}{16^2\omega_0^3} + \frac{81A^6\varepsilon^3}{8 \cdot 16\omega_0^5} + \ldots \tag{3.256}$$

Nayfeh [119] proposes another approach, where in a first approximation to the problem, the amplitude is treated as unknown. Hence, one gets

$$\omega = \omega_0 + \frac{3\alpha^2\varepsilon}{8\omega_0} - 15\frac{\alpha^4\varepsilon^4}{16^2\omega_0^3} + \ldots \tag{3.257}$$

and α is defined by the equation

$$A = \alpha + \frac{\alpha^3\varepsilon}{32\omega_0^2} - \frac{5\alpha^5\varepsilon^2}{16^2\omega_0^4} + \ldots \tag{3.258}$$

However, these two different methods are equivalent in the asymptotic sense.

The inverse form of (3.258) and application of the series

$$\alpha = A + \varepsilon\varphi_1(A) + \varepsilon^2\varphi_2(A) + \ldots \tag{3.259}$$

into equation (3.257) yields the equation (3.256). Observe that real and asymptotic errors represent aspects of the same problem. In what follows, we show [8], that the accuracy of the solution depends on the form of both assumed solution and of initial conditions (see also [61]). The results of numerical simulation are reported in Figure 3.7, where curves 1–3 correspond to the following solutions: solution (3.256); exact solution obtained through elliptic functions [119]; solution defined through (3.257) and (3.258). In the last case, first α has been found numerically from equation α and then this value has been substituted to equation (3.258), and (3.257). After the analysis of reported drawings, one may feel convinced that the method proposed by Nayfeh gives better results.

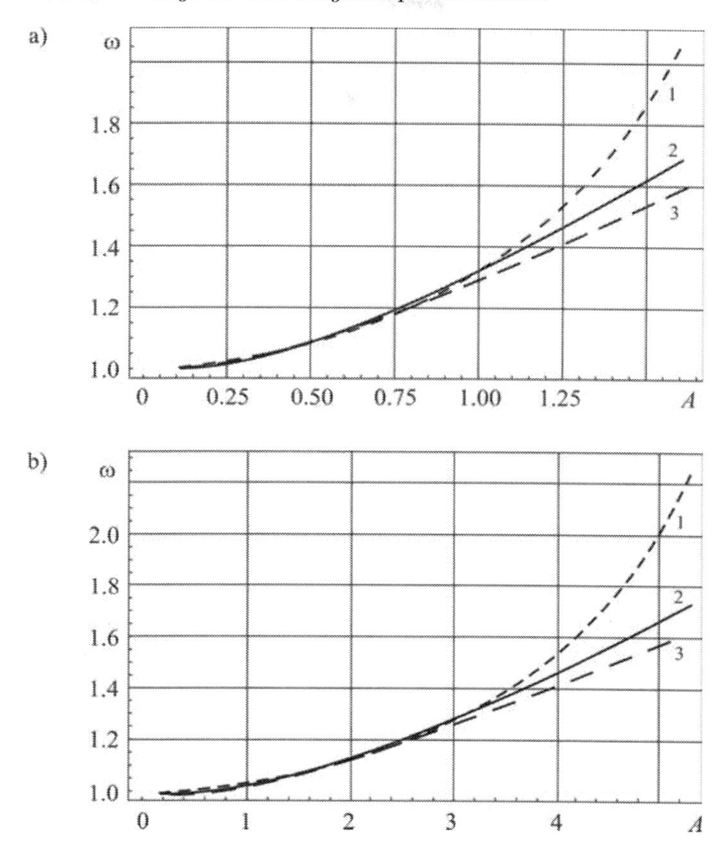

FIGURE 3.7: Comparison of results obtained through different analytical methods: a) $\omega_0 = 1$, $\varepsilon = 0.1$; b) $\omega_0 = \varepsilon = 1$.

As it has been already observed [166, 167, 168, 14, 29], an asymptotic series approximates a sought solution better when the structure of the problem is defined more exactly. For example, according to an asymptotic approach, the following two amplitude-frequency characteristics are equivalent

$$\omega = \omega_0 + \varepsilon\omega_1 + \varepsilon^2\omega_2 + \dots \, , \tag{3.260}$$

$$\omega^2 = \omega_0^2 + \varepsilon\gamma_1^2 + \varepsilon^2\gamma_2^2 + \dots \tag{3.261}$$

However, the quantitatively better results are obtained through the application of formula (3.261) [14, 29].

Consider now a pendulum vibrations governed by the following equations

$$\ddot{\theta} + \sin\theta = 0 \, , \tag{3.262}$$

$$\theta(0) = \theta_0 \, ; \quad \dot{\theta}(0) = 0 \, . \tag{3.263}$$

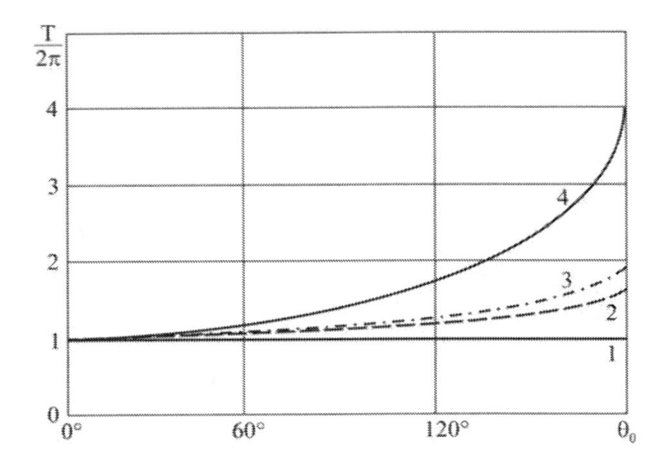

FIGURE 3.8: Period of vibrations of a mathematical pendulum vs initial deflection θ_0.

A classical method of solution to this problem relies on development of $\sin\theta$ into series with respect to θ, and next on the application of a perturbation technique.

However, the approach that is more effective relies on the application of the transformation

$$\sin\theta = \varphi \ . \tag{3.264}$$

The initial Cauchy's problem is defined by the equations (3.262), (3.263), defined by the following equations

$$\ddot{\varphi} + \frac{\varphi\dot{\varphi}^2}{1-\varphi^2} \dot{+} \varphi\sqrt{1-\varphi^2} = 0 \ , \tag{3.265}$$

$$\varphi(0) = \varphi_0 \equiv \sin\theta_0 \ ; \quad \dot{\varphi}(0) = 0 \ . \tag{3.266}$$

The solutions to problems (3.262), (3.263) and (3.265), (3.266) can be found using a perturbation method [135]. In the first case one gets

$$T = 2\pi(1 + \theta_0^2/16 + \ldots) \ . \tag{3.267}$$

whereas in the second one we have

$$T = 2\pi[1 + 0.25\sin^2(0.5\theta_0) + (9/64)\sin^4(0.5\theta_0) + \ldots] \ . \tag{3.268}$$

The computational results are shown in Figure 3.8. Curve 1 corresponds to a linear solution, whereas curves 2 and 3 correspond to solutions (3.267) and (3.268), respectively.

Curve 4 corresponds to an exact solution expressed by elliptic functions [110]. If an analytical solution structure is defined by the transformation (3.264), then the accuracy of the results increases. From the point of view of quantitative accuracy, the method described in reference [119], is better, since it better represents the characteristic features of the considered Cauchy's problem.

3.12 Analysis of strongly nonlinear dynamical problems

Asymptotic methods focusing on the analysis of strongly nonlinear dynamical systems are still not enough developed in comparison with the expectation of their application. It seems that a first step of progress is associated with the application of simple, but strongly nonlinear dynamical systems. This observation is motivated by the following remarks:

1. Simple dynamical systems can be characterized by complex nonlinear dynamics [123, 142, 164].
2. Fundamental properties of higher order dynamical systems can be described by simple dynamical systems.
3. The application of a concept of nonlinear modes often allows to reduce even a complex system to that of either one or few degrees of freedom [29, 142, 146, 147, 164].
4. Recently obtained results show that the approximate analytical methods can be successfully applied to the analysis of strongly nonlinear systems [3, 4, 29, 131, 164]. As a zero order approximation, a so called *vibro-impact model* is often used, and then either iteration methods [164, 131] or asymptotic techniques [3, 4, 29] are used. Unfortunately, as a result of those operations, a nonsmooth solution is obtained. In reference [7] it has been shown that one may also obtain smooth solutions.

Consider the following nonlinear differential equation

$$\ddot{x} + \gamma\dot{x} + \omega^2 x + \varepsilon x^n = 0 \ , \quad n = 3, 5, 7, \ldots \tag{3.269}$$

For $|\varepsilon| \ll 1$ and relatively small values of n ($n = 3, 5$), a solution to the equation (3.269) can be obtained by applying Lindstedt-Poincaré method, or various forms of averaging approaches [29, 121, 123, 125]. For large values of n a standard approach does not yield proper results. If a solution to the equation (3.269) is sought in the following form

$$x = x_0 + \varepsilon x_1 + \varepsilon^2 x_2 + \ldots \ , \tag{3.270}$$

then a nonlinear term can be approximated in the following way

$$\left(x_0 + \varepsilon x_1 + \varepsilon^2 x_2 + \ldots\right)^n = x_0^n + \varepsilon n x_0^{n-1} x_1 + \ldots \tag{3.271}$$

Observe that now ε plays a role of small parameter, instead of εn. If this change is not essential for $n = 3$ and $n = 5$, then it plays a key role for large values of n, and in particular for $n \to \infty$. In what follows, the following approximate formula is applied

$$x = x_0 \sqrt[n]{1 + \varepsilon n \frac{x_1}{x_0} + \ldots} \tag{3.272}$$

For small values of n formulas (3.270) and (3.272) are equivalent. For $n \to \infty$ a higher order singularity appears associated with the estimation of a n-th order root. Applying a quasi-linear method, a value of x_1 can be estimated through the analysis of the following equation [29, 121, 123, 125]

$$\ddot{x}_1 + \gamma \dot{x}_1 + \omega^2 x_1 = -x_0^n \, . \tag{3.273}$$

One may expect that the formula (3.272) can be applied in theory of oscillators in quantum mechanics, where a classical approach associated with a use of *quasi-linear series* does not always give proper results [141].

In order to estimate the application range of formulas (3.270) and (3.271) numerical simulations are carried out with results shown in Figure 3.9.

Graphically obtained results indicate that two formulas estimate the numerical solution of equation (3.269) (dashed curve) relatively well. It is worth noticing that already $x_0(t)$ [4] approximates the exact (numerical) solution reasonably well. The increase of n improves the accuracy of obtained results.

Similar observations are also associated with small values of γ and ε, although obtained results are not reported here. The decrease of values of γ and ε improves the approximation.

Similar computations are carried out for $\varepsilon = 1$ (the remaining parameters and initial conditions are the same), and the obtained results are reported in Figure 3.10.

One may easily get the interpretation similar as in the previous case, although now a convergence is slightly worse. The analysis of the problem yields a conclusion that large values of initial conditions may lead to large errors of approximation. In order to investigate this phenomenon, numerical computations are carried out with results shown in Figure 3.11. Also, in this case, the results associated with 1/4 period of oscillations are satisfactory.

Let us now consider a particular case of equation (3.269), i.e., when there is a lack of damping and linear stiffness ($\gamma = \omega^2 = 0$). Then, the equation (3.269) takes the form

$$\ddot{x} + \varepsilon x^n = 0 \, , \quad n = 3, 5, 7 \ldots \, , \tag{3.274}$$

which has been analyzed in references [3, 4, 29, 131, 164]. Notice that for large values of n, this equation approximates a vibro-impact process [44, 45].

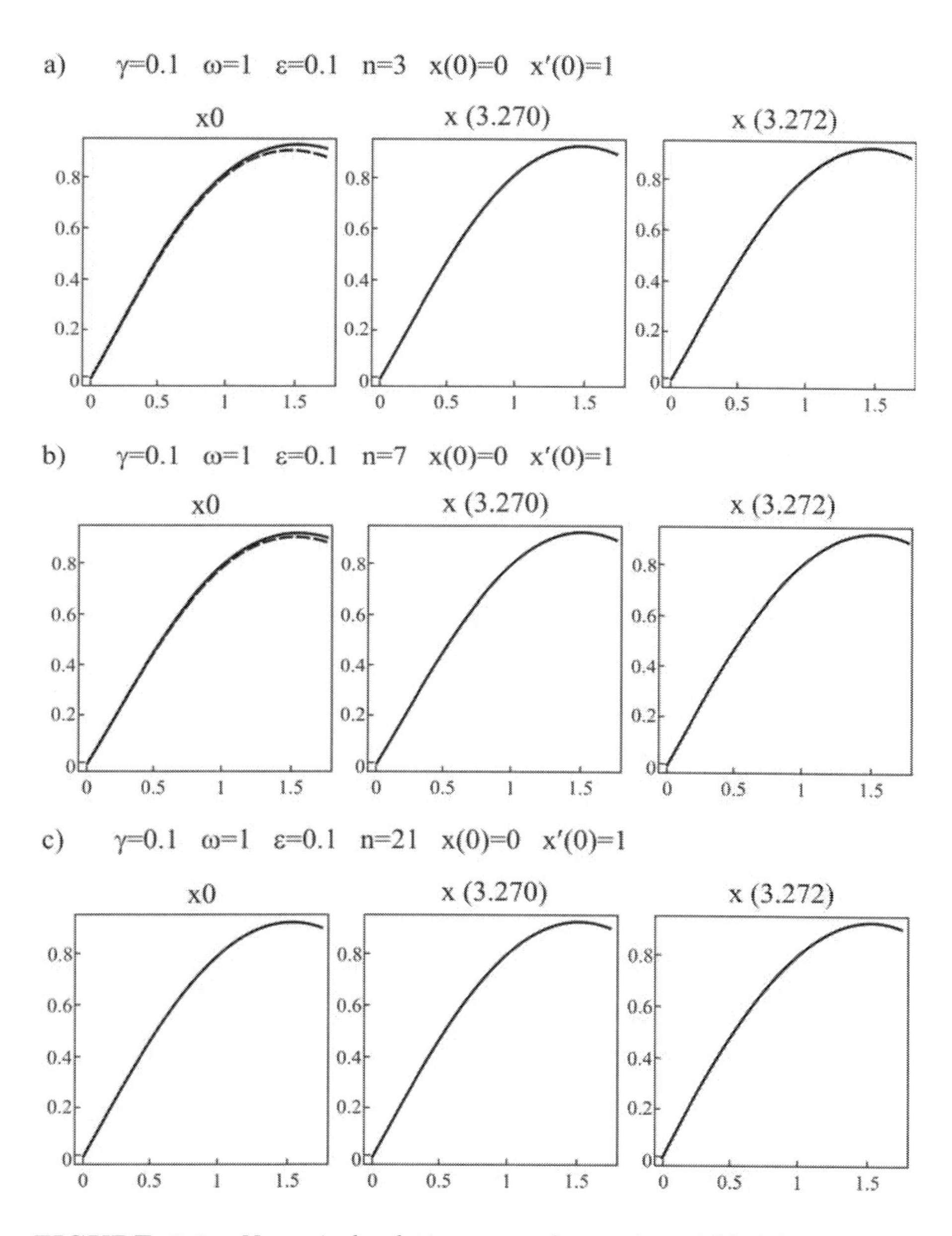

FIGURE 3.9: Numerical solution x_0 and x and x yielded by equation (3.270), and x obtained from equation (3.272) for $\gamma = 0.1$, $\omega = 1$, $\varepsilon = 0.1$, $x(0) = 0$, $x'(0) = 1$ and for different values of n: a) 3; b) 7; c) 21.

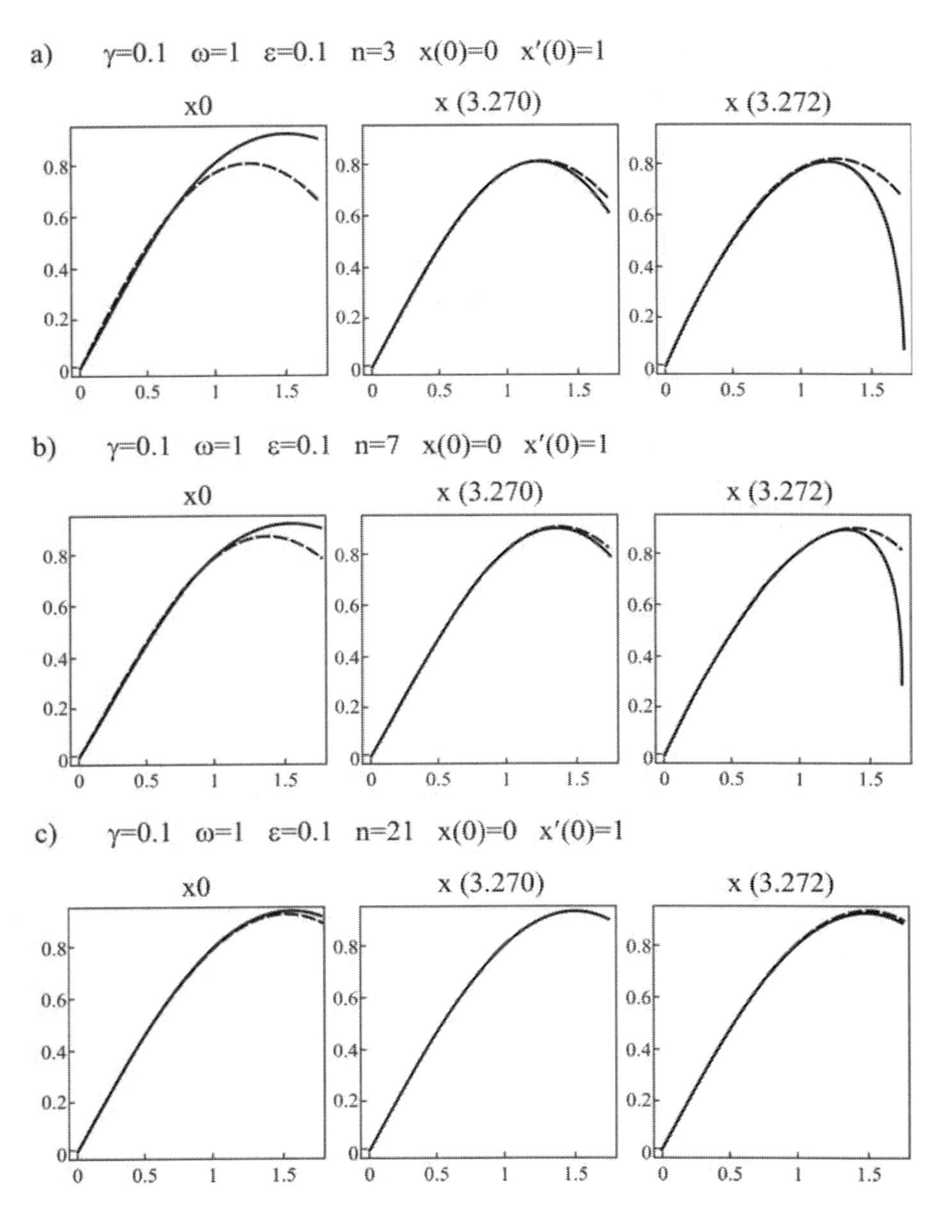

FIGURE 3.10: Numerical solution x_0 and x obtained from equation (3.270), and x obtained from equation (3.272) for $\gamma = 0.1$, $\omega = 1$, $\varepsilon = 1$, $x(0) = 0$, $x'(0) = 1$ and for different values of n: a) 3, b) 7, c) 21.

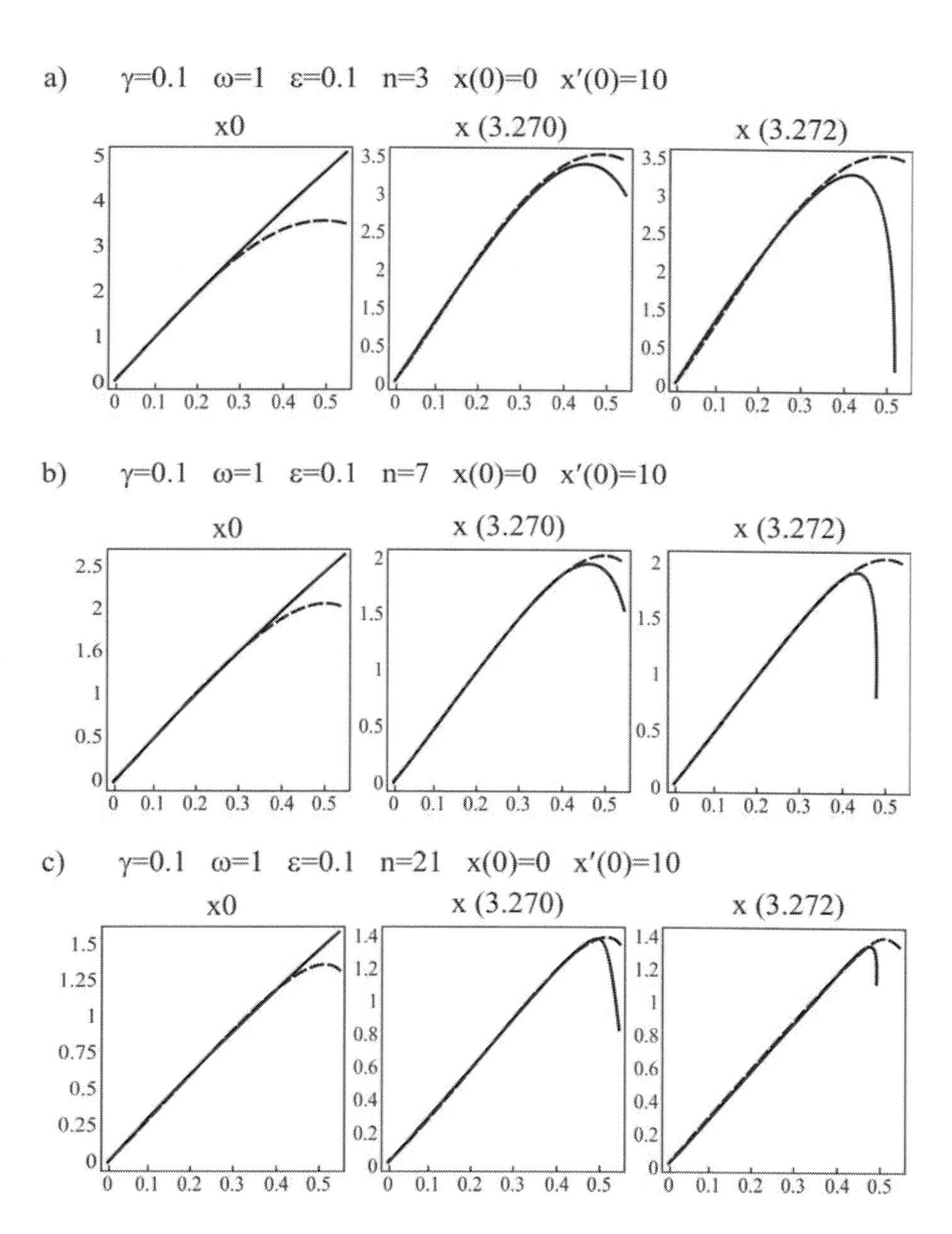

FIGURE 3.11: Numerical solution x_0 and x yielded by equation (3.270), and x obtained from (3.272) for $\gamma = 0.1$, $\omega = 1$, $\varepsilon = 1$, $x(0) = 0$, $x'(0) = 10$ and for different values of n: a) 3, b) 7, c) 21.

One may integrate this equation using special functions (cam, sam or Ateb functions) introduced by Rosenberg [147].

Since equation (3.274) is not a quasi-linear one, the application of parameter ε does not help during the solution and hence we take further $\varepsilon = 1$. Note that a parameter associated with power can be interpreted in two ways. For small n we have $n = 1 + \delta$, and one may further apply the so-called *small δ method* widely used in physics [17, 51]. On the other hand, for large n ($n \to \infty$),an appropriate asymptotic can be applied by the application of the parameter n^{-1}. In a limiting case, one deals with the vibro-impact process [3, 4, 29, 131, 164], but the obtained solution in nonsmooth.

Consider the following equation

$$\ddot{x} + x^n = 0 \,, \tag{3.275}$$

with the attached initial conditions

$$x(0) = 0 \,, \quad \dot{x}(0) = 1 \,, \tag{3.276}$$

in the case $n \to \infty$.

We are going to define a periodic solution of Cauchy problem (3.275), (3.276). First integral of this problem reads

$$p^2 = 1 - \frac{2x^{n+1}}{n+1} \,. \tag{3.277}$$

Introducing the change of variables $x = \sqrt[n+1]{0.5(n+1)\xi}$, and carrying out the integrations, we have

$$\sqrt[n+1]{\frac{2}{n+1}}\, t = \int\limits_0^{0 \le \xi \le 1} \frac{d\xi}{\sqrt{1 - \xi^{n+1}}} \,. \tag{3.278}$$

Notice that the implicit solution (3.278) after the change of variables $\xi = \sin^{2/(n+1)} \theta$ yields a formula with the explicitly given small parameter $\varepsilon = 2/(n+1)$ of the form

$$\varepsilon^{\varepsilon/2} t = \varepsilon \int\limits_0^{0 \le \theta \le \frac{\pi}{2}} \sin^{-1+\varepsilon} \theta d\theta \,. \tag{3.279}$$

Consider first the following formula (see [121])

$$\sin^{-1+\varepsilon} \theta = \theta^{-1+\varepsilon} \left(\frac{\theta}{\sin \theta} \right)^{1-\varepsilon} = \theta^{-1+\varepsilon} \left[\frac{\theta}{\sin \theta} - \varepsilon \ln \frac{\theta}{\sin \theta} + \ldots \right] . \tag{3.280}$$

Using the approximation

$$\frac{\theta}{\sin \theta} = 1 + \frac{\theta^2}{3!} + \ldots \tag{3.281}$$

and neglecting terms of ε order in the right hand side of equation (3.280), one gets

$$\sin^{-1+\varepsilon}\theta = \theta^{-1+\varepsilon} + \frac{\theta^2}{3!} + \ldots + 0(\varepsilon) \ . \tag{3.282}$$

The application of the package "Mathematica" yields the following approximations

$$\frac{\Theta}{\sin\Theta} = 1 + \frac{\Theta^2}{6} + \frac{7}{36}\Theta^4 + \frac{127}{604800}\Theta^8 + \frac{73}{3421440}\Theta^{10} + \frac{1414477}{653837184000}\Theta^{12} + \\ + \frac{8191}{37362124800}\Theta^{14} + \frac{16931177}{762187345920000}\Theta^{16} + \frac{574961557}{2554547108585472000}\Theta^{18} + \\ + \frac{91546277357}{401428831349145600000}\Theta^{20} + 0[\Theta]^{21} \ , \tag{3.283}$$

which is shown in Figure 3.12. Since during the integration, two first terms of series (3.283), play a crucial role, the first approximation has the form

$$\varepsilon^{\varepsilon/2}t \approx \theta^{\varepsilon} \ . \tag{3.284}$$

Since

$$\theta \approx \varepsilon^{1/2}t^{\varepsilon^{-1}} \ , \tag{3.285}$$

then, coming back into initial coordinates, one gets

$$x \approx \sqrt[n+1]{\frac{n+1}{2}} \sin^{\frac{2}{n+1}}\left(\sqrt{\frac{2}{n+1}}t^{\frac{n+1}{2}}\right) \ . \tag{3.286}$$

Although the sought solution holds in the interval of $1/4\ T$, t may be further extended to construct the solution in the whole period T interval.

The period of solution (3.286) reads

$$T = 4\left(\frac{\pi}{2}\sqrt{\frac{n+1}{2}}\right)^{\frac{2}{n+1}} \ . \tag{3.287}$$

It is worth noticing that for $\varepsilon = 1$ one gets a periodic solution: $x = \sin t$, $T = 2\pi$. If $n \to \infty$, then $T \to 4$, which coincides with our expectation.

If the expansion of solution (3.286) is carried out into a series with respect to t, and taking into account only one series term, a nonsmooth solution proposed by Pilipchuk [131, 164] is obtained.

In what follows, we estimate the accuracy of the obtained solution (3.287). For this purpose, a definition of *incomplete Beta* $B(\ldots,\ldots)$ function (see [152]) of the form is applied

$$\frac{2}{n+1}\int_0^{\frac{\pi}{2}} \sin^{-1+\frac{2}{n+1}}\theta d\theta = \frac{1}{n+1}B\left(\frac{1}{n+1},\frac{1}{2}\right) = A_1 \ . \tag{3.288}$$

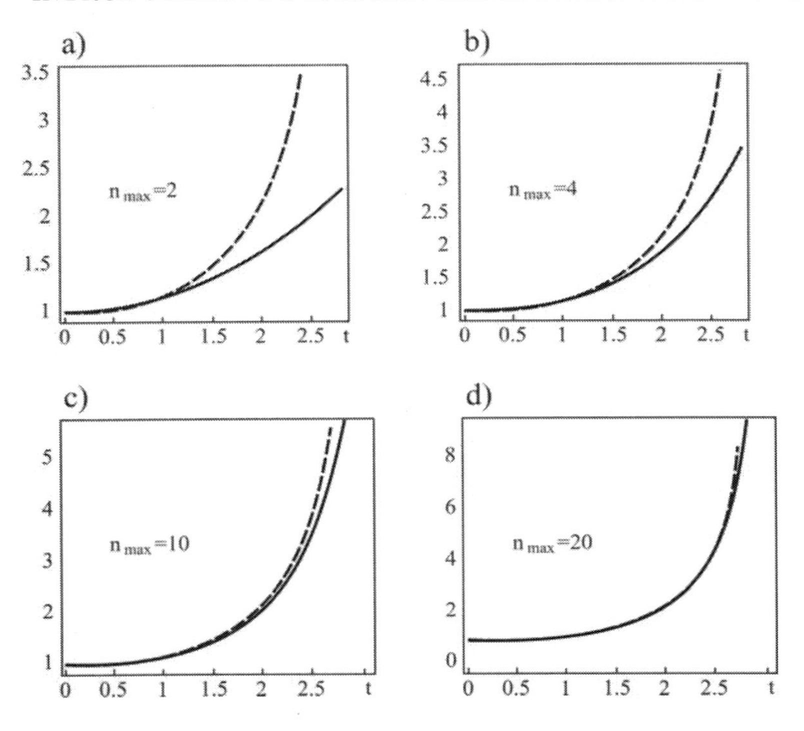

FIGURE 3.12: Approximation of $\Theta/\sin\Theta$ for various values of n: a) $n = 2$; b) $n = 3$; c) $n = 6$; d) $n = 11$.

The approximation to the integral standing in the left-hand side of formula (3.288) has the form

$$A_2 = \left(\frac{\pi}{2}\right)^{\frac{2}{n+1}} . \qquad (3.289)$$

Digital values of functions A_1 and A_2 and the occurring errors are reported in Table 3.1.

TABLE 3.1: The estimation of errors of values of A_1 and A_2

n	A_1	A_2	Error %
1	$\pi/2$	$\pi/2$	~ 0 %
3	1.30	1.25	~ 5 %
5	1.20	1.16	~ 3 %
...	...	...	...
∞	1	1	~ 0 %

The obtained results yield the following conclusion. Already the first approximation for $n \to \infty$ yields a reasonable accuracy applied in engineering (even for small values of n).

Notice that formulas (3.286), (3.287) generalize the opposite of the incomplete function Beta for $n = 1$ (sinus), and for $n \to \infty$ (linear function).

In reference [156], the following estimations of incomplete Beta function are given

$$B(\nu, 0.5, x) \approx \frac{x^\nu}{\nu} \quad \text{for small } x \tag{3.290}$$

and

$$B(\nu, 0.5, x) \approx \frac{\sqrt{\pi}\Gamma(\nu)}{\Gamma(0.5+\nu)} \quad \text{for } x \text{ close } 1 \ . \tag{3.291}$$

Observe that (using *asymptotic* $\Gamma(\nu) \sim \nu^{-1}$ for $\nu \to 0$ [156]), for small ν z he following *estimation for Beta function* holds

$$B(\nu, 0.5, x) \sim \frac{x^\nu}{\nu} \quad \text{dla} \;\; \nu \to 0 \ , \tag{3.292}$$

which testifies the earlier obtained results.

If the equation (3.275) is associated with the condition of the form (see [147])

$$x(0) = 0 \, ; \quad x = 1 \quad \Rightarrow \quad \dot{x} = 0 \ , \tag{3.293}$$

then asymptotic for $n \to \infty$ in the first approximation is defined by the formulas

$$x \approx \sin^{\frac{2}{n+1}} \left(\frac{2t^2}{n+1} \right)^{n+1} , \quad T = 4 \left(\frac{\pi}{2} \right)^{\frac{2}{n+1}} \left(\frac{n+1}{2} \right)^2 . \tag{3.294}$$

Consider the case $\omega^2 = 0$, assuming that a damping coefficient may take arbitrary values, i.e., the following equation is analyzed

$$\ddot{x} + \gamma \dot{x} + \varepsilon x^n = 0 \ . \tag{3.295}$$

Recall that a second order differential equation may be reduced to a second order differential equation without damping, through the change of variables [160]. However, in a nonlinear case, such change can be carried out only for strictly defined parameters [148]. In the considered case, for large values of n, equation (3.295) can be reduced asymptotically to equation (3.275). For this purpose, the following change of variables is introduced

$$\tau = \exp(-\gamma t) \ . \tag{3.296}$$

Hence, instead of equation (3.296), the following one is obtained

$$\gamma^2 \exp(2\gamma t) \frac{d^2 x(\tau)}{d\tau^2} + \varepsilon x^n(\tau) = 0 \ . \tag{3.297}$$

Changing variables through the formula

$$x = \exp\left(\frac{2\gamma y}{n}\right)\left(\frac{\gamma^2}{\varepsilon}\right)^{\frac{1}{n}} , \tag{3.298}$$

equation (3.297) is reduced to equation (3.275) with the accuracy of order $1/n$.

Consider now the case of damping lack, i.e., $\gamma = 0$, and in addition we take $\omega^2 = \varepsilon = 1$. Now equation (3.295) takes the form

$$\ddot{x} + x + x^n = 0 . \tag{3.299}$$

According to initial conditions (3.276), a solution to equation (3.299) has the following implicit form

$$t = \int_0^x \frac{dx}{\sqrt{1 - x^2 - \frac{2}{n+1}x^{n+1}}} . \tag{3.300}$$

Let us introduce the following change of variables

$$x^2 + \frac{2}{n+1}x^{n+1} = \sin^2\theta . \tag{3.301}$$

A solution to equation (3.301) with respect to x^2 has the following form

$$x = \sin\theta \sqrt[n-1]{1 + x_1} , \tag{3.302}$$

which gives

$$x_1 \approx \frac{-\sin^{n-1}\theta}{1 + \sin^{n-1}\theta} . \tag{3.303}$$

Finally, the formula (3.300) can be transformed to the form

$$t = \int_0^{0 \le \theta \le \pi/2} \left[\left(1 - \frac{\sin^{n-1}\theta}{1 + \sin^{n-1}\theta}\right)^{\frac{1}{n-1}} - \frac{\sin^{n-1}\theta}{1 + \sin^{n-1}\theta}\right] d\theta \tag{3.304}$$

with the accuracy of order $1/n$.

The expression standing under integral in (3.304) can be approximated by

$$\left(1 + \frac{\sin^n\theta}{1 + \sin^n\theta}\right)^{\frac{1}{n}} - \frac{\sin^n\theta}{1 + \sin^n\theta} = \frac{1}{1 + \sin^n\theta} + \theta(1/n) . \tag{3.305}$$

Consider the following approximation

$$\sin^n\theta \sim \begin{cases} \theta^n, & 0 \le \theta < \frac{1}{\sqrt[n]{n}} , \\ 1, & \theta = \pi/2 \end{cases} . \tag{3.306}$$

Applying a two-point Padé approximation [29], one gets

$$\sin^n \theta \sim \frac{\theta^n}{1+\theta^n} \; . \tag{3.307}$$

Therefore, one gets

$$\int\limits_0^\theta \frac{1+\theta^n}{1+2\theta^n} d\theta = \theta - \frac{1}{2n} \ln\left(1+2\theta^n\right) \tag{3.308}$$

and a solution is

$$t = \theta - \frac{1}{2n} \ln(1+2\theta^n) \; . \tag{3.309}$$

Furthermore, the sought function θ has the form

$$\theta = t\sqrt[n]{1+0.5\theta_1/t} \; . \tag{3.310}$$

The variable θ_1 is found from the following *transcendental equation*

$$\theta_1 - \ln[1+2t^n + t^{n-1}\theta_1] = 0 \; , \tag{3.311}$$

and its solution can be found numerically for each t.

Equations (3.302) and (3.304) define a sought asymptotic. In Figure 3.13, a comparison of results obtained by equation (3.299) and equations (3.302) and (3.304) is carried out. According to graphs, the accuracy increases with the increase of n, and for $n = 21$ it is already very high.

Let us focus briefly on the analysis of the considered equation in the case of small damping. Introducing the following change of variables

$$\tau = \exp(-\gamma t) \; ,$$

$$x = \exp\left(\frac{2\gamma y}{n}\right)\varepsilon^{-1/n} \; , \tag{3.312}$$

equation (3.269) is reduced to the following one

$$\gamma^2 \frac{d^2x(\tau)}{d\tau^2} + \frac{\omega^2 x(\tau)}{\tau^2} + x^n(\tau) = 0 \; . \tag{3.313}$$

Integration of the above equation can be carried out by two scaling and equivalent averaging methods, WKB [125] method, or other asymptotic methods concerning equations with small parameters [121, 123].

The introduction of a new *fast variable* $\tau_1 = \gamma^{-1}\tau$ yields an equation with formally of the form *frozen coefficients* of the form

$$\frac{d^2x(\tau_1)}{d\tau_1^2} + c^2\omega^2 x(\tau_1) + x^n(\tau_1) = 0 \quad c^2 = \tau^{-2} \; . \tag{3.314}$$

In order to integrate the last equation, one may apply one of the earlier introduced methods.

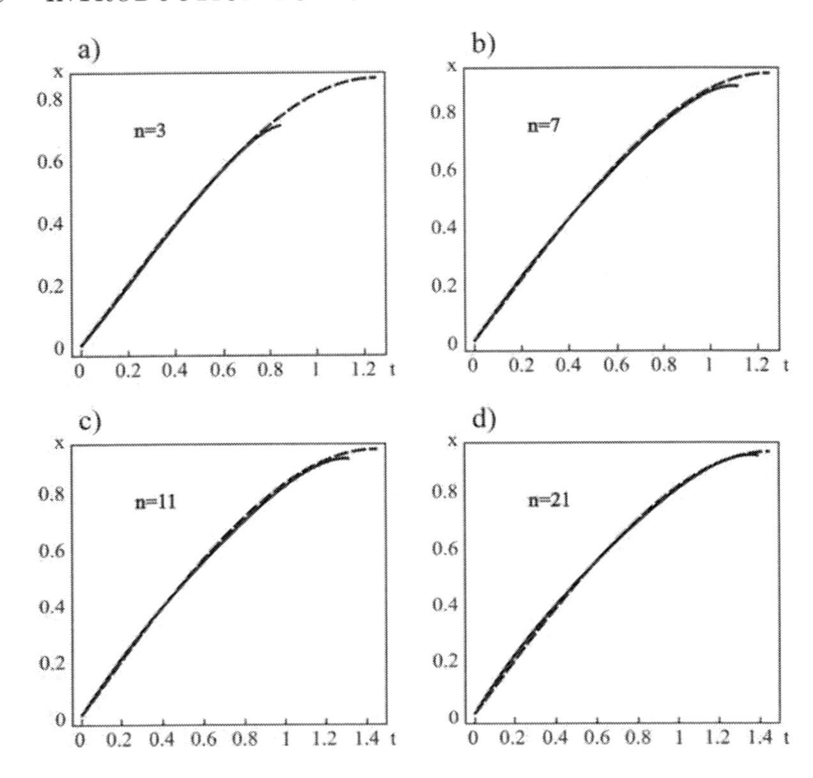

FIGURE 3.13: Comparison of solutions of equations (3.299) (dashed curve), and equations (3.302) and (3.304) for various values of n: a) $n = 3$, b) $n = 7$, c) $n = 11$, d) $n = 21$.

3.13 A few perturbation parameters

In classical monographs devoted to perturbation techniques (see Nayfeh [119, 120, 121, 122, 123], Hinch [84], Bush [57] and references cited there), mainly the applications of the asymptotic approaches associated with one parameter are considered. However, there is often a need to apply a few perturbation parameters.

There are many examples where the occurrence of two independent small parameters can not by equivalently substituted by only one small parameter. Some of such examples are given in works of Awrejcewicz [24, 25, 26, 27, 28], Awrejcewicz et al. [29, 32, 33]. In the considered works, one of perturbation parameters is associated with parametric excitation, whereas the other one characterizes either nonlinear stiffness, or time delay or friction.

Consider the following problem dealing with determination of ei- genvalues

$$(A + \varepsilon_1 B_1 + \varepsilon_2 B_2)\varphi = \lambda\varphi \ , \tag{3.315}$$

where λ, φ denote *frequency* and the associated *eigenfunction* [10]. In order to estimate the influence of perturbation parameters on a solution, one has to find the first zero order solution approximation($\varepsilon_1 = \varepsilon_2 = 0$) as well as associated independent corrections $\lambda_1^{(i)}$, separately (a subscript denotes approximation number, whereas a superscript defines a small parameter). The problem is reduced to the following:

$$A\varphi_0 = \lambda_0\varphi_0 \ , \tag{3.316}$$

$$A\varphi_1^{(1)} - \lambda_0\varphi_1^{(1)} = -B_1\varphi_0 + \lambda_1^{(1)}\varphi_0 \ , \tag{3.317}$$

$$A\varphi_1^{(2)} - \lambda_0\varphi_1^{(2)} = -B_2\varphi_0 + \lambda_1^{(2)}\varphi_0 \ . \tag{3.318}$$

Equations (3.317) and (3.318) yield corrections $\lambda_1^{(i)}$ (see Nayfeh [118, 119, 120, 121, 122, 123], Bush [57]); $\lambda_1^{(i)} = \frac{(B_i\varphi_0,\varphi_0)}{(\varphi_0,\varphi_0)}$, $i = 1, 2$, where symbol $(\cdot,\cdot)$ denotes a scalar product related to a corresponding norm.

Comparison of the values of $\varepsilon_1\lambda_1^{(1)}$ and $\varepsilon_2\lambda_1^{(2)}$ allows to estimate the relative influence of small parameters on the investigated system. Assume that $\varepsilon_1\lambda_1^{(1)} \approx \varepsilon_2\lambda_1^{(2)}$. Note that in this case we do not analyze the influence of the case of $(\varepsilon_1 B_1 \approx \varepsilon_2 B_2)$, a priori, but we focus on the investigation of the influence of these parameters on computational results. Furthermore, our main aim is to verify the often-applied hypothesis on superposition *of influence of perturbation parameters in higher approximations*.

A sought solution has the following form

$$\lambda \approx \lambda_0 + \lambda_1 + \lambda_2 \, ; \quad \varphi \approx \varphi_0 + \varphi_1 + \varphi_2 \, , \tag{3.319}$$

where:

$$\lambda_1 \approx \varepsilon_1\lambda_1^{(1)} + \varepsilon_2\lambda_1^{(2)} \, , \quad \varphi_1 \approx \varepsilon_1\varphi_1^{(1)} + \varepsilon_2\varphi_1^{(2)} \ . \tag{3.320}$$

Terms λ_2 and φ_2 are small corrections of the found values of $\lambda \approx \lambda_0 + \lambda_1$ and $\varphi \approx \varphi_0 + \varphi_1$. *The eigenvalue* λ_2 is obtained by the equation

$$A\varphi_2 - \lambda_0\varphi_2 = -B_1\varphi_1 - B_2\varphi_2 + \lambda_1\varphi_1 + \lambda_2\varphi_0 \ . \tag{3.321}$$

Multiplying both sides of this equation by φ_0, one gets

$$\lambda_2 = \frac{(B_1\varphi_1, \varphi_0) + (B_2\varphi_1, \varphi_0) - \lambda_1(\varphi_1, \varphi_0)}{(\varphi_0, \varphi_0)} \ . \tag{3.322}$$

In the case of $\varepsilon_1\lambda_1^{(1)} \gg \varepsilon_2\lambda_1^{(2)}$ (or $\varepsilon_1\lambda_1^{(1)} \ll \varepsilon_2\lambda_1^{(2)}$), the problem reduces to that of the analysis of one perturbation parameter, since the influence of a second one may be neglected.

Notice that knowledge of n-th eigenfunction allows to determine $2n + 1$ eigenfunctions (see Marchuk et al. [112]). In what follows, we show how to apply a procedure proposed by Marchuk, using only one perturbation parameter (in the case of a few perturbation parameters one may proceed similarly).

Consider the following set of perturbation parameters

$$A\varphi_0 = \lambda_0\varphi_0 \ , \tag{3.323}$$

$$A\varphi_1 - \lambda_0\varphi_1 = -B\varphi_0 + \lambda_1\varphi_0 \ , \tag{3.324}$$

$$A\varphi_2 - \lambda_0\varphi_2 = -B\varphi_1 + \lambda_1\varphi_1 + \lambda_2\varphi_0 \ , \tag{3.325}$$

$$A\varphi_3 - \lambda_0\varphi_3 = -B\varphi_2 + \lambda_1\varphi_2 + \lambda_2\varphi_1 + \lambda_3\varphi_0 \ . \tag{3.326}$$

Equations (3.323) and (3.324) yield λ_0, λ_1, and next φ_1, φ_2. The correction λ_2 is defined by the equation (3.325) and it reads

$$\lambda_2 = \frac{[(B-\lambda_1)\varphi_1, \varphi_0]}{(\varphi_0, \varphi_0)} \ . \tag{3.327}$$

Multiplying (3.326) by φ_0, one gets

$$\lambda_3 = \frac{[(B-\lambda_1)\varphi_2, \varphi_0] - \lambda_2(\varphi_1, \varphi_0)}{(\varphi_0, \varphi_0)} \ . \tag{3.328}$$

Since φ_2 is not known, the first term of numerator in (3.328) can be presented in the following form

$$[(B-\lambda_1)\varphi_2, \varphi_0] = [(B-\lambda_1)\varphi_0, \varphi_2] \ . \tag{3.329}$$

The equation (3.324) yields

$$(B-\lambda_1)\varphi_0 = -(A-\lambda_0)\varphi_1 \ , \tag{3.330}$$

and finally

$$\begin{aligned} [(B-\lambda_1)\varphi_0, \varphi_2] &= -\left[(A-\lambda_0)\varphi_1, \varphi_2\right] = \\ = -\left[(A-\lambda_0)\varphi_2, \varphi_1\right] &- [(-B\varphi_1 + \lambda_1\varphi_1 + \lambda_2\varphi_0), \varphi_0] = \\ &= [(B-\lambda_1)\varphi_1, \varphi_1] - \lambda_2(\varphi_0, \varphi_1) \ . \end{aligned} \tag{3.331}$$

Finally, one gets

$$\lambda_3 = \frac{((B-\lambda_1)\varphi_1, \varphi_1) - 2\lambda_2(\varphi_1, \varphi_1)}{(\varphi_0, \varphi_0)} \ . \tag{3.332}$$

Proceeding analogically, one obtains λ_0, λ_1, ..., λ_n, next φ_0, φ_1, ..., φ_n, and next λ_{n+1}, ..., λ_{2n+1}. Assuming that the perturbation equation has the form

$$A\varphi = \lambda(N + \varepsilon M)\varphi \ . \tag{3.333}$$

one gets after a splitting procedure

$$A\varphi_0 = \lambda_0 N\varphi_0 \ , \tag{3.334}$$

$$A\varphi_1 = \lambda_0 N\varphi_1 + \lambda_1 N\varphi_0 + \lambda_0 M\varphi_0 \ , \tag{3.335}$$

$$A\varphi_2 = \lambda_0 N\varphi_2 + \lambda_1 N\varphi_1 + \lambda_2 N\varphi_0 + \lambda_0 M\varphi_1 + \lambda_1 M\varphi_0 \ , \tag{3.336}$$

$$A\varphi_3 = \lambda_0 N\varphi_3 + \lambda_1 N\varphi_2 + \lambda_2 N\varphi_1 + \lambda_3 N\varphi_0 + \lambda_0 M\varphi_2 + \lambda_1 M\varphi_1 + \lambda_2 M\varphi_2 \ . \tag{3.337}$$

The obtained equations yield the determination of $\lambda_0, \lambda_1, \lambda_2$ as well as the determination of φ_0, φ_1. The formula serving for determination of λ_3 includes an unknown function φ_2:

$$\lambda_3 = \frac{[(\lambda_1 N + \lambda_0 M)\varphi_2, \varphi_0] + [(\lambda_2 N + \lambda_1 M)\varphi_1, \varphi_0] + \lambda_2(M\varphi_0, \varphi_0)}{(N\varphi_0, \varphi_0)} \ . \tag{3.338}$$

Notice that

$$[(\lambda_1 N + \lambda_0 M)\varphi_2, \varphi_0] = [(\lambda_1 N + \lambda_0 M)\varphi_0, \varphi_2] \tag{3.339}$$

and from equation (3.335), one gets

$$(\lambda_1 N + \lambda_0 M)\varphi_0 = (A - \lambda_0 N)\varphi_1 \ . \tag{3.340}$$

According to (3.340), formula (3.339) takes the form

$$[(\lambda_1 N + \lambda_0 M)\varphi_0, \varphi_2] = [(A - \lambda_0 N)\varphi_1, \varphi_2] = [(A - \lambda_0 N)\varphi_2, \varphi_1] \ . \tag{3.341}$$

Now, equation (3.336) yields

$$\begin{aligned} [(A - \lambda_0 N)\varphi_2, \varphi_1] &= [(\lambda_1 N\varphi_1 + \lambda_2 N\varphi_0 + \lambda_0 M\varphi_1 + \lambda_1 M\varphi_0), \varphi_1] = \\ & [(\lambda_2 N + \lambda_1 M\varphi_0)\varphi_0, \varphi_1] + [(\lambda_1 N + \lambda_0 M)\varphi_1, \varphi_1] = \\ & [(\lambda_2 N + \lambda_1 M\varphi_0)\varphi_1, \varphi_0] + [(\lambda_1 N + \lambda_0 M)\varphi_1, \varphi_1] \ , \end{aligned} \tag{3.342}$$

Finally, equations (3.338) and (3.342) yield

$$\lambda_3 = \frac{2\,[(\lambda_2 N + \lambda_1 M)\varphi_1, \varphi_0] + [(\lambda_1 N + \lambda_0 M)\varphi_1, \varphi_1] + \lambda_2(M\varphi_0, \varphi_0)}{(N\varphi_0, \varphi_0)} \ . \tag{3.343}$$

To sum up, the method proposed by Marchuk [112] can be generalized into a case of a few perturbation parameters.

Exercises

3.1. Analyze the following Cauchy problems for $\varepsilon \ll 1$:

$$\begin{cases} y' + y = \varepsilon y^2 \\ y(0) = 1 \end{cases} ; \quad \begin{cases} y' - 2y = \varepsilon y^3 \\ y(0) = 0 \end{cases} ; \quad \begin{cases} y'' + y = \varepsilon u^2 \\ y'(0) = 0 \\ y(0) = 1 \end{cases} ,$$

(a) determine, using a classical method of small parameter, three terms of asymptotic decompositions of the given problems;

(b) define a uniform suitability of the obtained decompositions;

(c) apply the methods of elongated parameters and rescaling and compare their efficiency.

3.2. Study the following Cauchy problem

$$\varepsilon y' + xy = -1\,, \quad y(0) = 1\,; \quad \varepsilon \ll 1\,.$$

(a) find first three terms of asymptotic decomposition using a small parameter method;

(b) define a dimension of decomposition uniformity;

(c) apply the methods of rescaling and matching the asymptotic decompositions, and compare their efficiency.

3.3. Consider the following Cauchy problem

$$(x + \varepsilon y)y' + y = 0\,, \quad y(1) = 1\,; \quad \varepsilon \ll 1\,,$$

and:

(a) determine three terms of asymptotic decomposition of the problem;

(b) define a space of nonuniformity;

(c) find an asymptotic decomposition of the following exact solution

$$y = -\frac{x}{\varepsilon} + \sqrt{\frac{x^2}{\varepsilon^2} + \frac{2}{\varepsilon} + 1}$$

for $\varepsilon \to 0$ and compare the obtained result with that of point (a).

(d) make a conclusion of a source of decomposition nonuniformity;

(e) construct uniformly suitable decomposition using one of the methods of singular perturbations.

3.4. Consider the following equation

$$y'' + \omega_0^2 y = \varepsilon y' y\,, \quad \varepsilon \ll 1\,,$$

and solve the following problems:

(a) construct the two-term asymptotic series of solution, using a classical perturbation technique;

(b) investigate its uniformity;

(c) transform it into a uniformly suitable series using the method of rescaling;

(d) transform it into a uniformly suitable series using the method of elongated parameters;

(e) construct a uniformly suitable series of first order using the multiple scales method;

(f) construct a uniformly suitable first order series using the averaging method;

3.5. Introduce to Duffing equation $u_{tt} + u + \varepsilon u^3 = 0$ a new independent variable of the form $\xi = \left(1 + \frac{3}{8}\varepsilon a^2\right) t$. Find two first terms of an asymptotic series of a solution. Is it uniformly suitable? What role do the independent variables play during construction of uniformly suitable asymptotic decompositions?

3.6. Consider the following boundary value problem for the function $y(x)$ defined by the following problem

$$\varepsilon y'' + xy' - y = 0\,, \quad x \in [0,1]\,, \quad \varepsilon \ll 1\,,$$

$$y(0) = 1/2\,, \quad y(1) = 1\,,$$

and:

(a) determine two first terms of an asymptotic series of a solution using the classical method of small parameter;

(b) construct an additive series using a method of matching of asymptotic decompositions;

(c) construct a graph and compare the obtained results for $\varepsilon = 0.1$.

Chapter 4

Wave-impact processes

Asymptotic methods may be useful for investigating even complex mathematical problems occurring during mathematical modelling of some impacting wave processes in fluids and gases. For didactic purposes, further solutions will be limited to first approximations, and comparison of the efficiency of results will be carried out with the use of various methods of the theory of singular perturbations.

First, the definition of a cylinder-like piston wave is introduced, and the vicinity of both piston and front of the impact wave is outlined. It is illustrated how an application of elongated coordinates and renormalization methods yields highly accurated results (note that up to now there is no efficient numerical approach to solve this problem appropriately). In the second part of Chapter 4, one-dimensional nonstationary nonlinear waves are analyzed. After mathematical formulation of the problem, irregularity of the solution is illustrated and discussed. Then it is shown how an application of the methods of renormalization, characteristics and multiple scales allows for overcoming the singularities to achieve the uniformly suitable solutions.

4.1 Definition of a cylinder-like piston wave

4.1.1 Defining the problem, its solution and analysis

Consider the following mathematical problems: find a function $f(x;\varepsilon)$ which in the interval $[\varepsilon; M]$ satisfies the following nonlinear differential equation:

$$\left[1-(\gamma-1)\left(f-xf_x+\frac{1}{2}f_x^2\right)\right]\left(f_{xx}+\frac{1}{x}f_x\right)=(x-f_x)^2 f_{xx} \tag{4.1}$$

with the following boundary conditions

$$f_x(\varepsilon)=\varepsilon\ , \tag{4.2}$$

$$f(M)=0\ , \tag{4.3}$$

$$f_x(M)=\frac{2}{\gamma+1}\left(M-\frac{1}{M}\right)\ , \tag{4.4}$$

(a subscript denotes differentiation with respect to x).

In the above, $\gamma > 1$ denotes a known constant, $\varepsilon \ll 1$ is a known small parameter, $M = 1 + \varepsilon_1$, $0 < \varepsilon_1 \ll 1$, ε_1 is the unknown number to be found.

Notice that the problem (4.1)–(4.4) occurs during mathematical modeling of a thin wire explosion in the air. As a result of such explosion, the cylinder-like piston expands with constant speed, and the front of a weak impacting cylindrical wave moves with constant speed before the piston.

Notice that the fundamental steps of mathematical model (4.1)–(4.4) (simplifying assumptions, a transition from dependent into independent variables) are given in the reference [118] (page 97).

It is worth noticing that the additional difficulty occurring during the solution of the nonlinear problem (4.1)–(4.4) is posed by the lack of full knowledge of the limit $x = M$ of the domain of determinacy of function $f(x;\varepsilon)$. It seems that the problem is overdetermined according to three boundary conditions (4.1)–(4.4) formulated earlier. The "unnecessary" boundary condition (4.4) serves, however, for the determination of the unknown quantity M.

According to the considerations reported in ([92], § 61) instead of the condition (4.4) the following one should be applied

$$f_x(M) = \frac{1}{\gamma M}\left\{\left[\frac{M}{M - f_x(M)}\right]^{\gamma} - 1\right\} . \tag{4.5}$$

Let us briefly discuss the physical interpretation of quantities occurring in (4.1)–(4.5): f, f_x denote, respectively, the velocity potential and velocity of a medium (between piston ($x = \varepsilon$) and the front of impacting wave ($x = M$), where M is *Mach number of the impacting wave* $M > 1$), and γ is the thermodynamic parameter characterizing a medium (for air $\gamma = 1.4$, whereas for water $\gamma \approx 7.15$).

After determining f, what becomes important from a physical point of view is pressure P in a medium between the piston and impacting wave front, which is found from the formula

$$(1 + \gamma P)^{(\gamma-1)/\gamma} = 1 - (\gamma - 1)\left(f - x f_x + \frac{1}{2} f_x^2\right) , \tag{4.6}$$

where $P = (p - p_o)/\gamma p_o$ is relative nondimensional needless pressure, and p_o is the pressure in medium undisturbed by explosion.

An exact analytical solution to problems (4.1)–(4.4) or to problems (4.1)–(4.3), (4.5) cannot be found, and also the application of numerical approaches poses some problems. In what follows, assuming that f, P, $\varepsilon_1 \ll 1$, classical perturbation methods are used to solve this problem.

Let us first apply the classical small parameter method:

$$f(x;\varepsilon) = f_1^*(x;\varepsilon) + f_2^*(x;\varepsilon) + \dots , \quad f_2^* \ll f_1^* \ll 1 ,$$

$$P(x;\varepsilon) = P_1^*(x;\varepsilon) + P_2^*(x;\varepsilon) + \dots , \quad P_2^* \ll P_1^* \ll 1 ,$$

$$M(\varepsilon) = 1 + M_1^*(\varepsilon) + M_2^*(\varepsilon) + \ldots \,, \qquad M_2^* \ll M_1^* \ll 1 \,. \tag{4.7}$$

Substituting (4.7) into (4.1), (4.6), one gets in the first approximation

$$\left(1 - x^2\right) f_{1xx}^* + \frac{1}{x} f_{1x}^* = 0 \,, \tag{4.8}$$

$$P_1^* = x f_{1x}^* - f_1^* \,. \tag{4.9}$$

A general solution to the obtained differential equation (4.8) with separable variables reads

$$f_{1x}^* = d_1 \frac{\sqrt{1-x^2}}{x} \,, \quad f_1^* = d_1 \left(\sqrt{1-x^2} - \ln \frac{1+\sqrt{1-x^2}}{x} \right) + c \,. \tag{4.10}$$

According to (4.9), (4.10) one gets

$$P_1^* = d_1 \ln \frac{1+\sqrt{1-x^2}}{x} + c \,, \quad d_1, c = const \,. \tag{4.11}$$

Arbitrary constants d_1, c are obtained from boundary conditions. From (4.2) one gets $f_{1x}^*(\varepsilon) = \varepsilon$, and next, one finds $d_1 = \varepsilon^2/\sqrt{1-\varepsilon^2}$, whereas (4.4), (4.3) and (4.10) yield $M = 1$, $c = 0$.

Finally, according to (4.10), (4.11) and taking into account the values of d_1, c, the asymptotic series (4.7) are found in the following form

$$f(x;\varepsilon) = d_1 f_1(x) + d_1^2 f_2(x) + o(d_1^2) \,, \tag{4.12}$$

$$P(x;\varepsilon) = d_1 P_1(x) + d_1^2 P_2(x) + o\left(d_1^2\right) \,, \quad d_1 = \frac{\varepsilon^2}{\sqrt{1-\varepsilon^2}} \,, \tag{4.13}$$

$$M(\varepsilon) = 1 + d_1 M_1 + d_1^2 M_2 + o\left(d_1^2\right) \,, \tag{4.14}$$

where M_1, M_2 are unknown numbers.

Notice that solutions for f_1, f_{1x}, P_1 are already found (see 4.10, 4.11) and they have the form

$$f_1 = \sqrt{1-x^2} - \ln \frac{1+\sqrt{1-x^2}}{x} \,, f_{1x} = \frac{\sqrt{1-x^2}}{x} \,, P_1 = \ln \frac{1+\sqrt{1-x^2}}{x} \,. \tag{4.15}$$

Therefore, according to (4.8) and (4.9), and taking (4.15) into account, one gets

$$f_{1xx} = -\frac{f_{1x}}{x\left(1-x^2\right)} = -\frac{1}{x^2\sqrt{1-x^2}} \,, \quad P_{1x} = x f_{1xx} = -\frac{1}{x\sqrt{1-x^2}} \,. \tag{4.16}$$

Let us analyze the results of (4.8) and (4.9) in the frame of a linear theory. According to (4.7), (4.12)–(4.15), functions f_1 , f_{1x}, P_1 are defined for $x \in [\varepsilon; 1]$, and functions f_{1xx}, P_{1x} for $x \in [\varepsilon; 1)$.

Let us investigate the behavior of the obtained equations in the vicinity of points $x = \varepsilon$ and $x = 1$. Equation (4.15) yields

$$\lim_{x\to\varepsilon}\frac{f_1}{f_{1x}} = \lim_{x\to\varepsilon}\frac{\left[\sqrt{1-x^2} - \ln(1+\sqrt{1-x^2}) + \ln x\right]x}{\sqrt{1-x^2}} =$$

$$\lim_{x\to\varepsilon}(1 - \ln 2 + \ln x)x = \lim_{x\to\varepsilon} x\ln x = 0 .$$

Proceeding analogically with (4.15) and (4.16), one gets

$$\lim_{x\to\varepsilon}\frac{f_{1x}}{f_{1xx}} = \lim_{x\to\varepsilon}\frac{P_1}{P_{1x}} = 0 .$$

Finally, for $x \to \varepsilon$, we have $f_1 \ll f_{1x} \ll f_{1xx}$, $P_1 \ll P_{1x}$, i.e., for $x \to \varepsilon$ the fundamental assumption of the small parameter method is violated, and hence a linear theory cannot be applied.

Consider now the point $x = 1$. For $x = 1$, one has f_1, f_{1x}, $P_1 \to 0$; f_{1xx}, $P_{1x} \to -\infty$, i.e., linear theory is not applicable in the vicinity of the point $x = 1$.

4.1.2 Nonlinear solution in the vicinity of the piston

By applying the method of matched asymptotic series, a solution uniformly suitable in the vicinity of the point $x = \varepsilon$ will be constructed (in the vicinity of the piston). Let us extend the external independent x and dependent f, P variables, introducing the following internal variables S, Φ, R:

$$x = \varepsilon S , \quad f(x;\varepsilon) = d_1\Phi(S) = d_1\Phi_1(S) + o(d_1) ,$$

$$P(x;\varepsilon) = d_1 R(S) = d_1 R_1(S) + o(d_1) ; \quad d_1 = \frac{\varepsilon^2}{\sqrt{1-\varepsilon^2}} . \tag{4.17}$$

Observe that

$$f_x = f_s S_x = \frac{d_1}{\varepsilon}\Phi_{1s} + o\left(\frac{d_1}{\varepsilon}\right) , \quad f_{xx} = \frac{d_1}{\varepsilon^2}\Phi_{1ss} + o\left(\frac{d_1}{\varepsilon^2}\right) . \tag{4.18}$$

Let us substitute (4.17), (4.18) into (4.1), (4.2), (4.6). Then, for an internal subspace, the following internal first order boundary value problem is formulated

$$\Phi_{1ss} + \frac{1}{S}\Phi_{1s} = 0 , \quad \Phi_{1s}(1) = 1 , \tag{4.19}$$

where

$$R_1 = S\Phi_{1s} - \Phi_1 - \Phi_{1s}^2/2 . \tag{4.20}$$

A general solution of the second order differential equation (4.19) contains two arbitrary constants e_1, e_2

$$\Phi_1 = e_1 \ln S + e_2 . \tag{4.21}$$

One of them, (e_1), is determined by the boundary condition (4.19), whereas the other one, (e_2), is found through matching of the internal (4.21) and external (4.15) solutions.

The equation (4.21) yields $\Phi_{1S} = e_1/S$, whereas the boundary condition (4.19) yields $e_1 = 1$. The internal solution reads

$$\Phi_1(S) = \ln S + e_2 \; . \tag{4.22}$$

Let us denote an external limit, i.e., the limit of the external solution (4.15) for $\varepsilon \to 0$. Since S is fixed, according to (4.17) one may observe that $x \to 0$.

According to the equation (4.15), for $x \to 0$, one gets

$$f^{(1)} = \ln x + 1 - \ln 2 + O(x^2) \; .$$

Let us describe the internal limit with the use of internal variables

$$\Phi_1(S) = \ln \varepsilon + \ln S + 1 - \ln 2 \; . \tag{4.23}$$

Comparing external and internal limits, i.e., comparing (4.22) and (4.23), one gets

$$e_2 = \ln \varepsilon + 1 - \ln 2 \; .$$

Finally, the internal solution holding in the vicinity of $x = \varepsilon$ (in the piston vicinity) described with the use of external variables takes (according to 4.17, 4.23, 4.20) the following form

$$f(x;\varepsilon) = d_1(\ln x + 1 - \ln 2) + o(d_1) \; ,$$

$$P(x;\varepsilon) = d_1\left(-\ln x + \ln 2 - \varepsilon^2/2x^2\right) \; . \tag{4.24}$$

According to equation (4.24), for $x = \varepsilon$, one gets

$$P(\varepsilon) = d_1\left(-\ln \varepsilon + \ln 2 - \frac{1}{2}\right) + o(d_1) \; . \tag{4.25}$$

The obtained result indicates that the linear solution (4.15), for which $P(\varepsilon) = d_1(-\ln \varepsilon + \ln 2) + o(d_1)$, increases the value of the pressure $P(\varepsilon)$ in comparison to the value of (4.25) defined through the nonlinear solution holding for $x = \varepsilon$ (inside of the piston), and in its vicinity $x = \varepsilon$.

To conclude, the application of both traditional methods of deformed coordinates and rescaling to investigate the solution in the vicinity of $x = \varepsilon$ does not yield satisfactory results.

4.1.3 Nonlinear solution in the vicinity of the front of the impact wave

Let us proceed now to constructing the solution in the vicinity of the point $x = M = 1 + \varepsilon_1$, where the unknown constant $\varepsilon_1 > 0$ is determined in the process of constructing a solution.

Let us remember that with the earlier use of a small parameter method, we have established that in the first approximation $\varepsilon_1 = 0$, the solutions (4.15), (4.16) are useless in the vicinity of $x = 1$.

Let us determine now, with the use of a small parameter method, the solution of the second order approximation f_2, P_2. Substituting (4.12), (4.13) into (4.1), (4.5) and equating the elements standing by d_1^2, to zero, we get the nonhomogenous linear equation, whose right-hand side depends on the first approximation for the determination of f_2, and the equations needed to determine P_2

$$\left(1-x^2\right) f_{2xx} + \frac{1}{x} f_{2x} = (\gamma - 1)\left(f_1 - x f_{1x}\right)\left(f_{1xx} + \frac{1}{x} f_{1x}\right) - 2x f_{1x} f_{1xx} \ , \tag{4.26}$$

$$P_2 = x f_{2x} - f_2 - \frac{P_1^2}{2} - \frac{f_{1x}^2}{2} \ . \tag{4.27}$$

It is convenient to present (4.26), using (4.9), in the shorter form

$$(1-x^2) f_{2xx} + \frac{1}{x} f_{2x} = -P_{1x}(2f_{1x} + (\gamma+1)xP_1 \ . \tag{4.28}$$

Integrating the equation (4.28) and satisfying the boundary condition (4.2) arising from $f_{2x}(\varepsilon) = 0$, we get

$$f_{2x} = \frac{1}{2x}\left[A\left(x\right) - \frac{\sqrt{1-x^2}}{\sqrt{1-\varepsilon^2}} A\left(\varepsilon\right)\right] \ ,$$

$$A\left(a\right) = \gamma + 3 + \frac{\left(\gamma+3\right)a^2 - 4}{\sqrt{1-a^2}} \ln\frac{1+\sqrt{1-a^2}}{a} \ , \tag{4.29}$$

where the auxiliary function A dependent on the formal parameter a is introduced. Let us notice that

$$A\left(a\right) = 4\ln a + \gamma + 3 - 4\ln 2 + O(a^2 \ln a) \qquad \text{for} \ \ a \to 0 \ ,$$

$$A\left(a\right) = 2(\gamma+1) - \frac{4}{3}(\gamma+5)(1-a) + o(1-a) \qquad \text{for} \ \ a \to 1 \ .$$

Integrating (4.29), we find f_2, and from (4.27) we will determine P_2:

$$f_2 = B(\varepsilon)\left(\sqrt{1-x^2} - \ln\frac{1+\sqrt{1-x^2}}{x}\right) -$$

$$-\frac{\gamma+3}{2}\sqrt{1-x^2}\ \ln\frac{1+\sqrt{1-x^2}}{x} + \left(\ln\frac{1+\sqrt{1-x^2}}{x}\right)^2 \ , \tag{4.30}$$

$$B(\varepsilon) = -\frac{\gamma+3}{2\sqrt{1-\varepsilon^2}} + \frac{\gamma+3}{2}\ln\frac{1+\sqrt{1-\varepsilon^2}}{\varepsilon} -$$

$$-\frac{\gamma-1}{2(1-\varepsilon^2)}\ln\frac{1+\sqrt{1-\varepsilon^2}}{\varepsilon}=-2\ln\varepsilon+2\ln 2-\frac{\gamma+3}{2}+O(\varepsilon^2\ln\varepsilon)\ ,$$

$$P_2=-\frac{1}{2x^2}-\frac{1}{2}\left(\ln\frac{1+\sqrt{1-x^2}}{x}\right)^2+$$

$$+\left(B(\varepsilon)+\frac{\gamma-1}{2\sqrt{1-x^2}}\right)\ln\frac{1+\sqrt{1-x^2}}{x}+\frac{\gamma+4}{2}\ . \tag{4.31}$$

Eventually, substituting (4.15), (4.29)–(4.31) into (4.12), (4.13), and using the classical small parameter method, we will get two approximations determinig the functions f, f_x, P.

By prolonging this procedure, it is possible to obtain the third approximation (it is not given here, because of its complexity).

However, it turns out, that constructing the first and further approximations does not improve the situation in the vicinity of the point $x=1$, where the asymptotic series (4.12), (4.13) remain unequally useful (irregular).

Let us construct the solution holding in the vicinity of the point $x=1$ with the use of the method of matched asymptotic series. Let us introduce $\lambda=1-x$, and then $\lambda\to 0$ for $x\to 1$. We will represent the series (4.12), (4.13) by distributing (4.15), (4.29)–(4.31) into the series for $\lambda\to 0$ $(x\to 1)$

$$f(x;\varepsilon)=d_1\left[-\frac{2}{3}\sqrt{2}\lambda^{3/2}+O(\lambda^2)\right]+$$

$$+d_1^2\left[-(\gamma+1)\lambda+\frac{\sqrt{2}}{3}(4\ln\varepsilon-4\ln 2+\gamma+3)\lambda^{3/2}+O(\lambda^2)\right]+o(d_1^2)\ ,$$

$$d_1=\frac{\varepsilon^2}{\sqrt{1-\varepsilon^2}}\ . \tag{4.32}$$

It is plainly visible that the series (4.12), (4.13) are irregular in the vicinity of the point $x=1$, because for $x\sim 1-d_1^2$, i.e., for $\lambda\sim d_1^2$, their second elements are of the same order as the first ones. Moreover, the functions f, P are, respectively, of d_1^4, d_1^2 order. Those observations allow to introduce in the internal region (in the vicinity of the point $x=1$) internal independent variable δ and dependent variables (whose magnitude order in this region is equal to one) in the following way

$$x=1-d_1^2\delta\,,\quad (\lambda=1-x=d_1^2\delta)\ , \tag{4.33}$$

$$f\left(x;\varepsilon\right)=-\frac{2}{\gamma+1}d_1^4\left[\Phi\left(\delta\right)+o\left(1\right)\right]\,,\quad P\left(x;\varepsilon\right)=\frac{2}{\gamma+1}d_1^2\left[\Pi\left(\delta\right)+o\left(1\right)\right]\,, \tag{4.34}$$

(coefficient $2/(\gamma+1)$ in (4.34) was introduced to simplify the computations).

Substituting (4.33), (4.34) into the exact equations (4.1), (4.6) and comparing the elements standing by the same powers of d_1 we will get the following inner equations for higher elements of the series (4.34)

$$2\left(\Phi_\delta+\delta\right)\Phi_{\delta\delta}-\Phi_\delta=0\,,\quad \Pi=\Phi_\delta\,. \tag{4.35}$$

Inner equations (4.35) are simpler than initial equations (4.1), (4.6) and they lead to the determination of two exact solutions from $h=\pm1$, including the arbitrary constant c_1, c_2 in the form of

$$\Phi_\delta=\Pi=\left(1+h\sqrt{1+\delta c_1}\right)\Big/\,c_1\,, \tag{4.36}$$

$$\Phi=\delta/c_1+2h\left(1+\delta c_1\right)^{3/2}\Big/\left(3c_1^2\right)+c_2\,,\quad h\pm1\,. \tag{4.37}$$

Let us combine the internal and external series for the function of pressure P (combining of the function f or f_x gives the same result).

The unary external series, according to (4.17), gets the following form

$$P=\sqrt{2}d_1\sqrt{\lambda}\,.$$

Then, in internal variables, we have

$$P=\sqrt{2}d_1^2\sqrt{\delta}\,. \tag{4.38}$$

We get *the unary internal series* from (4.36), (4.34) taking into account that for the established external variable x for $\varepsilon\to0$, the internal variable $\delta\to\infty$:

$$P=\frac{2}{\gamma+1}hd_1^2\sqrt{\delta/c_1}\,. \tag{4.39}$$

Comparing (4.38) to (4.39), one gets

$$c_1=2/\left(\gamma+1\right)^2\,. \tag{4.40}$$

Let us determine now *the variable x_* of the front of the impacting wave.* Since according to (4.33) $x=1-d_1^2\delta$, then x_* is searched in the form of

$$x_*=M=1-d_1^2\delta_*+o\left(d_1^2\right)\,. \tag{4.41}$$

Let us represent the boundary condition (4.4) in the form of internal variables. Since

$$f_x=f_\delta\delta_x=-\frac{2}{(\gamma+1)}d_1^4\Phi_\delta\left(-\frac{1}{d_1^2}\right)=\frac{2d_1^2\Phi_\delta}{(\gamma+1)}$$

and taking the fact that $1/M=1+d_1^2\delta_*+o(d_1^2)$ into account, then from (4.4) one gets

$$\Phi_\delta\left(\delta_*\right)=-2\delta_*\,. \tag{4.42}$$

Substituting (4.36) into (4.42) and solving this equation, we get

$$\delta_* = -3/(4c_1) = -3(\gamma+1)^2/8 \ .$$

Then, from the equation (4.41), we get

$$x_* = M = 1 + 3(\gamma+1)^2 d_1^2/8 + o\left(d_1^2\right) \ . \tag{4.43}$$

The obtained result was given for the first time in [103], and the next elements of the series (4.43) exacting the position of the front of the impacting wave is given in the work [117].

Let us now consider, instead of (4.4) the boundary condition (4.5). It is convenient to present it as

$$\gamma x_* f_x = (1 - f_x/x_*)^{-\gamma} - 1 \ . \tag{4.44}$$

Let us use the relation (2.3). Given that $f_x = O(d_1^2)$, $x_* = O(1)$, after equating in (4.44) higher elements of asymptotic series, the formula (4.42), is obtained. It means that the positions of the impacting wave front $x_* = M$, occurring as a result of explosion of a thin wire in air and in water overlap with the accuracy of up to $O(d_1^2)$ inclusive (therefore, let us notice that there is an error in the work [136]).

In order to finish the process of constructing the inner solution, the boundary condition (4.3) has to be satisfied, which is expressed for inner variables in a following way

$$\Phi\left(\delta_*\right) = 0 \ . \tag{4.45}$$

Substituting (4.37) into (4.45) and taking the fact that $\delta_* c_1 = -3/4$ into account, we get

$$h = 1 \, , \quad c_2 = 2/\left(3c_1^2\right) = (\gamma+1)^4/6 \ .$$

Eventually, the position of the front of impacting wave is determined by the relation (4.43), and parameters associated with it are determined by internal variables, with the use of the following relations

$$\Phi(\delta) = \delta/c_1 + 2\left[1 + (1+\delta c_1)^{1/2}\right] \Big/ \left(3c_1^2\right) \, , \tag{4.46}$$

$$\Pi(\delta) = \Phi_\delta(\delta) = \left[1 + (1+\delta c_1)^{1/2}\right] \Big/ c_1 \, , \quad c_1 = 2/(\gamma+1)^2 \, ,$$

that should be substituted into (4.34).

Let us study now the problem (4.1)–(4.4) with the use of traditional method of deformed coordinates.

4.1.4 Methods of strained coordinates and renormalization

Let us now deform the independent and dependent variables, forgetting the results obtained earlier with the use of nonlinear theory.

Basing on the small parameter method (4.12)–(4.14), we will introduce new variables:

$$x = S + d_1 x_1 (S) + d_1^2 x_2 (S) + o\left(d_1^2\right) , \tag{4.47}$$

$$f(x;\varepsilon) = V(S;\varepsilon) = d_1 V_1(S) + d_1^2 V_2(S) + o(d_1^2) . \tag{4.48}$$

Let us introduce new variables S, V into the problem (4.1)–(4.4). In order to do that, one has to determine the derivatives f_x, f_{xx}, eg. $f_x = V_s S_x$. There are two ways of finding S_x. Firstly, from (4.47) one gets

$$x_s = 1 + d_1 x_{1s} + d_1^2 x_{2s} + \dots ,$$

and applying the method of differentiating the inverse function, we have

$$S_x = 1/x_s = 1 - d_1 x_{1s} + d_1^2 (x_{1s}^2 - x_{2s}) + \dots$$

Secondly, we can construct the function

$$g(x;S) = x - S - d_1 x_1 - d_1^2 x_2 + \dots = 0 .$$

Let us notice that the total differential $dg = g_x dx + g_s dS$, which yields $S_x = -g_x/g_s$, and we get the same result. Eventually, we have

$$f_x = d_1 V_{1s} + d_1^2 \left(V_{2s} - x_{1s} V_{1s}\right) + o\left(d_1^2\right) , \tag{4.49}$$

and then we find $f_{xx} = (f_x)_s S_x$.

In order to determine the problem which would lead to finding V_1, V_2, etc., one should transform the asymptotic series in problem (4.1)–(4.4) according to the rules of chapter 2 and equate to zero the coeeficients with the same powers of d_1.

As a result, we get:

1. The equation used for determining $V_1(S)$ with the accuracy as for the accepted notations overlaps with (4.8), and its solution (see 4.15, 4.16) has the following characteristics

$$V_1 (S) = \sqrt{1 - S^2} - \ln \frac{1 + \sqrt{1 - S^2}}{S} = -\frac{2}{3}\sqrt{2}\,(1 - S)^{3/2} + o\,(1 - S)^{3/2} , \tag{4.50}$$

$$V_{1s}(S) = \sqrt{1 - S^2}/S = \sqrt{2(1 - S)} + o(\sqrt{1 - S}) , \tag{4.51}$$

$$V_{1ss}(S) = -1/(S^2\sqrt{1 - S^2}) = -1/\sqrt{2(1 - S)} + O(\sqrt{1 - S}) , \tag{4.52}$$

$$V_1 (1) = V_{1s} (1) = 0 , \quad V_{1ss} \to -\infty \text{ for } S \to 1 . \tag{4.53}$$

2. The equation used for determining V_2 has the form of (4.8)

$$\left(1-S^2\right)V_{2ss}+V_{2s}/S=k_1V_{1ss}+R_1\ , \tag{4.54}$$

where:

$$k_1(S)=2(1-S^2)x_{1s}-(\gamma+1)SV_{1s}+(\gamma-1)V_1+2Sx_1\ , \tag{4.55}$$

$$R_1(S)=(1-S^2)x_{1ss}V_{1s}+\frac{\left[x_{1s}+\frac{x_1}{S}+(\gamma-1)(V_1-SV_{1s})\right]V_{1s}}{S}\ . \tag{4.56}$$

Since V_{1SS} has singularity for $S=1$ (see 4.53), then, it arises from (4.51) that in order for the singularity in V_2 not to exceed the order of singularity for V_1 (Lighthill law, chapter 3.7.4), for $S\to1$ the following condition should be introduced

$$k_1(1)=0\ . \tag{4.57}$$

It arises form the equations (4.56), (4.50), (4.51) that in R_1, elements are gathered that do not cause the increase of the order of singularity in V_2.

From (4.55), (4.53) we have

$$k_1\,(1)=-\,(\gamma+1)\,V_{1s}\,(1)+(\gamma-1)\,V_1\,(1)+2x_1\,(1)=2x_1\,(1)\ . \tag{4.58}$$

After taking (4.57)into account, this leads to the condition:

$$x_1=0\ , \tag{4.59}$$

which means that the deformation in the first approximation is not required.

Let us determine the asymptote $\tilde{V}_2(S)$ of the solution $V_2(S)$ of the equation (4.54) for $S\to1$ nd for the condition (4.59). Let us express the equation (4.54) for $S\to1$ in the following form

$$2(1-S)\tilde{V}_{2s}=\tilde{k}_1\tilde{V}_{1ss}+\tilde{R}_1\ , \tag{4.60}$$

(a tilde above the quantities means that higher elements of the asymptotic series for these quantities are taken into account for $S\to1$).

It arises from (4.55), (4.56), (4.59), (4.51) and (4.50) that

$$\tilde{k}_1=-(\gamma+1)\sqrt{2(1-S)}\,,\qquad \tilde{R}_1=O(1-S)\ .$$

Eventually, taking into consideration (4.52), from (4.60), we get the following equation

$$2\,(1-S)\,\tilde{V}_{2ss}+\tilde{V}_{2s}=\gamma+1\ .$$

Let us notice that its solution

$$\tilde{V}_2=-\,(\gamma+1)\,(1-S)+o\,(1-S)\ ,\quad \tilde{V}_{2s}=\gamma+1\ , \tag{4.61}$$

for $S\to1$ is not more singular than V_1.

Substituting (4.50), (4.51) and (4.61) into (4.48), one gets (for $S \to 1$):

$$f(x;\varepsilon) = d_1(-\frac{2}{3}\sqrt{2}(1-S)^{3/2}+\ldots)+d_1^2(-(\gamma+1)(1-S)+\ldots)+o(d_1^2) \quad (4.62)$$

$$f_x(x;\varepsilon) = d_1(\sqrt{2(1-S)}+\ldots)+d_1^2(\gamma+1+\ldots)+o(d_1^2)\ . \quad (4.63)$$

For $1-S=O(d_1^2)$, the second elements of these series are of the same order as the first elements. Hence, in order for these series to be equally useful up to the order $O(d_1^2)$, the independent variable $x_2(S)$ should be deformed with the accuracy to $O(d_1^2)$, i.e., in the series (4.47) should be determined. To do it, we have derived the equation for $V_3(S)$, i.e., for the next element of the asymptotic series (4.48):

$$(1-S^2)V_{3ss}+\frac{1}{S}V_{3s} = k_1V_{2ss}+k_2V_{1ss}+R_2\ , \quad (4.64)$$

where $k_2(S)$ for $x_1=0$ is (see (4.59)):

$$k_2\,(S) = 2x_{2s}+(\gamma-1)\,V_2+(\gamma+1)\,SV_{2s}+(\gamma+1)\,V_{1s}^2/2-2S^2x_{2s}+2Sx_2\ ,$$

(the expression for $R_2(S)$ is not provided here because of its complexity, and because it does not cause the increase of the order of singularity in V_3 for $S \to 1$).

Since the condition (4.57) is already satisfied, then, in accordance with Ligthill's law (concerning the fact that the order of singularity does not increase in successive approximations) one should require $k_2(1)=0$. Since

$$k_2(1) = (\gamma-1)V_2(1)-(\gamma+1)V_{2s}(1)+(\gamma+1)V_{1s}^2(1)/2+2x_2\ ,$$

then, taking (4.61) into consideration, we get

$$x_2 = (\gamma+1)^2/2\ .$$

Eventually, *the deformation* has the following form

$$x = S+d_1^2\,(\gamma+1)^2/2+o\left(d_1^2\right)\ , \quad (4.65)$$

and the asymptotic series (4.62), (4.63) taking (4.65) into consideration, are equally useful with the accuracy of up to $O(d_1^2)$ inclusive.

In order to fully solve the problem, it is enough now to satisfy the condition (4.4) and determine Mach number M from this condition. According to the assumption (4.14), we have

$$x = M = 1+d_1M_1+d_1^2M_2+o\left(d_1^2\right)\ ,$$

and from (4.65), we get

$$1-S = -d_1M_1+d_1^2\left[\frac{(\gamma+1)^2}{2}-M_2\right] = d_1^2\left[\frac{(\gamma+1)^2}{2}-M_2\right]\ , \quad (4.66)$$

since it arises from the left-hand side (4.4) taking (4.63) into consideration, that $M_1=0$.

Substituting (4.63) into (4.4) and taking (4.66) and (4.14) into consideration, we get the equation

$$d_1^2 = \left\{2\left[\frac{(\gamma+1)^2}{2} - M_2\right]\right\}^{1/2} + d_1^2(\gamma+1) = \frac{4d_1^2 M_2}{\gamma+1} ,$$

which we can easily solve. As a result, we get $M_2 = 3(\gamma+1)^2/8$ and

$$x = M = 1 + \frac{3(\gamma+1)^2 d_1^2}{8} + o\left(d_1^2\right) , \quad d_1 = \frac{\varepsilon^2}{\sqrt{1-\varepsilon^2}} . \tag{4.67}$$

The results (4.43) and (4.67) determining the position of the impacting wave front, obtained with the use of the method of matched asymptotic series and the method of elongated coordinates, overlap with the accuracy of up to $O(d_1^2)$.

It should not seem strange, because otherwise one of the methods would lead to less accurate results [106, p. 206]. It should be emphasized that during the study of complex problems of the interaction of impacting waves with obstacles, the method of matched asymptotic series is most often used (see also the results following the formula 4.25).

What is surprising is the fact that the principal differences with respect to ideology and technique of applying methods nevertheless lead to the same results for f_x. Transforming (4.63), and taking (4.65) into consideration, we get

$$f_x(x;\varepsilon) = d_1\sqrt{2\left[1 - x + \frac{d_1^2(\gamma+1)^2}{2}\right]} + d_1^2(\gamma+1) + o\left(d_1^2\right) . \tag{4.68}$$

We obtain the same result by transforming (4.34), and taking (4.46) and (4.33) into consideration

$$f_X(x;\varepsilon) = \frac{2d_1^2}{\gamma+1}\left[1 + \sqrt{1 + \frac{2(1-x)}{d_1^2(\gamma+1)}}\right]\frac{(\gamma+1)^2}{2} =$$

$$= d_1^2(\gamma+1) + d_1\sqrt{d_1^2(\gamma+1)^2 + 2(1-x)} .$$

Let us apply now the modification of the method of strained variables, i.e., the renormalization method into the problem discussed earlier (4.1)–(4.4).

According to the chapter 3, section 3.5, in the renormalization method, one should first construct the first few elements of the asymptotic series of the searched function with the use of the classical method of small parameter. Therefore, we will search for the solution in the form of asymptotic series with respect to the powers of the function $d_1(\varepsilon) = \varepsilon^2/\sqrt{1-\varepsilon^2}$ of the form:

$$f(x;\varepsilon) = d_1 f_1(x) + d_1^2 f_2(x) + d_1^3 f_3(x) + o\left(d_1^3\right) , \tag{4.69}$$

$$x = M + d_1 M_1 + d_1^2 M_2 + o\left(d_1^2\right) \ . \tag{4.70}$$

Let us notice that the form of those series was discussed with the introduction of (4.12)–(4.14). Then, we get

$$f_x = d_1 f_{1x} + d_1^2 f_{2x} + d_1^3 f_{3x} + o(d_1^2) \ ,$$

$$f_{xx} = d_1 f_{1xx} + d_1^2 f_{2xx} + d_1^3 f_{3xx} + o(d_1^3) \ . \tag{4.71}$$

Substituting (4.69), (4.71) into (4.1)–(4.4) we obtain the equations needed for determining f_1, f_2, f_3. The solutions f_1, f_2 and their derivatives were found earlier (see (4.15), (4.29), (4.30)). It has also been proved that the series (4.69), (4.71) are irregular in the vicinity of the point $x = 1$.

One should notice that determining f_2, and all the more f_3, poses certain computational difficulties. For example, the equation of the third approximation has the following form

$$\left(1 - x^2\right) f_{3xx} + \frac{1}{x} f_{3x} = (\gamma - 1)\left[f_1 f_{2xx} + f_2 f_{1xx} - x\left(f_{1x} f_{2xx} + f_{2x} f_{1xx}\right) + \right.$$

$$\left. + \frac{1}{2} f_{1x}^2 f_{1xx} + \frac{1}{x}\left(f_1 f_{2x} + f_2 f_{1x} + \frac{1}{2} f_{1x}^3\right) - 2 f_{1x} f_{2x}\right] -$$

$$-2x\left(f_{1x} f_{2xx} + f_{2x} f_{1xx}\right) + f_{1x}^2 f_{1xx} \ . \tag{4.72}$$

In order to avoid those difficulties, a different method will be proposed. Since we want to construct the solution in the vicinity of the point $x = 1$, instead of searching for the exact expressions for the function f_1, f_2, (that have already been found) and f_3, it is enough to reduce computations to determine their asymptotics (higher elements of the series) for $x \to 1$.

It should be emphasised that such an approach significantly simplifies solving the problem. If, in order to shorten the notation, we include $\lambda = 1 - x$, then from (4.15), (4.16), (4.29), (4.30) we can determine the order of quantities we are interested in. Proceeding like this, we get

$$f_1 \sim \lambda^{3/2} \ , \quad f_{1x} \sim \lambda^{1/2} \ , \quad f_{1xx} \sim \lambda^{-1/2} \ ,$$

$$f_2 \sim \lambda + \lambda^{3/2} \ , \quad f_{2x} \sim 1 + \lambda^{1/2} \ , \quad f_{2xx} \sim \lambda^{-1/2} \ .$$

The final asymptotic representation of the solution (4.72) for $x \to 1$ has the form

$$2(1 - x)\tilde{f}_{3xx} + \tilde{f}_{3x} = -(\gamma + 1)\tilde{f}_{1xx}\tilde{f}_{2x} \ , \tag{4.73}$$

(a tilde above the symbols means that the elements of higher order of the asymptotic series of those quantities are taken into account for $x \to 1$).

From the equation (4.30) we get $\tilde{f}_{1xx} = -1/\sqrt{2(1-x)}$, $\tilde{f}_{2x} = \gamma + 1$ and the equation (4.73) transforms into the equation

$$2(1 - x)\tilde{f}_{3xx} + \tilde{f}_{3x} = (\gamma + 1)^2/\sqrt{2(1 - x)} \ ,$$

whose solution has the form of

$$\tilde{f}_{3x} = (\gamma+1)^2 \left[2\sqrt{2(1-x)}\right] .$$

Taking into consideration the form of equations (4.8), (4.28), (4.72), (4.73), it is convenient to apply the renormalization method in order to construct the uniformly useful asymptotic series for f_x. Its asymptotic series for $x \to 1$, in accordance with the earlier results, is of the form

$$\tilde{f}_x(x;\varepsilon) = d_1\tilde{f}_{1x} + d_1^2\tilde{f}_{2x} + d_1^3\tilde{f}_{3x} + o(d_1^3) =$$

$$= d_1\sqrt{2(1-x)} + d_1^2(\gamma+1) + d_1^3(\gamma+1)^2/(2\sqrt{2(1-x)}) + o(d_1^3) . \quad (4.74)$$

It is clear that for $x \sim 1 - d_1^2$ all the elements in (4.74) have the same order $O(d_1^2)$, i.e., the series (4.74) is irregular.

Let us introduce the following *deformation*

$$x = S + d_1 x_1(S) + d_1^2 x_2(S) + o\left(d_1^2\right) ,$$

$$f_x(x;\varepsilon) = W(S;\varepsilon) = d_1 W_1(S) + d_1^2 W_2(S) + d_1^3 W_3(S) + o\left(d_1^3\right) . \quad (4.75)$$

Let us use the formulas (3.79)–(3.81) derivated in chapter 3.5.1. They have the following form

$$W_1(S) = \tilde{f}_{1x}(S) . \quad (4.76)$$

$$W_2(S) = \tilde{f}_{2x}(S) + x_1\tilde{f}_{1xx}(S) . \quad (4.77)$$

$$W_3(S) = \tilde{f}_{3x}(S) + x_1\tilde{f}_{2xx}(S) + x_2\tilde{f}_{1xx}(S) + x_1^2\tilde{f}_{1xxx}/2 . \quad (4.78)$$

Taking (4.74) and (4.75) into consideration, from (4.76), (4.77) we get

$$W_1(S) = \sqrt{2(1-S)}, \quad W_2(S) = \gamma + 1 - x_1/\sqrt{1-S} .$$

In order to get rid of singularity occurring in W_2, we will require

$$x_1 = 0 . \quad (4.79)$$

We also get $W_2(S) = \gamma + 1$. Then (4.78) taking (4.79), (4.74), (4.75) into consideration, has the following form

$$W_3(S) = (\gamma+1)^2/\left[2\sqrt{2(1-S)}\right] - x_2(S)/\sqrt{2(1-S)},$$

and in order to get rid of singularity occuring in W_3, we will require satisfying the following condition

$$x_2(S) = (\gamma+1)^2/2 .$$

To finish our considerations in this section, let us notice that the deformation of coordinates has been introduced, and we obtained the results (4.65), (4.63), (4.62) with the accuracy of up to $O(d_1^2)$, overlapping with earlier results obtained with the use of the method of elongated coordinates and the method of matched asymptotic series.

4.1.5 Effectiveness of various asymptotic methods

Comparing asymptotic methods applied in solving the considered problem leads to the following conclusions.

1. Applying the classical method of small parameter yields unreliable results both in the vicinity of a piston and in the vicinity of the impacting wave front.

2. Constructing the effective solution in the vicinity of the point $x = \varepsilon$ (in the vicinity of a piston) is possible only with the use of the method of matched asymptotic series. Applying both the method of elongated coordinates and the renormalization method here yields unreliable results.

3. It was visible during the construction of the nonlinear solution in the vicinity of the point $x = 1$ in the first approximation, that the methods of matched asymptotic series and of elongated coordinates are uniform with respect to the complexity of application, and the renormalization method is the most effective one. For constructing further approximations the method of matched asymptotic series is more effective than the elongated coefficients method, and applying the renormalization method enables to conduct the analysis of the problem of determination of function $f_4(x)$ from the asymptotic series (4.7).

4.2 One-dimensional nonstationary nonlinear waves

4.2.1 Formulation of the problem and its solution

Let us consider the following mathematical problem. Find the solution of the nonlinear partial differential equation of the form

$$\Phi_{tt} - a_0^2 \Phi_{xx} = -2\Phi_x \Phi_{xt} - \Phi_x^2 \Phi_{xx} - (\gamma - 1)\Phi_t \Phi_{xx} - \frac{\gamma - 1}{2} \Phi_x^2 \Phi_{xx} \, , \quad (4.80)$$

satisfying the condition

$$\Phi(0, t) = \varepsilon(t) \, , \quad (4.81)$$

(lower indices denote differentiation with respect to adequate variables).

Above, $\Phi(x, t; \varepsilon)$ denotes the velocity potential, i.e., $\Phi_x = u$, where $u(x, t; \varepsilon)$ is the velocity of the medium, $a_0 = const$ is the acoustic velocity in a steady medium, x is the distance covered by the wave, and t is time ($0 \leq t < \infty$). The boundary condition (4.81) is determined by the law of propagation of waves. A small dimensionless parameter ε ($\varepsilon \ll 1$) is determined during the formulation of a particular physical problem. For example, during motion of a weak impacting wave, ε characterizes its intensity, and during propagation of a wave with finite amplitude, ε characterizes its amplitude, etc.

It is impossible to obtain the exact analytical solution of the equation (4.80), so we apply the small parameter method. We will search for the solution in the form of the following power series:

$$\Phi(x,t;\varepsilon) = \varepsilon\Phi_1(x,t) + \varepsilon^2\Phi_2(x,t) + \mathrm{O}(\varepsilon^3) \ . \tag{4.82}$$

Substituting (4.82) into (4.80) and (4.81), and equating to zero the elements with the same powers of ε, we get the following problems for the first approximation

$$\Phi_{1tt} - a_0^2\Phi_{1xx} = 0 \ , \tag{4.83}$$

$$\Phi_1(0,t) = g(t) \ , \tag{4.84}$$

and for the second approximation

$$\Phi_{2tt} - a_0^2\Phi_{xx} = -2\Phi_{1x}\Phi_{1xt} - (\gamma - 1)\Phi_{1t}\Phi_{1xx} \ , \tag{4.85}$$

$$\Phi_2(0,t) = 0 \ . \tag{4.86}$$

The equation (4.83) is a well-known *wave equation*, and the equation (4.85) is *the nonhomogeneous wave equation*, and its right-hand side depends on Φ_1. Both equations are hyperbolic.

According to the lectures concerning the analysis of the equations of mathematical physics, in order to solve the equation (4.83) it is convenient to introduce instead of x, t new independent variables ξ, η of the form

$$\xi = t - x/a_0 \, , \quad \eta = t + x/a_0 \ . \tag{4.87}$$

Let us notice that instead of the suggested change of coordinates (4.87) the following one is often used

$$\varphi = x - a_0 t \, , \quad \psi = x + a_0 t \ . \tag{4.88}$$

Independent variables ξ, η or φ, ψ allow to reduce the equation to the canonical form, and the sets of straight lines $\xi = \mathit{const}$, $\eta = \mathit{const}$ or $\varphi = \mathit{const}$, $\psi = \mathit{const}$ are called *equations of characteristic*.

For example, let us consider the exchange of variables (4.87). In this case, the velocity potential Φ_1 will be the function of variables ξ, η, i.e., $\Phi_1 = \Phi_1(\xi,\eta)$.

We will express the equation (4.83) in the form of new variables. In order to do that, we determine the derivatives of the function Φ_1, using the exchange (4.87), obtaining according to the differential law

$$\Phi_1 = \Phi_{1\xi}\xi_x + \Phi_{1\eta}\eta_x = \frac{1}{a_0}(-\Phi_{1\xi} + \Phi_{1\eta}) \ ,$$

$$\Phi_{1t} = \Phi_{1\xi}\xi_t + \Phi_{1\eta}\eta_t = \Phi_{1\xi} + \Phi_{1\eta} \ .$$

After further differentiation, we get

$$\Phi_{1xx} = \frac{1}{a_0^2}(\Phi_{1\xi\xi} - 2\Phi_{1\xi\eta} + \Phi_{1\eta\eta}) \, ,$$

$$\Phi_{1tt} = \Phi_{1\xi\xi} + 2\Phi_{1\xi\eta} + \Phi_{1\eta\eta} \, . \tag{4.89}$$

Substituting (4.89) into (4.83), we get the canonical form of the wave equation

$$\Phi_{1\xi\eta} = 0 \, . \tag{4.90}$$

Differentiating the equation (4.90) twice, we get its general solution of the form

$$\Phi_1 = F(\xi) + H(\eta) = F(t - x/a_0) + H(t + x/a_0) \, , \tag{4.91}$$

where F, H are arbitrary functions determining the raveling waves directed, respectively to the right and to the left.

In our further considerations we will focus on the example of the wave running to the right, i.e., in (4.91) we assume

$$H(t + x/a_0) = 0 \, .$$

Substituting $\Phi_1 = F(t - x/a_0)$ into (4.85) and satisfying the boundary conditions (4.84), (4.86), we find

$$\Phi_1(x, t) = g(t - x/a_0) \, , \tag{4.92}$$

$$\Phi_2(x, t) = -\frac{\gamma + 1}{4a_0^3} x g'^2 (t - x/a_0) \, . \tag{4.93}$$

The correctness of the formula (4.93) can be easily proved by the direct substitution of (4.93) into (4.85), (4.86).

Eventually, according to (4.92), (4.93) and (4.82), we get the solution determining the form of *the velocity potential*

$$\Phi(x, t; \varepsilon) = \varepsilon g(\xi) - \varepsilon^2 \frac{\gamma + 1}{4a_0^3} x g_\xi^2 + O(\varepsilon^3) \, . \tag{4.94}$$

By differentiating (4.94), we will determine *the velocity of molecules of a medium*

$$u(x, t; \varepsilon) = \Phi_x = \Phi_\xi \xi_{xx} = -\frac{\varepsilon}{a_0} g_\xi - \varepsilon^2 \frac{\gamma + 1}{4a_0^3} g_\xi^2 + \varepsilon^2 \frac{\gamma + 1}{2a_0^4} x g_\xi g_{\xi\xi} +$$

$$+ O(\varepsilon^3) = \varepsilon \, u_1(x, \xi) + \varepsilon^2 u_2(x, \xi) + O(\varepsilon^3) \, . \tag{4.95}$$

It is clear that the asymptotic series (4.94), (4.95) are nonuniformly useful for various values of x, because the elements proportional to x appear for them (which is the analogy for secular elements occuring in the asymptotic series (3.27)). For $x = O(1/\varepsilon)$ the second elements in the series (4.94), (4.95) have the order of the first elements, and for $x > 1/\varepsilon$ they are higher than this. Hence, the typical perturbation of singular problems occurs.

In order to construct the solution holding for large values of x the method of singular perturbations is further used.

4.2.2 Renormalization method and singularities

Let us deform the dependent and independent variable in the form of

$$\xi = t - x/a_0 = K + \varepsilon\mu_1(x, K) + O(\varepsilon^2) \ , \tag{4.96}$$

$$\Phi(x, t; \varepsilon) = \phi(x, K; \varepsilon) = \varepsilon\phi_1(x, K) + \varepsilon^2\phi_2(x, K) + O(\varepsilon^3) \ , \tag{4.97}$$

where μ_1 is the so far unknown function of deformation; K, ϕ are new dependent and independent variables. Using the relations (3.79), (3.80), we get

$$\phi_1(x, K) = g(K) \ , \tag{4.98}$$

$$\phi_2(x, K) = -\frac{\gamma+1}{4a_0^3} x g_\xi^2(K) + \mu_1(x, K) g_\xi(K) =$$

$$= g_\xi(K)\left[\mu_1(x, K) - \frac{\gamma+1}{4a_0^3} x g_\xi(K)\right] \ . \tag{4.99}$$

In order to get rid of the secular element in (4.99), we will require

$$\mu_1(x, K) = \frac{\gamma+1}{4a_0^3} x g_\xi(K) \ . \tag{4.100}$$

Eventually, the asymptotic series for the function Φ with the accuracy of up to $O(\varepsilon)$ assumes the form of

$$\Phi(x, t; \varepsilon) = \varepsilon g(K) + O(\varepsilon^3) \ , \tag{4.101}$$

where

$$\xi = K + \varepsilon\frac{\gamma+1}{4a_0^3} x g_\xi(K) + O(\varepsilon^2) \ . \tag{4.102}$$

It turns out, however, that the solution (4.101), (4.102) is not the uniformly useful series for the vicinity of the medium molecules. Since $u = \Phi_x = \Phi_\xi \xi_x$, from (4.102) we get

$$u = -\frac{\varepsilon}{a_0} g_\xi(K)\left[1 + \frac{\gamma+1}{4a_0^2}\varepsilon\, g_\xi(K)\right] / \left[1 + \frac{\gamma+1}{4a_0^3}\varepsilon\, x g_{\xi\xi}(K)\right] =$$

$$= -\frac{\varepsilon}{a_0} g_\xi(K) + \varepsilon^2\frac{\gamma+1}{4a_0^3}\left[g_\xi\,(K) - \frac{x}{a_0} g_{\xi\xi}\,(K)\right] + O(\varepsilon^3) \ . \tag{4.103}$$

The occurrence of a secular element proportional to x in the series (4.103) results in the fact that this series is nonuniformely useful for $x \sim O(1/\varepsilon)$,which will be further proved during the analysis with the use of the method of characteristics and the method of multiple scales.

Let us apply the renormalization method for the function $u(x, t; \varepsilon)$.

$$\xi = K + \varepsilon\mu_1(x, K) + O(\varepsilon^2) \ ,$$

$$u(x,t;\varepsilon) = U(x,K;\varepsilon) = \varepsilon U_1(x,K) + \varepsilon^2 U_2(x,K) + O(\varepsilon^3) \ , \tag{4.104}$$

where μ_1 is the unknown function of deformation, K, U are new variables (dependent and independent).

According to the renormalization method (3.79) and (3.80), we have

$$U_1(x,K) = u_!(x,K) \ , \tag{4.105}$$

$$U_2(x,K) = u_2(x,K) + \mu_1 u_1(K) \ , \tag{4.106}$$

where $u_{1x} = u_{1\xi}\xi_x = -u_{1\xi}/a_0$.

Substituting the expression arising from (4.95) into (4.104)–(4.106), we get

$$u(x,t;\varepsilon) = U(x,K;\varepsilon) =$$

$$= -\frac{\varepsilon}{a_0} g_\xi(K) - \varepsilon^2 \frac{\gamma+1}{4a_0^3} g_\xi^2(K) + \varepsilon^2 \frac{\gamma+1}{2a_0^4} x g_\xi(K) g_{\xi\xi}(K) - \mu_1 \frac{\varepsilon^2}{a_0} g_{\xi\xi}(K) =$$

$$= -\frac{\varepsilon}{a_0} g_\xi(K) - \varepsilon^2 \frac{\gamma+1}{4a_0^3} g_\xi^2(K) + \frac{\varepsilon^2}{a_0} g_{\xi\xi}(K) \left(\frac{\gamma+1}{2a_0^3} x g_\xi(K) - \mu_1 \right) \ . \tag{4.107}$$

It arises from the above equation that the choice of

$$\mu_1 = \frac{\gamma+1}{2a_0^3} x g_\xi(K)$$

allows to get rid of a secular element. Eventually, the uniformly applied asymptotic series for $u(x,t;\varepsilon)$ assumes the form of

$$u(x,t;\varepsilon) = -\frac{\varepsilon}{a_0} g_\xi(K) - \varepsilon^2 \frac{\gamma+1}{4a_0^3} g_\xi^2(K) + O(\varepsilon^3) \ , \tag{4.108}$$

where

$$\xi = K + \varepsilon \frac{\gamma+1}{2a_0^3} x g_\xi(K) + O(\varepsilon^2) \ . \tag{4.109}$$

Let us notice that the deformations (4.109) and (4.102), and the expressions (4.108) and (4.103) for the velocity u differ from one another. The possibility of such a result, though during the analysis of other problem, was earlier mentioned in references [166] and [119].

The deformation of the velocity potential Φ instead of the deformation of velocity Φ, yielded wrong results. This proves that the deformations should be conducted for "primitive" variables, such as velocity, pressure, etc.

4.2.3 Analytical method of characteristics

Let us derive the characteristic equations of the differential equation (4.80). In the case of one-dimensional nonstationary stream determined by equation (4.80) the following relation occurs

$$a^2 = a_0^2 - (\gamma - 1)\left(\Phi_t + \frac{1}{2}\Phi_x^2 \right) \ , \tag{4.110}$$

where a is the acoustic velocity of the moving medium under consideration.

Grouping (4.80) the elements for Φ_{xx} and using (4.110), we get from (4.80) the equivalent equation

$$\Phi_{xx}(a^2 - \Phi_x^2) - 2\Phi_x\Phi_{xt} - \Phi_{tt} = 0 \ . \tag{4.111}$$

Let us remember that the characteristics of a second order differential equation

$$a_{11}\Phi_{xx} + 2a_{12}\Phi_{xt} + a_{22}\Phi_{tt} = 0$$

are determined by equations

$$\left(\frac{dx}{dt}\right)_{1,2} = \frac{a_{12} \pm \sqrt{a_{12}^2 - a_{11}a_{22}}}{a_{11}} \ . \tag{4.112}$$

It arises form the formula (4.112) that equation (4.111) has two families of characteristics

$$\left(\frac{dx}{dt}\right)_{1,2} = \Phi_x \pm a = u \pm a \ . \tag{4.113}$$

In the case of the waves directed to the right, we have

$$dx/dt = \Phi_x - a \quad \text{lub} \quad dt/dx = 1/(\Phi_x - a) \ . \tag{4.114}$$

In order to find a, we will use (4.110):

$$a = a_0\left[1 - \frac{\gamma - 1}{a_0^2}\left(\Phi_t + \frac{1}{2}\Phi_x^2\right)\right]^{1/2} \ . \tag{4.115}$$

Let us express $u = \Phi$, $v = \Phi_t$ in the form of the following asymptotic series

$$u = \Phi_x(x,t;\varepsilon) = \varepsilon\Phi_{1x}(x,t) + \varepsilon^2\Phi_2(x,t) + O(\varepsilon^3) \ , \tag{4.116}$$

$$v = \Phi_t(x,t;\varepsilon) = \varepsilon\Phi_{1t}(x,t) + \varepsilon^2\Phi_{2t}(x,t) + O(\varepsilon^3) \ . \tag{4.117}$$

Let us expand (4.115) the series according to the formula (2.3) of the form

$$a = a_0\left[1 - \frac{\gamma - 1}{2a_0^2}\left(\Phi_t + \frac{1}{2}\Phi_x^2\right) - \frac{(\gamma-1)^2}{8a_0^4}\left(\Phi_t + \frac{1}{2}\Phi_x^2\right)^2 + \ldots\right] \ . \tag{4.118}$$

Let us substitute the series (4.116) and (4.117) into (4.118), and let us focus only on the elements of the order ε. As a result, we get

$$a = a_0 - \varepsilon\frac{\gamma - 1}{2a_0}\Phi_{1t} + O(\varepsilon^2) \ . \tag{4.119}$$

Substituting (4.116) and (4.119) into (4.114), we obtain the following *characteristic equation* in the second approximation

$$\frac{dx}{dt} = -a_0 + \varepsilon\left(\Phi_{1x} + \frac{\gamma - 1}{2a_0}\Phi_{1t}\right) + O(\varepsilon^2) \ . \tag{4.120}$$

Observe that it is not necessary, since in sections 2.1, 2.2 the solutions for the first approximation have been constructed only.

Introducing

$$u = \Phi\,, \quad v = \Phi_t\,, \tag{4.121}$$

$$u_t = v_x\,, \tag{4.122}$$

from (4.80), (4.121), we obtain

$$v_t - a_0^2 u_x = -2uu_t - (\gamma - 1)vu_x - \frac{\gamma - 1}{2}u^2 u_x\,, \tag{4.123}$$

whereas from (4.120), we obtain

$$\frac{dt}{dx} = -\frac{1}{a_0} - \frac{u}{a_0^2} + \frac{\gamma - 1}{2a_0^3}v + O(\varepsilon^2)\,. \tag{4.124}$$

Solutions of the equations (4.122)–(4.124) are searched in the form of asymptotic series

$$u = \varepsilon u_1(\xi, x) + O(\varepsilon^2)\,, \quad v = \varepsilon v_1(\xi, x) + O(\varepsilon^2)\,, \quad t = t_0(\xi, x) + O(\varepsilon^2)\,. \tag{4.125}$$

Assuming that

$$t(\xi, 0) = \xi\,, \tag{4.126}$$

and substituting (4.125) into (4.122)–(4.124) (and then equating to zero the coefficients standing by ε), integrating the obtained equations, and satisfying the conditions (4.81), (4.126), uniformly useful solution of the first order is determined

$$u = -\varepsilon g'(\xi)/a_0 + O(\varepsilon^2)\,, \quad v = \varepsilon g'(\xi) + O(\varepsilon^2)\,, \tag{4.127}$$

where:

$$t = x/a_0 = \xi + \varepsilon(\gamma + 1)xg'(\xi)/2a_0^3 + O(\varepsilon^2)\,. \tag{4.128}$$

The formulas (4.127), (4.128) overlap with the formulas (4.108), (4.109), obtained with the use of the renormalization method applied for the velocity $u = \Phi_x$.

The analytical method of characteristics for the two-dimensional problem is professionally called *the method of precised characteristics*. The solution uniform with the accuracy of up to $O(\varepsilon)$, is obtained from the linear solution, found with the use of the classical small parameter method (if we exchange the linearized characteristices for precised characteristics with accuracy of up to $O(\varepsilon)$ inclusive).

4.2.4 Multiple scales method

According to the method of multiple scales, we will search for the solution in the equation (4.80), (4.81) of the form

$$\Phi(x,t;\varepsilon) = \Phi(\varphi,\psi,X;\varepsilon) = \varepsilon\Phi_1(\varphi,\psi,X) + \varepsilon^2\Phi_2(\varphi,\psi;X) + O(\varepsilon^3)\ , \quad (4.129)$$

where:

$$\varphi = t - x/a_0\,, \quad \psi = t + x/a_0\,, \quad X = \varepsilon x\,. \quad (4.130)$$

In the above, $\varphi = const$, $\psi = const$ are the equations of characteristics moving, respectively, to the left and to the right; x is "the slow," and X is "the fast" coordinate.

Let us put the integral equation (4.80) in the form of new variables. For this purpose, the derivatives corresponding to the rule of the differentiation of a complex function are determined, i.e.,

$$\Phi_x = \Phi_\varphi\varphi_x + \Phi_\psi\psi_x + \Phi_X X_x = -\frac{1}{a_0}\Phi_\varphi + \frac{1}{a_0}\Phi_\psi + \varepsilon\Phi_X\ ,$$

$$\Phi_t = \Phi_\varphi\varphi_t + \Phi_\psi\psi_t + \Phi_X X_t = \Phi_\varphi + \Phi_\psi\ .$$

Here and further, a letter in the lower index denotes a partial derivative determined with respect to that variable. We determine Φ_{xx}, Φ_{xt}, Φ_{tt} in a similar way.

Let us substitute the expressions for the derivatives into the equation (4.80) and let us distribute them with respect to the powers of ε using (4.129). Equating to zero the coefficients for ε, we obtain the equation for the first approximation of the form

$$\Phi_{1\varphi\psi} = 0\ ,$$

whose general solution $\Phi_1(\varphi,\psi,X) = q(\varphi,X) + h(\psi,X)$ is known (q, h are arbitrary functions).

We will focus our further considerations to the wave moving to the right. Then we have

$$\Phi_1 = \Phi_1(\varphi,X)\ , \quad (4.131)$$

and from the condition (4.81) we obtain

$$\Phi_1(t,0) = g(t)\ . \quad (4.132)$$

The equation for the second approximation (elements in 4.80 in ε^2 wih respect to 4.131) has the form of

$$\Phi_{2\varphi\psi} = -2a_0\Phi_{1\varphi X} - \frac{\gamma+1}{a_0^2}\Phi_{1\varphi}\Phi_{1\varphi\varphi}\ . \quad (4.133)$$

In order to obtain the uniformly useful asymptotic series for Φ, one has to remove the right-hand side in (4.133). This yields

$$\left[\frac{\partial}{\partial X} + \frac{\gamma+1}{2a_0^3}\Phi_{1\varphi}\frac{\partial}{\partial\varphi}\right]\Phi_{1\varphi} = 0\ . \quad (4.134)$$

In what follows we show that the following function is the solution of the equation (4.134), which satisfies the condition (4.132)

$$\Phi_1 = g(\xi) + \left[(\gamma+1)/\left(4a_0^3\right)\right] X g'^2(\xi) \ , \qquad (4.135)$$

where:

$$\varphi = \xi + [(\gamma+1)/(2a_0^3)] X g'(\xi) \ . \qquad (4.136)$$

Therefore, we will find ξ_φ i ξ_X. Differentiating (4.136) with respect to φ, we have

$$1 = \xi_\varphi + (\gamma+1) X g'' \xi_\varphi / (2a_0^3)$$

and we find

$$\xi_\varphi = 1/\left[1 + (\gamma+1) X g''/(2a_0^3)\right] \ . \qquad (4.137)$$

Differentiating (4.136) with respect to X, we have

$$0 = \xi_X + (\gamma+1) g'/(2a_0^3) + (\gamma+1) X g'' \xi_X/(2a_0^3)$$

and therefore

$$\xi_X = -\frac{(\gamma+1) g'/(2a_0^3)}{1 + (\gamma+1) X g''/(2a_0^3)} \ . \qquad (4.138)$$

Then, we differentiate (4.135) with respect to φ:

$$\Phi_{1\varphi} = g' \xi_{\,\varphi} + \frac{\gamma+1}{4a_0^3} X \cdot 2 g' g'' \xi_{\,\varphi} = \xi_\varphi g' \left(1 + \frac{\gamma+1}{2a_0^3} X g''\right) \ ,$$

Hence, using (4.137), we have

$$\Phi_{1\varphi} = g' \ . \qquad (4.139)$$

Next, using (4.139) and (4.137), (4.138), it is easy to prove that the left-hand side of (4.134) is equal to zero. From (4.134), we obtain

$$\frac{\partial}{\partial X}(g') + \frac{\gamma+1}{2a_0^3} g' \frac{\partial}{\partial \varphi}(g') = g'' \left(\xi_X + \frac{\gamma+1}{2a_0^3} g' \xi_\varphi\right) = 0 \ ,$$

which was to be proved. Substituting (4.135) into (4.129) and taking the fact that $X = \varepsilon x$, into consideration, in the case of the waves running to the right, we obtain

$$\Phi(x,t;\varepsilon) = \varepsilon\, \Phi_1(\varphi, X) = \varepsilon\, g(\xi) + \varepsilon^2 \left(\frac{\gamma+1}{4a_0^3} x g'^2(\xi)\right) \ , \qquad (4.140)$$

where:

$$\varphi = t - x/a_0 = \xi + \varepsilon(\gamma+1) x g'(\xi)/(2a_0^3) \ . \qquad (4.141)$$

The obtained results prove that the transformation (4.141) of independent variables overlaps with (4.128) and (4.109). The expression determining the velocity potential (4.140) overlaps with (4.101), and by the differentiation of the first approximation in (4.140) with respect to x we get (4.127).

Let us emphasise that all three methods of the theory of perturbations give the same results.

Comparing the method of renormalization, the analytical method of characteristics, and the method of multiple scales proves that the method of renormalization is easiest in application. However, in order to avoid errors, it should be applied not for the basic function, but for its derivative with respect to the variable that is the source of nonuniform usefulness of the asymptotic series (in the considered example, it should be used not for the velocity potential $u = \Phi_x$, but for the longitudinal component x because for large values of x, the asymptotic series becomes nonuniformly useful due to cumulation of the effect of localization of perturbations). The analytical method of characteristics is useful only for obtaining the first order nonlinear solution. The method of multiple scales, although complicated in use, has the following advantage: One element of the asymptotic series obtained with the use of the method of multiple scales corresponds to two elements of the asymptotic series obtained with the use of the renormalization method.

Comparisons of these, and other, methods of the theory of singular perturbations are presented also in Chapter 3.

Chapter 5

Padé approximations

Recent years in the field of asymptotic methods have brought a growing interest in Padé approximations [16, 19, 143, 144, 145]. The detailed study of the applications of Padé approximations can be found in the works [19, 47, 48].

This chapter deals with Padé approximation and their applications in the analysis of various problems of applied mathematics. It should be emphasized that recently an increase of interest in applying Padé approximations to mechanical problems is observed. First, basic relations and characteristics of Padé approximations are introduced, and then a few simple examples are followed by the analysis of a supersonic flow around a thin cone in a circumsonic regime. The example provided emphasizes important features and advantages of applying Padé approximations, especially for practical use. It is worth noticing that modeling of burning of the fuel in a combustion engine, various biological models, buckling of constructions, etc. are all associated with an occurrence of the sudden blow-up phenomenon. This problem is addressed among others, and solved efficiently with the use of Padé approximations.

5.1 Determination and characteristics of Padé approximations

Let the function $f(z)$ be determined as the power series

$$f(z) = \sum_{n=0}^{\infty} c_n z^n , \tag{5.1}$$

where c_n are the known constants.

What we call *Padé approximation (AP)* of the function $f(z)$ of the order $[L/M]$ is the rational function of the form

$$[L/M]_f = \frac{P_{L/M}(z)}{Q_{L/M}(z)} = \frac{a_0 + a_1 z + \ldots + a_L z^L}{b_0 + b_1 z + \ldots + b_M z^M} , \tag{5.2}$$

where coefficients of the polynomial $P_{L/M}(z)$ and $Q_{L/M}(z)$, dependent on c_n, are chosen in such a way that the decomposition of the fraction (5.2) into Taylor series overlaps with the series (5.1) as long as it is possible.

To be more precise, the initial classical presentation of PA was formulated as follows [47]: the rational fraction $P_{L/M}(z)/Q_{L/M}(z)$, where $P_{L/M}(z)$, $Q_{L/M}(z)$ are the polynomials of respective degrees L and M, is called PA of the order $[L/M]$ of the function $f(z)$, presented in the form of a series (5.1), if the following relation holds:

$$Q_{L/M}(z)f(z) - P_{L/M}(z) = O\left(z^{L+M+1}\right) . \tag{5.3}$$

Let us notice that polynomials P and Q satisfying (5.3) always exist, and for the demanded L and M there is one PA of the order $[L/M]$ of the function $f(z)$. Let us assume now that for the searched function $f(x;\varepsilon)$, PA is obtained in the form of

$$f(x;\varepsilon) = \sum_{n=0}^{N} \varepsilon^n f_n(x) + O(\varepsilon^{N+1}) = f_0(x) + \varepsilon f_1(x) + \varepsilon^2 f_2(x) + \ldots \tag{5.4}$$

On this basis, we will construct PA for the function f, applying the definition introduced above. For the demanded L/M Padé approximation $[L/M]$ is computed from the formulas

$$[L/M]_f = \frac{\Delta\{\xi_0^{(L)}, \xi_1^{(L)}, \ldots, \xi_M^{(L)}\}}{\Delta\{1, \varepsilon, \ldots, \varepsilon^M\}} ,$$

and

$$\xi_j^{(L)} = \sum_{k=j}^{L} f_{k-j}\varepsilon^k ; \quad \Delta\{x_0, \ldots, x_M\} = \begin{vmatrix} f_{L_1-M} & f_{L+2-M} & \cdots & f_{L+1} \\ f_{L+2-M} & f_{L+3-M} & \cdots & f_{L+2} \\ \cdots & \cdots & \cdots & \cdots \\ f_L & f_{L+1} & \cdots & f_{L+M} \\ x_M & x_{M-1} & \cdots & x+0 \end{vmatrix} ,$$

where

$$f_{j=0} \ \text{for} \ j < 0 , \quad f_j = \sum_{k=j}^{L} = 0 \ \text{for} \ j > L .$$

Now, we shall provide the most popular computation formulas for PA, but only of the low orders, since the increase of order leads to the sharp increase of the complexity of formulas, and the computations have to be conducted with the use of a computer. We have especially

$$[1/1]_f = \frac{f_1 f_0 + \left(f_1^2 - f_0 f_2\right)\varepsilon}{f_1 - f_2\varepsilon} , \tag{5.5}$$

$$[1/2]_f = \frac{f_0\left(f_2 f_0 - f_1^2\right) + \left[f_1\left(f_2 f_0 - f_1^2\right) - f_0\left(f_3 f_0 - f_1 f_2\right)\right]\varepsilon}{\left(f_2 f_0 - f_1^2\right) - \left(f_3 f_0 - f_1 f_2\right)\varepsilon + \left(f_1 f_3 - f_2^2\right)\varepsilon^2} , \tag{5.6}$$

$$[2/1]_f = \frac{f_2 f_0 + \left(f_2 f_1 - f_0 f_3\right)\varepsilon + \left(f_2^2 - f_1 f_3\right)\varepsilon^2}{f_2 - f_3\varepsilon} , \tag{5.7}$$

$$[2/2]_f = \frac{A}{B}\,, \tag{5.8}$$

where:

$$A = \left(f_2^2 - f_1 f_3\right) f_0 + \left[f_1\left(f_2^2 - f_1 f_3\right) + f_0\left(f_1 f_4 - f_2 f_3\right)\right]\varepsilon +$$

$$+ \left[f_2\left(f_2^2 - f_1 f_3\right) + f_1\left(f_1 f_4 - f_2 f_3\right) + f_0\left(f_3^2 - f_2 f_4\right)\right]\varepsilon^2\,,$$

$$B = f_2^2 - f_1 f_3 + \left(f_1 f_4 - f_2 f_3\right)\varepsilon + \left(f_3^2 - f_2 f_4\right)\varepsilon^2\,.$$

We have enumerated the most important features of Padé approximations. If PA is presented in the form of *Padé table*

$$\begin{array}{cccc} [0/0] & [0/1] & [0/2] & \ldots \\ [1/0] & [1/1] & [1/2] & \ldots \\ [2/0] & [2/1] & [2/2] & \ldots \\ \bullet & \bullet & \bullet & \ldots \end{array}\,,$$

the left-hand side of this table will overlap with partial sums of the series (5.1), and the approximations in the form of {[1/0], [1/1], [2/1], [2/2], ...} will present the sequence of fractions appearing on the decomposition of the series (5.1) into the adequate complex fraction [19, 47].

From the point of view of possible applications in the problems of applied mathematics, PA have the following advantages:

1. Contrary to the partial sums of the series (5.1) in general, PAs have poles, which allows to apply them in the approximations of analytical functions.

2. From all possible fractionally rational approximations, PA, in a sense, "corresponds best" with problems of the linear theory of perturbations.

3. The series (5.1) determining the function $f(z)$ may be *divergent* or very slowly *convergent*, but regardless of that, the application of PA in many cases leads to very good computational results.

4. The PA sequence constructed for the series (5.1) may have the radius of convergence larger than that of the series (5.1).

There are a lot of books devoted to PA characteristics (see [19]).

The systematic introduction to the theory of Padé approximations is given in the monograph [47]. This work points, among others, at the relation between PA and sequence transformations.

5.2 Application of Padé approximations

PA may be treated as the instrument for the construction of effective approximations in various problems of applied mathematics, physics, and chemistry.

It occurs that PA approximate functions better than partial sums of the series [16, 19, 143, 144, 145] used for their determination. Works [144, 145] present how accurate approximations of many nonlinear problems of acoustics were constructed with the use of PA. Works [16, 143] show that during the analysis of the problem of the subsonic flow round an elliptic profile and during the supersonic flow round a thin wing, the results obtained with the use of deformed coordinates and PA practically overlap with each other, and with precise computations.

The examples given in [19, 47] illustrate the characteristic very important for practical applications of PA. The PA sequence constructed for the first series may have the radius of convergence larger than that of the initial series.

Many other examples of advantages coming from applying PA in various problems of applied mathematics are given in the works [16, 19, 47, 143, 144, 145] and others. Below, we consider some of the examples in more detail.

5.2.1 Simple examples

PAs belong to the most effective instruments leading to *the acceleration of the convergence* of the series. For example, the first ten elements of the slowly convergent Gregory series

$$\pi = 4 - \frac{4}{3} + \frac{4}{5} - \frac{4}{7} + \dots \tag{5.9}$$

yield $\pi = 3.0418396$, whereas $[5/4]_\pi$ after applying the formula (5.1), (5.2) gives the quantity $\pi \approx 3.1415925$. In other words, after using the first ten elements of the series (5.9) we obtain seven accurate digits of the sign π.

Moreover, PAs can be applied for *summing of divergent series.* For example, using the first ten elements of the series for $\ln(1+\varepsilon)$ divergent for $\varepsilon = 2$, we obtain $\ln(3) = [5/4]_{\ln(3)} = 1.098626\dots$ (with the accuracy of up to $O(10^{-6})$ of the value $\ln(3) = 1.098612$).

Let us consider one more example. Find $f(\infty)$ of the function

$$f(x) = \sqrt{(1+2x)\,/\,(1+x)}\ .$$

Let us notice that

$$\lim_{x\to\infty} f(x) = \lim_{x\to\infty}\left(\frac{x(2+1/x)}{x(1+1/x)}\right)^{1/2} = \sqrt{2} \approx 1.4142\dots$$

and let us distribute $f(x)$ into the following Taylor series

$$f(x) = 1 + \frac{x}{2} - \frac{5}{8}x^2 + \frac{13}{16}x^3 - \frac{141}{128}x^4 + \ldots \equiv \sum_{h=0}^{\infty} a_n x^n . \tag{5.10}$$

Partial sums of the series (5.10) for $x = \infty$ take alternately the value $\pm\infty$ and do not allow to find $f(\infty)$. However, $[1/1]_f = \frac{1+7x/4}{1+5x/4}$ yields the value $f(\infty) = 7/5 = 1.4$.

Further computations show that $[5/5]_f$ allows to obtain $f(\infty)$ with the accuracy of $O(10^{-8})$.

5.2.2 Supersonic flow round a thin cone in a circumsonic regime

Let us consider *the circumsonic regime of the symmetric flow round a thin cone* with the apex angle 2ε by the supersonic stream of perfect gas with Mach number $M = U/a$ for the conditions $M > 1$, in $M - 1 \ll 1$. Let us assume that the cone is thin enough for the stream occurring around it to be also a cone for the demanded M. Basic features of the analyzed process are determined by the solution of the boundary problem. This solution includes the equation determining small circumsonic perturbations for the velocity potential, and boundary conditions on the surface of a cone and on the sudden conical change of rigidity of the wing attached to the vertex of the cone.

Mathematical difficulty of the mentioned problem is formulated in the work [19]. The solution of this problem with the use of the classical method of small parameter [19] cannot be applied in the vicinity of the sudden change of rigidity of the wing and on the surface of a cone.

The authors of the works [103, 104] managed to construct the solution equally useful with the use of the method of deformed coordinates. The element of this solution that is most important for applications is the expression determining the damping coefficient $C_p = 2(p - p_\infty)/(\rho_\infty U_\infty^2)$ on the surface of a cone, which, according to the circumsonic law of similarity takes the form of

$$Y = \frac{C_p}{\varepsilon^2} + 2\ln\left(\varepsilon\sqrt{M^2-1}\right) = 2\ln 2 - 1 + \varepsilon_1 + \left(\frac{\pi^2}{12} - \frac{1}{4}\right)\varepsilon_1^2 + O(\varepsilon_1^3) , \tag{5.11}$$

where $\varepsilon_1 = (\gamma+1)\varepsilon^2/(M^2-1)$ is *the circumsonic parameter of similarity.*

The formula (5.11) allows to obtain the results close to the results of numerical computations [166] for $\varepsilon_1 \leq 0.4$. However, for large ε_1 the error in the determination of Y becomes more and more significant.

Applying PA [1/1] in order to obtain the asymptotic decomposition (5.11) yields

$$[1/1]_Y = \frac{2\ln 2 - 1 + \left[1 - (2\ln 2 - 1)\left(\frac{\pi^2}{12} - \frac{1}{4}\right)\right]}{1 - \left(\frac{\pi^2}{12} - \frac{1}{4}\right)\varepsilon_1} . \tag{5.12}$$

The obtained result corresponds to the numerical solution in the wider range of changes of the similarity parameter (Figure 5.1; curves 1, 2, 3 denote respectively the first, second, and third approximation obtained from (5.11); points characterize the quantities of numerical computations [166]).

5.2.3 Damping of the ball-shaped waves of pressure in a free space and in a tube

Nonlinear effects occurring during the propagation of *a sinusoidal wave of pressure* in a perfect medium lead to the occurrence of the explosion of pressure within the range of any length of the wave.

In many works (see [145]), the following relations have been obtained

$$\frac{d(1/\delta)}{d(x/\lambda)} = \frac{6\gamma}{R\delta^3}\Delta S , \tag{5.13}$$

that determine the relation of the amplitude of a ball-shaped wave to the distance it has covered ($\delta = (p_2 - p_1)/p_1$; p_2, p_1 denote the pressure vertices of a ball-shaped wave; x is the distance covered by the wave; λ is the length of the wave; γ is the adiabate exponent; R is the universal gas constant; ΔS denotes the entropy gain).

The exact formula determining *the entropy gain* has the following form:

$$\Delta S = \frac{R}{\gamma - 1}\ln\left[(1+\delta)\left(\frac{\delta\,(\gamma+1) + 2\gamma}{\delta\,(\gamma-1) + 2\gamma}\right)^{-\gamma}\right] . \tag{5.14}$$

However, in the case of the widely used approximation for weak shock waves [145] it is substituted by the first component of the decomposition ΔS into the series for $\delta < 1$ in the form of

$$\Delta S = \frac{\gamma+1}{12\gamma^2}R\delta^3\left(1 - \frac{3}{2}\delta + \frac{3\,(11\gamma^2+1)}{20\gamma^2}\delta^2 + \ldots\right) + O\left(\delta^6\right) . \tag{5.15}$$

In this case, equation (5.13) assumes the form of

$$\frac{d(1/\delta)}{d(x/\lambda)} = \frac{\gamma+1}{2\gamma} .$$

Its solution (for the initial conditions $\delta = \delta_0$, $x = 0$) has the form of

$$\frac{x}{\lambda} = \frac{2\gamma}{\gamma+1}\left(\frac{1}{\delta} - \frac{1}{\delta_0}\right) , \tag{5.16}$$

which qualitatively agrees with data obtained from the experiment.

After the substitution of the left-hand side (5.15) with PA [1/1], equation (5.13) is transformed into the form of

$$\frac{d\,(1/\delta)}{d\,(x/\lambda)} = \frac{\gamma+1}{2\gamma}\frac{10\gamma^2 - \left(4\gamma^2 - 1\right)\delta}{10\gamma^2 + \left(11\gamma^2 + 1\right)\delta} . \tag{5.17}$$

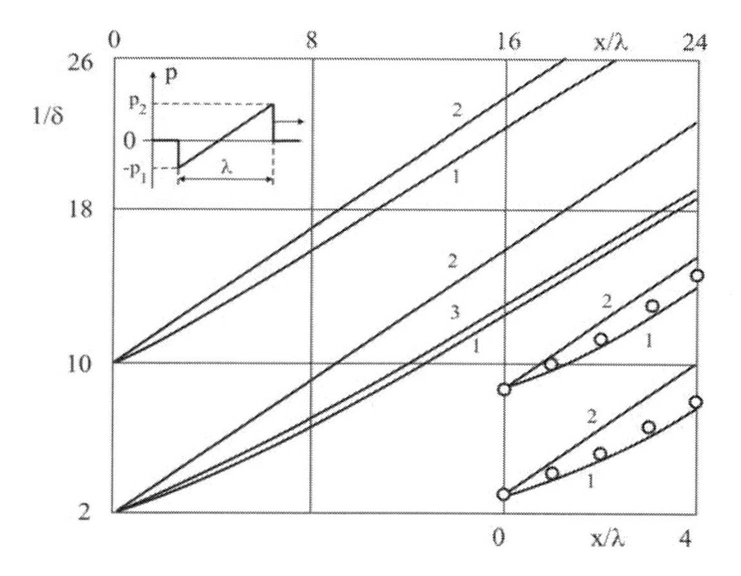

FIGURE 5.1: Comparison of the results of numerical computations, the application of PA and the experimental research (see the text).

Integrating (5.17), with the initial conditions $\delta = \delta_0$, $x = 0$, one gets

$$\frac{x}{\lambda} = \frac{2\gamma}{\gamma+1}\left(F\left(\delta\right) - F\left(\delta_0\right)\right) ,$$

$$F\left(\delta\right) = \frac{1}{\delta} + \frac{3}{2}\ln\left|\frac{1}{\delta} - \frac{4\gamma^2-1}{10\gamma^2}\right| . \tag{5.18}$$

In Figure 5.1, curves 1 determine the results obtained from numerical integration (5.13) with the right-hand side exact; curves 2 and 3 correspond to the results obtained from the formulas (5.16) and (5.18), respectively. In the left bottom corner of the figure in the enlarged scale, the comparison of results of the computation with the experiment made (circles).

Taking the influence of the tube walls into consideration and applying PA [1/1] for ΔS, the equation (5.13) is transformed to the form of [144]:

$$\frac{d\,(1/\delta)}{d\,(x/\lambda)} = \frac{\gamma+1}{2\gamma}\,\frac{10\gamma^2-(4\gamma^2-1)\,\delta}{10\gamma^2+(11\gamma^2+1)\,\delta} + \frac{3}{2}\alpha\lambda\left(\frac{1}{\delta}+\frac{1}{2}\right). \qquad (5.19)$$

The second element occurring in (5.19) determines the acoustic approximation of the velocity of dissipation of wave energy on the walls of the tube determined by *the coefficient* α of the form

$$\alpha = \frac{1}{r}\left(1+\frac{\gamma-1}{(Pr)^{1/2}}\right)\left(\frac{\omega\nu}{2a^2}\right)^{1/2},$$

where r is the radius of the tube, ω is frequency, ν is the kinematic viscosity of gas, Pr is Prandtl number characterizing gas, and a denotes the velocity of propagation of sound in gas.

Equation (5.19) allows to obtain the solution corrresponding to the experimental data in wider range of parameter changes than the solution obtained with the use of the approximation based on the conception of weak shock waves [144].

In Figure 5.2, curve 1 determines the results obtained with the use of the approximation based on *the conception of constant shock waves*, curve 2 denotes the result obtained from numerical integration (5.19) for $\gamma = 1.4$, $\alpha\lambda = 0.046$, $r = 0.955\,cm$, $\omega/(2\pi) = 540\,Hz$, $\nu = 0.15\,cm^2/s$, $\lambda = 63\,cm$, $Pr = 0.72$ for the initial condition $\delta = 0.04$, $x = 0$, which corresponds to the data obtained in the experiment; circles denote the results obtained from the experiment (see [144]).

It is shown in the work [143] on the basis of many practically important examples that the sequence of Padé approximations constructed for a certain series can have the radius of convergence larger than the initial series.

TABLE 5.1: Exact solutions and relative errors.

δ	Exact solution $\Delta S/C_V$	Relative error (%) of various approximations $\Delta S/C_V$		
		$(\gamma^2-1)\delta^3/(12\gamma^2)$	[1/1]	[2/2]
0.01	$0.40 \cdot 10^{-7}$	1.505	0.000	0.000
0.1	$0.35 \cdot 10^{-4}$	15.53	0.015	0.000
0.5	$0.27 \cdot 10^{-2}$	88.25	1.302	0.003
1.0	$0.13 \cdot 10^{-1}$	203.80	8.060	0.047

The advantages of applying *Padé approximations* are illustrated in Table 5.1. In the second column of the table, the exact value of $\Delta S/C_V$ is given, and

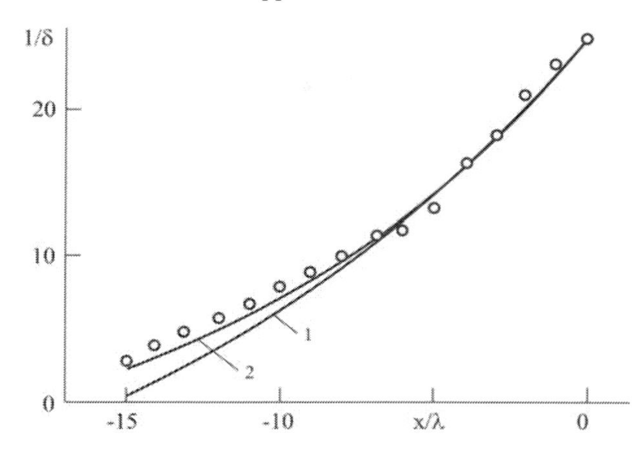

FIGURE 5.2: Results of computations obtained from the approximation of constant shock waves (curve 1), numerical computations (curve 2) and the experiment (circles) – see text.

in other columns the relative error is given that is admissible in the computations of $\Delta S/C_V$, with the use of approximation associated with the conception of constant shock waves, Padé approximations [1/1] and [2/2] according to [144] (C_V is the heat volume characteristic for the constant volume).

The example provided emphasizes the feature of Padé approximations, especially important for practical applications.

Finally, let us notice that in many works the generalizations of Padé approximations are considered that can be applied in case of the analysis of the problems with several perturbations [5].

5.2.4 Analysis of the "blow-up" phenomenon

The analysis of the sudden *blow-up* phenomenon is very important both in theory and in practice. Let us enumerate only modeling of the phenomenon of burning of the blend in a combustion engine, various biological models (e.g., the description of population growth), buckling of construction, etc. [87, 111, 139, 114, 149, 176]. In order to understand and explain the mentioned phenomena, various asymptotical methods are applied, especially *the method of asymptotical matching* [87, 111], the so-called *geometrical asymptotics* [114], *Witham method* [107, 176] or *the method of boundary effect* [173, 174].

According to the work [139], the phenomenon of a sudden jump can be determined with the use of rational functions. This remark brings the idea of the direct application of Padé approximation (AP) [6, 29, 47]. PA was successfully used in the theory of solitons. It also contributed to the rise of a new notion of "a padeon" [97, 98, 99, 108, 162]. Moreover, PA can be effectively applied in the cases where the so-called phenomenon of the localization of solution is observed.

In order to construct PA, one only has to solve the systems of algebraic equations. Comparing the expressions standing next to the coefficients ε^0, ε^1, ε^2, ..., ε^{m+n}, the mentioned system of algebraic equations is generated by nonlinearities.

Let us notice that if PA is convergent to a certain function, the roots of expressions in denominators tend to singular points. This allows us to identify singularity, and then to conduct further computations. For example, if the "clipped" series has the form of

$$\left(c_0 + c_1\varepsilon + \ldots + c_{m+n}\varepsilon^{m+n}\right)\left(1 + \sum_{i=1}^{n} b_i\varepsilon^i\right) = \sum_{i=0}^{m} a_i\varepsilon^i \ , \tag{5.20}$$

then the diagonal form of PA is determined by the formula

$$AP[1,1] = \frac{a + (a^2 - b)\varepsilon}{a - b\varepsilon} \ . \tag{5.21}$$

Leaving only two elements of the series, we get

$$AP[0,1] = \frac{1}{1 - a\varepsilon} \ . \tag{5.22}$$

Let us now consider the following boundary problem

$$y'' - y + 2y^3 = 0 \ , \tag{5.23}$$

$$y(0) = 1 \ , \tag{5.24}$$

$$y(\infty) = 0 \ , \tag{5.25}$$

which has the following exact solution

$$y = \cosh(x) \ . \tag{5.26}$$

This solution can be presented in the form of the following Dirichlet series

$$y = Ce^{-x}\varphi(x) \, , \quad C = const \, , \tag{5.27}$$

where:

$$\varphi(x) = 1 - \frac{1}{4}C^2e^{-2x} + \frac{1}{16}C^4e^{-4x} + \ldots \tag{5.28}$$

Substituting (5.21) into the series (5.28) and assuming that

$$a = \frac{1}{4}C^2e^{-2x} \, ; \quad b = \frac{1}{16}C^4e^{-4x} \, ; \quad \varepsilon = e^{-2x} \, ; \tag{5.29}$$

from the equation (5.27), one gets

$$y = \frac{4C}{4e^x + C^2e^{-x}} \ . \tag{5.30}$$

Let us notice that the equation (5.30) fulfills the conditions (5.25). From the condition (5.24) we find $C = 2$ and the solution (5.30) overlaps with the exact solution (5.25).

In order to illustrate the application of PA in the problem of a "jump," let us consider the following problem

$$\frac{dx}{dt} = \alpha x + \varepsilon x^2 \,, \quad x(0) = 1 \,, \tag{5.31}$$

where $0 < \varepsilon \ll \alpha \ll 1$ (see [11]).

The exact solution of the boundary condition has the form of

$$x(t) = \frac{\alpha \exp(\alpha t)}{\alpha + \varepsilon - \varepsilon \exp(\alpha t)} \,. \tag{5.32}$$

In the case $t \to \ln[(\alpha + \varepsilon)/\varepsilon]$, the solution tends to infinity, i.e., the "blow-up" phenomenon occurs. Applying regular asymptotics in the form of

$$x(t) = \exp(\alpha t)\Psi(t), \tag{5.33}$$

where:

$$\Psi(t) = 1 - \varepsilon\alpha^{-1}\left[1 - \exp(\alpha t)\right] + \dots \,, \tag{5.34}$$

does not allow for the satisfactory description of the phenomenon.

Padé approximation (5.22) applied in the series (5.34) with the substitution of $a = \alpha^{-1}[1 - \exp(\alpha t)]$ allows to obtain the exact solution (5.32).

As it was mentioned, the main focus of our consideration is the process of burning of a blend in a combustion engine, determined by a differential equation

$$\dot{y} = y^2(1 - y) \,, \tag{5.35}$$

$$y(0) = \varepsilon \,. \tag{5.36}$$

This problem will be further analyzed, with the assumption that $\varepsilon \ll 1$.

The results of a numerical simulation of the equations (5.35) and (5.36) with the use of Runge-Kutta method of the fourth order are illustrated in Figure 5.3 for different values of ε. It is easy to notice that at the moment close to t^* a sudden jump of the solution occurs, which is characterized by two features. Firstly, on decreasing the value of ε the shape of of the solution $y(t)$ assumes the form of a scalar function. Secondly, the function $y(t)$ reaches zero almost horizontally for $t < t^*$. Similar behavior can be observed for $t > t^*$, i.e., the function is horizontal and it is close to 1 for sufficiently small values of ε (see Fig. 5.3c).

The aim of our analysis is to determine *the value of a jump* at the moment $t = t^*$, and to determine the analytical solutions for $t < t^*$ (area I) and for $t > t^*$ (area II) through their coupling in the point $t = t^*$.

Firstly, we will apply the following regular series

$$y = \sum_{i=0}^{\infty} \varepsilon^i y_i \,. \tag{5.37}$$

By substituting (5.37) into the equations (5.35) and (5.36), the problem is reduced to the analysis of the following one

$$\dot{y}_0 = y_0^2(1 - y_0)\,; \tag{5.38}$$

$$y_0 = 1\,; \tag{5.39}$$

$$\dot{y}_1 = 0\,; \quad y_1 = C \equiv const\,; \tag{5.40}$$

$$\dot{y}_2 = y_1^2\,; \quad y_2 = C^2 t\,; \tag{5.41}$$

$$\dot{y}_3 = y_1^3 + 2y_0y_2\,; \quad y_3 = C^3(t-1)t\,; \tag{5.42}$$

$$\dot{y}_4 = -3y_1^2y_2 + y_2^2 + 2y_1y_3\,; \quad y_4 = 5C^4(t-1)t\,; \tag{5.43}$$

$$\dot{y}_5 = -3y_1y_2^2 - 3y_1^2y_3 + 2y_2y_3 + 2y_1y_4\,;$$

$$y_5 = C^5t(14t^2 - 21t + 3)\,; \tag{5.44}$$

$$\dot{y}_6 = -y_2^3 - 6y_1y_2y_3 + y_3^2 - 3y_1^2y_4 + 2y_2y_4 + 2y_1y_5\,;$$

$$y_6 = 14C^6t^3(3t^2 - 6t + 2)\,. \tag{5.45}$$

The solution was obtained with the use of the program "Mathematica." Equation (5.37) has the solution $y_0 = 1$, which corresponds to the value of the searched function for $t \to \infty$. This part of the solution will be applied in further considerations. We will determine the constant C from the initial condition (5.36)

$$y_1 + \varepsilon y_2 + \varepsilon^2 y_3 + \ldots = 1 \quad \text{for} \quad t = 0\,, \tag{5.46}$$

which yields $C = 1$.

Let us now apply Padé approximation in order to detect the mentioned phenomenon of a sudden "blow-up" of the solution $y(t)$ at the moment $t = t^*$. Firstly, let us conduct computations "manually," and then let us compute higher approximations with the use of the program "Mathematica." Function y is approximated with the use of the following rational function:

$$y_1 + \varepsilon y_2 + \varepsilon^2 y_3 + \varepsilon^3 y_4 + \ldots = \frac{y_1 + a_1\varepsilon + a_2\varepsilon^2 + a_3\varepsilon^3 + \ldots}{1 + b_1\varepsilon + b_2\varepsilon^2 + b_3\varepsilon^3 + \ldots}\,, \tag{5.47}$$

where a_i and b_i are the unknown coefficients.

According to the construction of PA, we present the solution (5.47) as

$$(y_0 + \varepsilon y_1 + \varepsilon^2 y_2 + \ldots)(1 + b_1\varepsilon + b_2\varepsilon^2) = y_0 + a_1\varepsilon + a_2\varepsilon^2 + \ldots \tag{5.48}$$

Comparing in (5.48) the terms standing next to the same powers ε, one gets

$$\varepsilon^0: \quad y_1 = y_1\,;$$

$$\varepsilon^1: \quad y_1b_1 + y_2 = a_1\,;$$

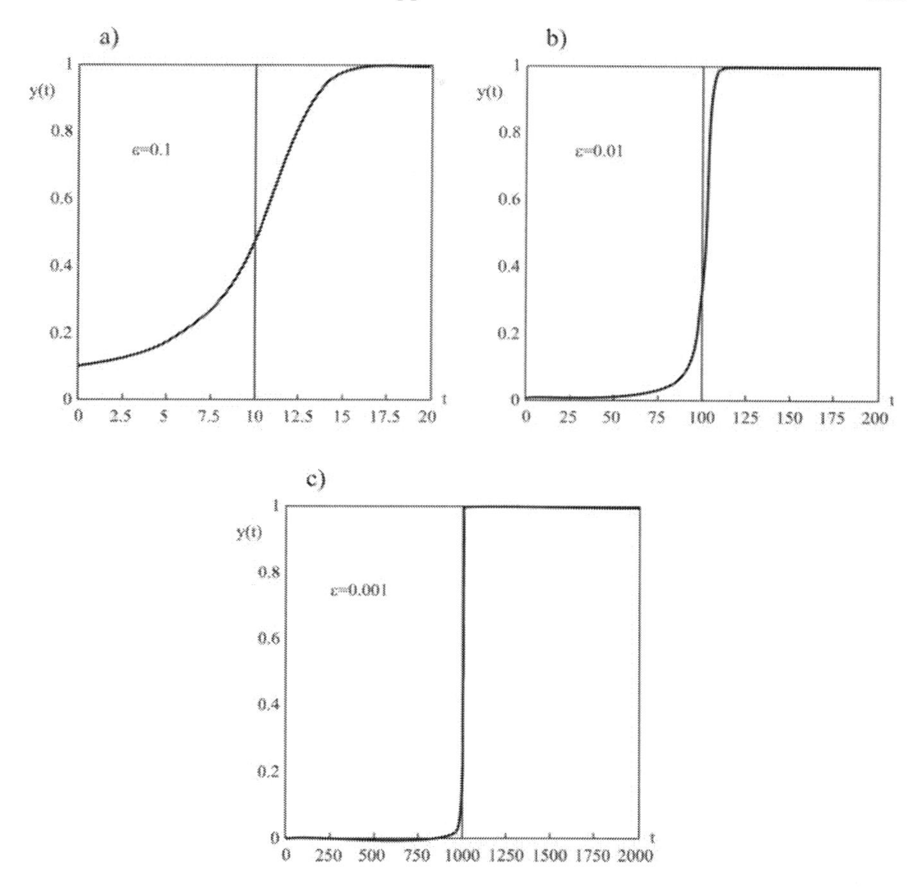

FIGURE 5.3: The numerical solution of equation (5.20) for various initial conditions: a) $\varepsilon = 0.1$; b) $\varepsilon = 0.01$; c) $\varepsilon = 0.001$.

$$\begin{aligned} &\varepsilon^2 : \quad y_1 b_2 + y_2 b_1 + y_3 = a_2 \, ; \\ &\varepsilon^3 : \quad y_1 b_3 + y_2 b_2 + y_3 b_1 + y_4 = a_3 \, . \end{aligned} \tag{5.49}$$

Let us notice that on the right-hand side of the equation (5.47) we have got the number of unknown $a_1, \ldots, a_n$, $b_1, \ldots, b_m$ equal to $m + n$. Hence, only $m + n$ equations are applied from the infinite number of equations generated by the series (5.37).

From equation (5.49), one gets

$$a_1 = C \, ; \quad a_2 = C^2 \, ; \quad b_1 = -C(t-1) \tag{5.50}$$

and $PA[2, 1]$ has the form of

$$AP[2, 1] = C\varepsilon \frac{1 + C\varepsilon}{1 - C(t-1)\varepsilon} \, . \tag{5.51}$$

The symbol $PA[m,n]$ denotes PA that has $(m+1)$ elements in a numerator and $(n+1)$ elements in a denominator. This result has been verified with the use of the program "Mathematica," where the constant $C = 1$ has also been determined.

PAs of higher orders can be determined in a similar way. They have the following form

$$AP[2,1] = \frac{\varepsilon(-2+\varepsilon t)}{2+t\varepsilon\left[-3+(2+t)\varepsilon\right]} \ ;$$

$$AP[3,1] = \frac{\varepsilon\left\{2+\left[-2-\varepsilon(3+(2+t)\varepsilon)\right]\right\}}{2+t\left[-2+(-5+2t)\varepsilon\right]} \ . \tag{5.52}$$

The threshold value t^* is determined from the condition of neutralization of denominators. The following values of t^* have been determined:

$$t^*[2,1] = -1 + \frac{1}{\varepsilon} \ , \tag{5.53}$$

$$t^*[2,2] = \frac{3-2\varepsilon+\sqrt{1+4\varepsilon(\varepsilon-3)}}{2\varepsilon} \ , \tag{5.54}$$

$$t^*[3,1] = \frac{2+5\varepsilon+\sqrt{4+\varepsilon(4+25\varepsilon)}}{4\varepsilon} \ . \tag{5.55}$$

Numerical values t^* for three different values of ε are given in Table 5.2.

TABLE 5.2: Numerical values t^* obtained with the use of PA.

Number of equation	ε 0.1	0.01	0.001
(5.39)	11.0	101.0	1001.0
(5.40)	x	102.08518	1002.008
(5.41)	11.640965	101.51492	1001.5015

Let us notice that in the case $\varepsilon = 0.1$ the imaginary value $t^*[2,2]$ occurs, which corresponds to equation (5.54). Moreover, for $AP[2,2]$ and $AP[3,1]$, equations (5.54) and (5.55) represent only one solution each (whereas, in fact, there are two real solutions for each equation).

Summing up the obtained results, we have to emphasise that applying Padé approximations allows to explain the phenomenon of the "jump" of the solution $y(t)$ (see Fig. 5.3).

Now, we will construct PA in the area II ($t > t^*$). Since in this area, the solution is close to the value of 1, we will introduce the following exchange of coordinates

$$y = 1 + \tilde{x} \ , \tag{5.56}$$

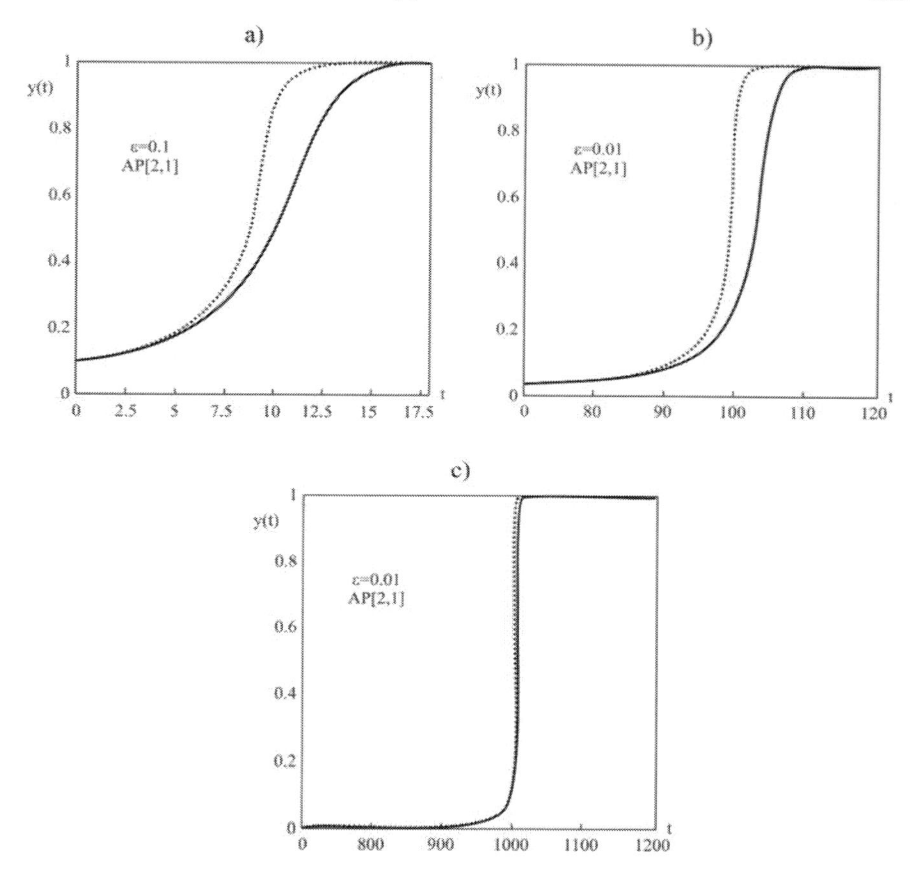

FIGURE 5.4: The numerical, asymptotic, and PA solution for various initial conditions: a) $\varepsilon = 0.1$; b) $\varepsilon = 0.01$; c) $\varepsilon = 0.001$ ($AP[2,1]$).

and from equation (5.35) one gets

$$\dot{\tilde{x}} = -\tilde{x}(1+\tilde{x})^2 \ , \tag{5.57}$$

where $\tilde{x} \ll 1$. Next, we proceed analogically as in the case of the area I. Let us introduce a so-called *formal small parameter* δ. Firstly, we apply it for the asymptotical splitting, and then we assume that $\delta = 1$.

In the next step, we will apply the classical method of perturbations in determining the solution

$$\tilde{x} = \delta x_0 + \delta^2 x_1 + \delta^3 x_2 + \dots \tag{5.58}$$

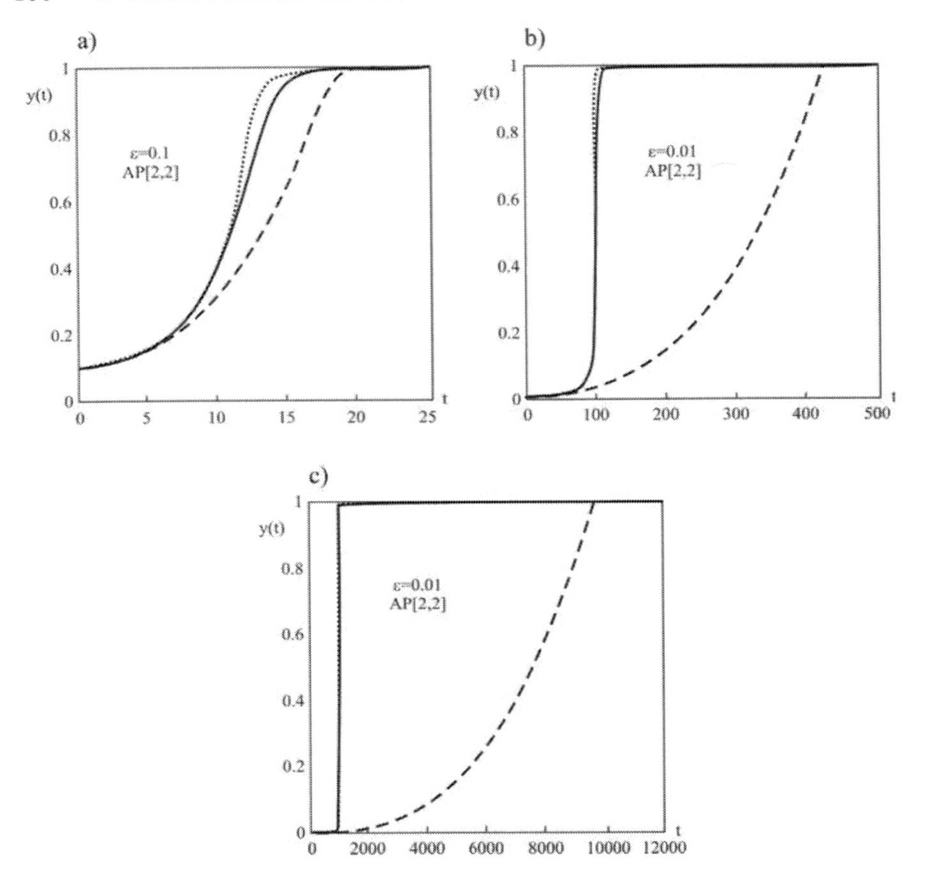

FIGURE 5.5: The numerical, asymptotic, and PA solution PA for various initial conditions: a) $\varepsilon = 0.1$; b) $\varepsilon = 0.01$; c) $\varepsilon = 0.001$ ($AP[3,1]$).

Substituting (5.58) into (5.57), we get the following sequence of differential equations

$$\frac{dx_0}{dt} = -x_0 \quad x_0 = e^{-(t-t_0)};$$

$$\frac{dx_1}{dt} = -x_1 - 2x_0\,, \quad x_1 = 2e^{-2(t-t_0)};$$

$$\frac{dx_2}{dt} = -x_2 - x_0^3 - 4x_0x_1\,, \quad x_2 = \frac{9}{2}e^{-3(t-t_0)};$$

$$\frac{dx_3}{dt} = -x_3 - 3x_0^2x_1 - 2x_1^2 - 4x_0x_2 \quad x_3 = \frac{32}{8}e^{-4(t-t_0)};$$

$$\frac{dx_4}{dt} = -x_4 - 3x_0x_1^2 - 3x_0^2x_2 - 4x_1x_2 - 4x_0x_3\,, \quad x_4 = \frac{625}{24}e^{-5(t-t_0)}. \quad (5.59)$$

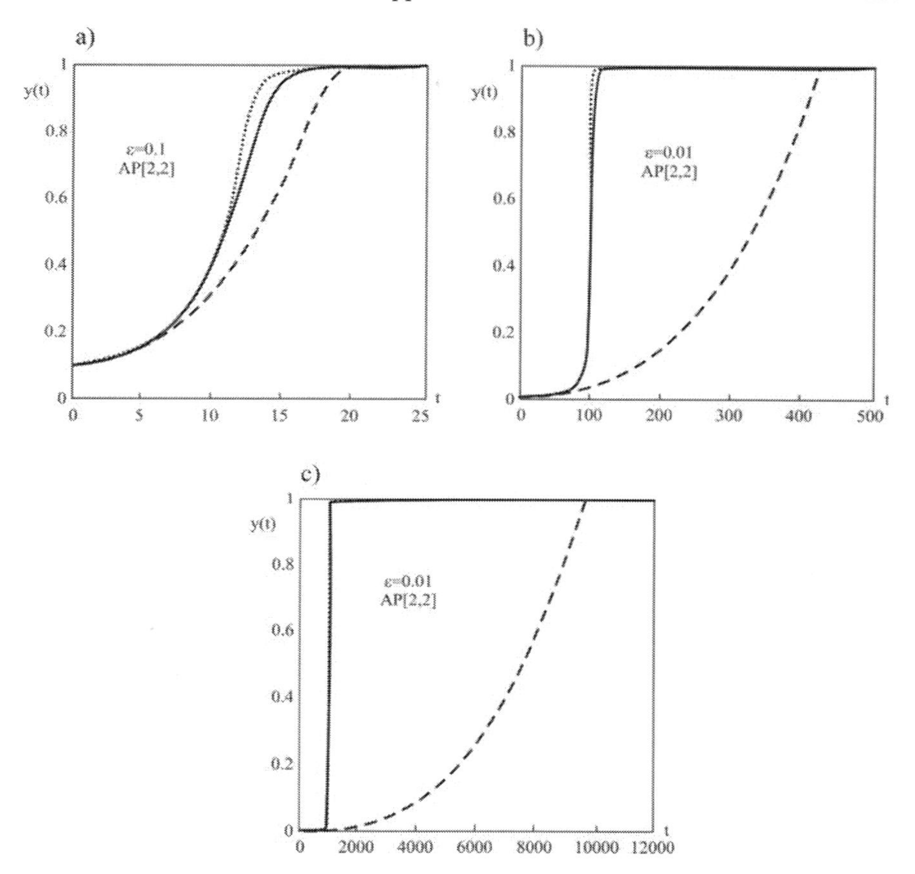

FIGURE 5.6: The numerical, asymptotic, and PA solution for various initial conditions: a) $\varepsilon = 0.1$; b) $\varepsilon = 0.01$; c) $\varepsilon = 0.001$ ($AP[2,2]$).

The value of a function at the moment $t = t_0$ determines the value of the constant of integration which will be determined in the process of matching of solutions. The values t^* and t_0, determined with the use of Padé approximations are given in the Table 5.3.

Two solutions $\bar{y}$ and $\tilde{y}$ valid for the areas I and II are then matched. The conditions of matching have the following form:

$$\bar{y} = \tilde{y}\,; \quad \frac{d\bar{y}}{dt} = \frac{d\tilde{y}}{dt} \quad \text{dla} \quad t = t_0\,. \tag{5.60}$$

The results of computations are given in the figures (Fig. 5.4-5.6), where (I) constant curve denotes the exact (numerical) solution; (II) discontinuous curve denotes the solution approximated by the asymptotic series; (III) solid curve corresponds to Padé approximation for $\varepsilon = 0.1$; 0.01; 0.001.

TABLE 5.3: The values t^* and t_0 obtained with the use of PA.

Padé approximations	Time constants	ε 0.1	0.01	0.001
PA[2, 1]	t^*	24.945	945.722	31110.408
	t_0	24.213	944.100	31107.647
PA[3, 1]	t^*	16.514	424.669	9641.779
	t_0	16.299	424.188	9640.571
PA[2, 2]	t^*	16.514	424.669	9641.779
	t_0	16.299	424.188	9640.571

Figure 5.4 presents $PA[2,1]$, Figure 5.5 presents $PA[3,1]$, whereas Figure 5.6 represents $PA[2,2]$.

We leave drawing the conclusion from the presented figures to the reader.

5.2.5 Homoclinic orbits

Iteration procedures are often used in determining *the parameters of homoclinic orbits* both with the use of numerical [46, 59, 60, 82, 83, 101, 150] or analytical methods [165]. It seems that applying Padé approximations yields much more economical results, in a sense of correcting *the convergence of iteration procedures* [29, 113, 12].

In order to illustrate the application of Padé approximations, the results of computations associated with the determination of analytical approximation of homoclinic orbits occurring in Lorenz system have been used

$$\dot{x} \equiv \frac{dx}{dt} = \delta(y - x) ,$$

$$\dot{y} \equiv \frac{dy}{dt} = \rho x - y - zx ,$$

$$\dot{z} \equiv \frac{dz}{dt} = xy - bz , \tag{5.61}$$

for the following parameters $\delta = 10$, $b = 8/3$ i $\rho = 13.926\ldots$ In the works [29, 113] the local analysis of the local structure of homoclinical solutions for $t \to \pm 0$ and $t \to \pm\infty$, is conducted, and then, the global solution is constructed with the use of *quasi-fractional approximations*. The accuracy of approximations is iteratially corrected. Since Lorenz system is autonomous, one can freely choose the initial conditions representing the orbits of the form $x(0) = y(0) = y_0$ and $z(0) = z_0$. Taking parameters $\gamma = 10$, $b = 8/3$ as constant, each iteration determines certain approximation to the homoclinical orbit and to the value ρ. Further iterations are presented in Table 5.4. The accuracy of computations is controlled on the basis of the parameter ρ.

In the works [29, 48] the possibility of applying Padé approximations in accelerating the iteration procedures is pointed. Let us consider the following

TABLE 5.4: Further iterations.

No. of iteration	y_0	z_0	ρ
0	15.08854	28.92802	17.63087
1	11.55446	19.02569	13.45975
2	11.65734	19.27001	13.91258

iteration procedure

$$T(u_0) = 0\,,$$
$$u_n = T(u_{n-1})\,,$$
$$u = 1, 2, 3, \ldots\,, \tag{5.62}$$

where T is a certain operator. We introduce the formal small parameter ε $(0 \leq \varepsilon \leq 1)$ and let the solution $u \approx u_n$ be searched in the form of the powers of the parameter ε in the following form

$$u \approx u_0 + (u_1 - u_0)\varepsilon + (u_2 - u_1)\varepsilon^2 + \ldots + (u_n - u_{n-1})\varepsilon^n\,. \tag{5.63}$$

For $\varepsilon = 0$ we have $u = u_0$, whereas for $\varepsilon = 1$ we get $u = u_n$. The series (5.63) can be approximated with the use of a rational function, according to Padé scheme of the form

$$\frac{u_0 + \sum\limits_{i=1}^{m} a_i \varepsilon^i}{1 + \sum\limits_{j=1}^{\ell} \beta_j \varepsilon^j} - [u_0 + (u_1 - u_0)\varepsilon + \ldots + (u_n - u_{n-1})\varepsilon^n] = 0(\varepsilon^{n+1})\,, \tag{5.64}$$

where $m + l = n$.

For $\varepsilon = 1$, we have

$$u \approx \frac{u_0 + \sum\limits_{i=1}^{m} a_i}{1 + \sum\limits_{j=1}^{l} b_j}\,. \tag{5.65}$$

For $m = 1$, we obtain *the diagonal Padé approximation.* Various examples given in the works [29, 48] show high effectiveness of the discussed method.

The results of computations show that applying the approximations of zero order leads to significant errors associated with the use of the iteration technique. The approximation based on applying the following equation is more effective:

$$u \approx u_1 + (u_2 - u_1)\varepsilon\,. \tag{5.66}$$

The approximation for $\varepsilon = 1$ yields

$$u \approx \frac{u_1^2}{2u_1 - u_2}\,, \tag{5.67}$$

TABLE 5.5: The results obtained with the use of Padé approximations.

y_0	z_0	ρ
11.65734	19.27001	13.9283

and the results are given in Table 5.5.

Applying the approach proposed in the work [165], one can estimate the accuracy of the determination of parameter ρ. Hassard and Zhang [82] apply the method of guessing to estimate the values of ρ, for which there is a homoclinical orbit. They show that in this case, for $\gamma = 10$ and $b = 8/3$, $\rho \in [13.9265, 13.9270]$.

According to Table 5.4, ρ has the error of 0.013%, whereas the third iteration has the error of 0.096%. Applying Padé approximations in determining homoclinical orbits in Lorenz system has improved the accuracy by one order of magnitude.

5.2.6 Vibrations of nonlinear system with nonlinearity close to sign (x)

Let us consider the equation (see also [9])

$$\ddot{x} + x^{1/(2n+1)} = 0 \tag{5.68}$$

with the initial conditions

$$x(0) = 0 \quad \dot{x}(0) = A\,, \tag{5.69}$$

for $n \to \infty$ [35].

Let us notice that similar mathematical model can be used in the theory of vibro-impact systems [44]. The solution of equation (5.68) can be found on the basis of *special functions Ateb* proposed by Rosenberg [147], which are the opposites of incomplete *Beta functions.* In further considerations, though, we will apply the asymptotic approach, basing on the application of *the method of small* δ [29, 51, 13].

For $n \to \infty$, the following equations is obtained

$$\ddot{x}_0 + \mathrm{sign}\,(x_0) = 0\ , \tag{5.70}$$

where

$$\mathrm{sign}\,(x) = \begin{cases} +1, & \text{dla} \;\; x > 0\ , \\ -1, & \text{dla} \;\; x < 0\ , \end{cases} \tag{5.71}$$

which is analyzed with the use of various methods in the works [14, 44, 29]. The analytical solution of equation (5.70) can be presented as *Fourier series* [77], *functions fragmentally continuous* [105] or *saw-shaped functions* [131].

In order to construct the solution of equation (5.68) we will apply the method of so-called small δ [51, 13, 29]. Substituting $\delta = (2n+1)^{-1}$ and assuming that $\delta \ll 1$, we will apply the following approximation

$$x^{\delta} = 1 + \delta \ln|x| + \ldots \tag{5.72}$$

We will consider only the case $x > 0$, since the solution for $x < 0$ can be obtained with the use of symmetrical mapping.

For equation (5.68) and with the initial conditions (5.71), we search for the following solution

$$x = x_0 + \delta x_1 + \ldots \tag{5.73}$$

After substituting (5.73) into the equations (5.68) and (5.69) we get the following equation:

$$\ddot{x}_1 = -\ln|x_0| \ . \tag{5.74}$$

Let us consider the solution determined in the quarter of a period [105]

$$x(t) = \begin{cases} -\dfrac{t}{2}(t-2A) \ , & 0 \le t \le 2A \ , \\ \dfrac{t^2}{2} - 3At + 4A^2 \ , & 2A \le t \le 4A \ , \end{cases} \tag{5.75}$$

$$x(t+nT) = x(t) \quad T = 4A \ .$$

The first approximation in the range $0 \le t \le A$ yields

$$\ddot{x}_1 = -\ln\left(tA - \frac{t^2}{2}\right) \ , \tag{5.76}$$

$$x_1(0) = 0 \ , \quad \dot{x}(0) = 0 \ . \tag{5.77}$$

Differentiating twice (5.76) and taking into consideration the equation (5.77) we obtain (during computations, program "Mathematica"):

$$x_1(t) = 2At - 3t^2 - t^2\ln 2 - 4A(A-t)\ln|-2A| + t^2\ln|2A-t| +$$

$$+t^2\ln t + 4A^2\ln|-2A+t| - 4At\ln|-2A+t| \ . \tag{5.78}$$

Although the solution can be prolonged only to the terms of the order δ^2, δ^3, ..., we will focus here only on the term of zero and first order, getting

$$x \approx x_0 + \delta x_1 \ . \tag{5.79}$$

Padé approximations can be applied for enlarging the area of application of the solution (5.68) [29, 51]. In this case, we obtain

$$x \approx \frac{x_0^2}{x_0 - \delta x_1} \ . \tag{5.80}$$

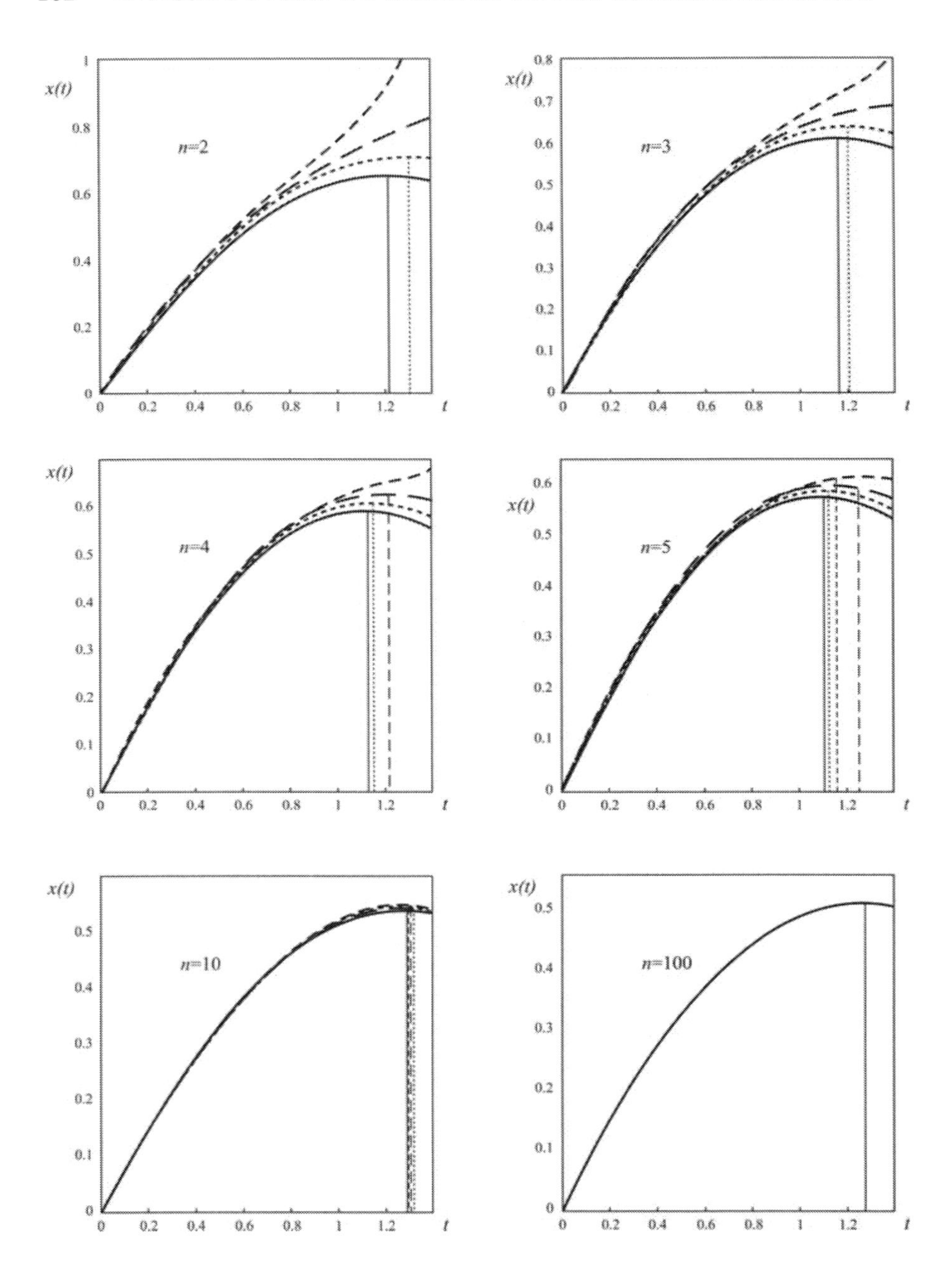

FIGURE 5.7: The solution of Cauchy problem (5.68), (5.69) for $A = 1$ with the use of Runge-Kutta method (——), and the approximations: (5.79) (------); (5.80) (- - - - - -); (5.81) (– – –) for different values of n.

TABLE 5.6: The numerical estimation $1/4T_i$, $i = 1, 2, 3$ where: T_1, T_2, T_3 are the periods associated with adequate approximations (5.79), (5.80) and (5.81).

	$n = 2$	$n = 3$	$n = 4$	$n = 5$	$n = 10$	$n = 100$
T_{num}	1·21981	1·16507	1·13200	1·10991	1·05976	1·00684
T_1	1·30571	1·20769	1·15748	1·12686	1·06436	1·00652
$\Delta_1(\%)$	7·04	3·66	2·25	1·53	0·434	1·00473
T_2	—	—	—	1·25367	1·07934	1·00664
$\Delta_2(\%)$	—	—	—	12·95	1·85	0·0160
T_3	—	—	1·22309	1·16116	1·07097	1·00658
$\Delta_3(\%)$	—	—	8·05	4·62	1·06	0·0103

As a result of exponential approximation, we obtain [51]

$$x \approx x_0 \exp(\delta x_1 / x_0) \ . \tag{5.81}$$

Some results of numerical computations for $A = 1$ are given in Fig. 5.7. Moreover, the numerical estimation with the use of Runge-Kutta method of the periods is given, together with errors, in Table 5.6.

For some n, the period values are not given. It is because of the observation of curves in Fig. 5.7, where, in some cases, curves corresponding to the approximations (5.80), (5.81) do not have the minimum. It has to be emphasised that the periods obtained with the use of the equations (5.79)–(5.81) are generally higher than their "exact values" obtained from numerical computations with the use of Runge-Kutta method for the small values of n. Moreover, this equation gives the best approximation of the values of the searched periods. Relative errors decrease quickly with the increase of n. In the analyzed case, it seems that exponential approximation is better than Padé approximation. However, with the increase of n all the analyzed equations are suitable for estimating the searched periods of vibrations.

Exercises

5.1. Applying PA, determine the value of $\sin 90°$. For this purpose, one has to take the first five elements of the asymptotic series (2.9) and, using the formulas (5.5)–(5.8), calculate successively PA of the order [1/1], [1/2], [2/1], [2/2]. Conduct the analysis of obtained results and compare them to the exact value $\sin 90° = 1$.

5.2. Determine $\ln(2.5) = \ln(1 + 1.5)$. Let us notice that the asymptotic series (2.11) is in this case divergent. Transforming this divergent series with the use of PA, however, allows to obtain the value $\ln(2.5)$ with the demanded degree of accuracy (one of the most important characteristics of PA). Check it.

5.3. Using the formula (2.10), construct the asymptotic decomposition of function $\cos(\varepsilon t)$, ($\varepsilon \ll 1$, $0 \leq t < \infty$). Prove that it is nonuniformly useful. On the basis of the formula (2.10) and the formulas (5.5)–(5.8) compute the values $\cos(\varepsilon t)$ and conduct the analysis of obtained results.

5.4. Realize the actions enumerated in ex. 3 for the function $e^{-\varepsilon t}$ applying the formula (2.7) (you can use figures 5.2, 5.3 given in the work [51]).

5.5. Consider the example associated with the real problem of aerodynamics. It is known ([15], p. 77), that the dimensionless quantity V_*, allowing to determine the velocity of gas and pressure on the surface of the profile with the elliptic intersection (aeroplane's wing) during the flow round of gas stream is determined by the following asymptotic decomposition

$$V_* = \varepsilon - \frac{1}{2}\varepsilon^2 \frac{x^2}{1-x^2} - \frac{1}{2}\varepsilon^3 \frac{x^2}{1-x^2} + \frac{3}{8}\varepsilon^4 \frac{x^4}{(1-x^2)^2} + \ldots$$

Find PA adequate for this decomposition, of orders [1/1], [1/2], [2/1]. Compare obtained results to the exact solution

$$V_* = (1+\varepsilon)/\sqrt{1+\varepsilon^2[x^2/(1-x^2)]} - 1$$

for $x = 0.5$ and $\varepsilon = 0.5,\ 0.1$.

5.6. For the series (3.61), construct PA of the order [1/1] and compare obtained results to the exact solution (3.60) obtained for $x = 0.5,\ 0.75$.

Chapter 6

Averaging of ribbed plates

This chapter contains examples of application of averaging to analyze ribbed plates. Averaging procedure is illustrated, as well as Kantorovich-Vlasov method and its modifications are discussed, among others. Examples include applications of the described methods to both static and dynamic problems.

6.1 Averaging in the theory of ribbed plates

Since many theories of supported shells have appeared, let us focus on the mathematical model of the problem.

The study of different variants of equations of supported shells is included in works [1, 2, 178]. Differences between the analyzed equations obtained by various scientists are associated with the members dependent on torsional rigidity of ribs and their resistance to bending, the factors which usually do not significantly affect the image of the condition of distortion stress and which can be omitted in many practical problems.

Behavior of the ribbed shell in the area of ribs, and all the more of frames, can be properly described with the use of a three-dimensional theory of elasticity. Beyond those small areas, results obtained even with the use of various calculation schemes for the case of rather narrow ribs (the rib's width is the same as the shell's breadth) practically overlap. For further considerations, we have assumed the scheme of contact between a rib and a shell along the line and we have taken into account the resistance of ribs to bending and stretching-compression and a small distortion associated with their position in relation to the average sheating area. The behavior of ribs is studied within the frames of Kirchhoff-Klebsch theory.

First, let us consider a simple example allowing to obtain an accurate solution. Other, more complicated examples may be found in the monograph [36]. Let a homogeneous infinite beam on parallel elastic bearings be uniformly loaded with the load P (Fig. 6.1).

Equations of the curve of *elastic deflection of a beam* have the following form in this case

$$EI\frac{d^4w}{dy^4} + c\sum_{k=-\infty}^{\infty}\delta(y-kl)w = P\ , \tag{6.1}$$

where δ is Dirac's function,

$$w = w_0 + w_1 = \frac{P}{c_0} + \frac{P}{24EI}F(y)\ , \tag{6.2}$$

$c_0 = c/l$, $F(y)$ is a periodic function with the period l, which for $0 \leq y \leq l$ has the form of: $F(y) = y^2(l-y)^2$.

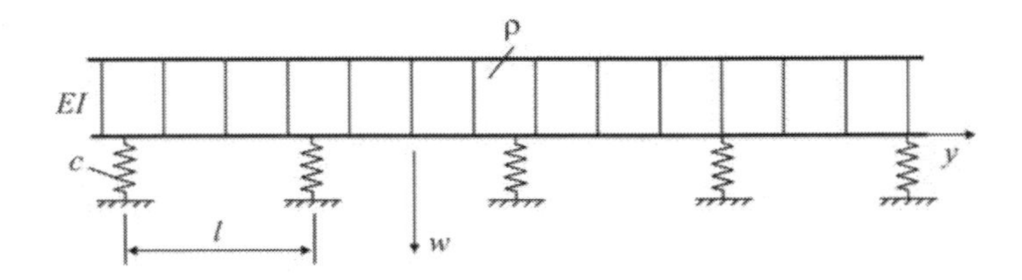

FIGURE 6.1: A homogeneous beam supported discretely with parallel bearings.

Let the distance between the supports approach zero, and their reduced rigidity c_0 remain constant. The first component of dislocation (w_0) is a constant, and the other component (w_1) approaches zero. The period of a function w_1 also approaches zero (Fig. 6.2).

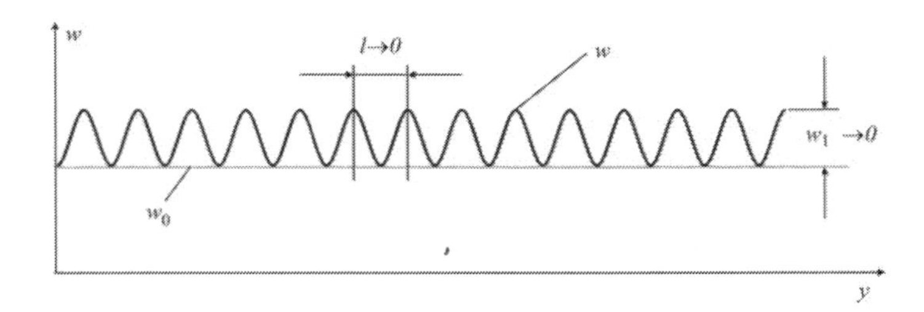

FIGURE 6.2: Dependencies w, w_0, w_1 on y.

Hence, for $l \to 0$, the actual displacement w differs form the constant component w_0 by small, although fast changing (with the period approaching zero for $l \to 0$) w_1 corrections. The function w_1 sharply (l^{-1} times) increases during differentiation (such functions are usually called *unstable* with respect to differentiation). The occurrence of the component w_1 results in significant difficulties during the numerical analysis of supported plates and shells. Moreover, due to the sharp increase in the value of a function caused by differentiation, the mentioned component may significantly affect the general image of the stress-strain condition, especially at the moments of force.

The situation described above is typical and it is suitable for the analysis with the use of the methods mentioned earlier. For that purpose, *the slow* (averaged or structurally ortothropic) and *the fast* (with the period equal to the distance of ribs) [1, 2] components are introduced. For our further considerations, we will use the averaging method, whose main aim will be to separate the slow and the fast component of the solution.

In order to demonstrate the application of the described procedure, let us come back to equation (6.1). On averaging the coefficient changing with respect to y, we get

$$EI\frac{d^4 w_{00}}{dy^4} + c_0 w_{00} = P \; . \tag{6.3}$$

From the physical point of view, the substitution of equation (6.1) by equation (6.3) means "blurring" the concentrated rigidities of the ribs and substitution of the initial discrete elastic support by a continuous support of *Winkler's type* (see also a similar approach given in the monograph [29]).

Now, we will present the initial dislocation w in the form of the sum of the slow (w_{00}) and the fast (w_{11}) parts of the solution $w = w_{00} + w_{11}$.

Coefficient w_{11} is determined from the following equation

$$EI\frac{d^4 w_{11}}{dy^4} + c\sum_{k=-\infty}^{\infty}\delta(y-kl)w_{11} = -c_0 w_{00}\left[l\sum_{k=-\infty}^{\infty}\delta(y-kl) - 1\right] \; . \tag{6.4}$$

Let us notice that equation (6.4) is not simpler than the initial one (it is even more complex with respect to the complexity of its right part). It can be significantly simplified, though, with the assumption that function w_{11} is periodic, with the period l. Equation (6.4) with the length of the period $0 \le y \le 1$, can be substituted by the equation

$$EI\frac{d^4 w_{11}}{dy^4} = c_0 w_{00} \; , \tag{6.5}$$

and conditions of symmetry and periodicity for $y = 0, l$ yield

$$\frac{dw_{11}}{dy} = 0 \, , \quad w_{11} = 0 \; . \tag{6.6}$$

The solutions of boundary conditions (6.3), (6.5), (6.6) have the form

$$w_{00} = \frac{P}{c_0} = w_0 \, , \quad w_{11} = \frac{P}{24EI}y^2(y-l)^2 = w_1 \, , \quad (0 \le y \le 1) \, .$$

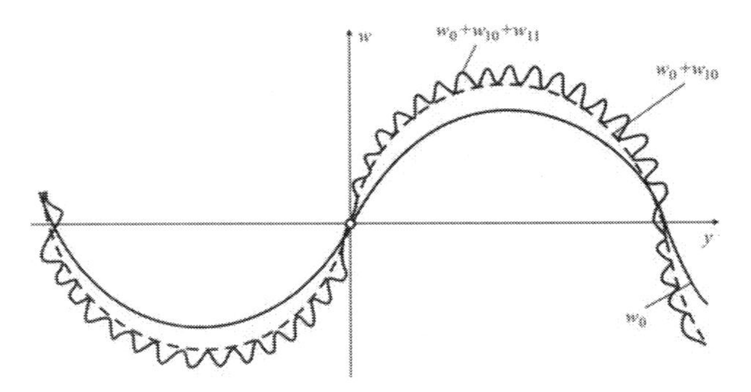

FIGURE 6.3: Relations between the accurate solution and the averaged one.

This means that we managed to obtain the accurate solution in this case.

Let us now consider more complex problem, when together with the support, a Winkler-type support with rigidity k also occurs, and the external load changes with coordinate y ($P = P_0 \sin(2\pi/L)$, $P_0 = const$).

The beam deflection equation has the form of

$$EI\frac{d^4w}{dy^4} + kw + c\sum_{k=-\infty}^{\infty} \delta(y - kl)w = P_0 \sin\frac{2\pi y}{L}\ , \tag{6.7}$$

whereas the averaged equation is expressed by the relation

$$EI\frac{d^4w_0}{dy^4} + (k + c_0)w_0 = P_0 \sin\frac{2\pi y}{L}\ . \tag{6.8}$$

Let us consider the case when the distance between ribs is much smaller than the external load, i.e.

$$\varepsilon = l/L \ll 1\ . \tag{6.9}$$

The accurate and averaged solutions for this case are presented in Fig. 6.3.

If the dislocation w is presented in the form of $w_0 + w_1$, conditions (6.6) for additional dislocation are not satisfied. Contrary to the example considered earlier, the correction made in the averaged solution has two components: the slow (w_{10}) and the fast one (w_{11}). In order to determine the fast component w_{11} we can again consider the additional equation on the period length $0 \le y \le l$ in the form of

$$EI\frac{d^4w_{11}}{dy^4} + kw_{11} = w_0 \tag{6.10}$$

with boundary conditions (6.6).

Let us notice that equation (6.10) allows to introduce further simplifications. Since function w_{11} is quick-changing with respect to y, it undergoes significant changes during differentiation process. Therefore, $EI|d^4w_{11}/dy^4| \gg k|w_{11}|$, and element kw_{11} in (6.10) can be omitted. Secondly, the slow-changing function w_0 (which changes slightly along length l) can be treated as the constant during differentiation of equation (6.10)[1].

Eventually, to determine w_{11}, we obtain the following equation

$$EI\frac{d^4 w_{11}}{dy^4} = w_0 \ , \tag{6.11}$$

with boundary conditions (6.6).

We will present the solution of the boundary equation (6.11), (6.6) in the form of

$$w_{11} = \frac{w_0}{24}F(y) \ .$$

In order to determine w_{10}, if this is demanded, we may apply the averaging method again. In practical computations, however, the approximation $w \approx w_0 + w_{11}$ is enough. Taking the fast component w_1 is essential for the process of the accurate description of the image of the stress – strain condition, since it can bring a significant correction to the determinaton of stresses. The element w_{10} introduces a minor correction with respect to stresses and dislocations, and that is why it can be omitted.

The above considerations are similar to those conducted earlier (*frame method* of Prokopov [134], *method of successive approximations* of Riabov [140] and Grebnia [78], or *the similar method* of Zarutsky [1, 2]).

It should be emphasized that during the analysis of plates and shells, a series of difficulties occurs, which cannot be overcome only by the use of intuition. First of all, those are various boundary problems generalized as for the dynamic problems, or the assessment of the error of basic approximations. It is not clear which of the elements should be left, and which of them should be dismissed in initial equations in order for approximations to be succesive (i.e., so that the dismissed elements were really smaller in comparison to those included in further considerations). Those difficulties can be overcome by constructing the asymptotic procedure basing on a physical interpretation factors discussed above.

6.2 Kantorovich-Vlasov-type methods

Let us consider the operational equation of the form

$$AU(x,y) = q(x,y) \, , \quad (x,y) \in \Omega = (a,b) * (c,d) \, , \tag{6.12}$$

[1]The solution of the equation (6.11) may be searched in the form of the product of the fast $\varphi(l^{-1}y)$ and slow $f(y)$ function of the form: $w_{11} = \varphi(l^{-1}y)f(y)$. Then $\frac{dw_{11}}{dy} = l^{-1}\frac{d\varphi}{d(l^{-1}y)}f + \frac{df}{dy}\varphi$ and the second element is small ($\sim l$) in comparison to the first element.

where A is a certain operator of the boundary problem including both the differential operator (differential operators system) and boundary conditions, $q(x, y)$ is the known function (the vector function for the system of equations); $u(x, y)$ is the demanded function (the vector function for the system of equations).

Let us further consider *Vlasov-Kantorovich projection methods*, often used in technical issues, and their modifications on the basis of the introduced equation (6.12).

6.2.1 Kantorowich-Vlasov method (KVM)

According to this method, we search for solution of the equation (6.12) in the form of

$$U_N(x, y) = \sum_{i=1}^{N} \varphi_i(x)\psi_i(y) , \tag{6.13}$$

where $\varphi_i(x)$ are the given functions, $\psi_i(y)$ are the demanded functions determined on the basis of the following system of projection equations

$$(AU_N - q, \varphi_j(x))_{H(a,b)} = 0 , \quad (j = 1, 2, \ldots, N) , \tag{6.14}$$

where $(AU_N - q, \varphi_j(x))$ is the conventional notation of the vector product in the space $H(a, b)$, $H = H(a, b) \times H(c, d)$ is the space in which there is an accurate solution U_0 of the equation (6.12). The procedure (6.14) with the use of Bubnov-Galerkin method with respect to one coordinate, reduces the demanded system of partial equations to the system of ordinary differential equations.

Disadvantages: If the expected equation should be characterized by the uniform symmetry with respect to all variables, it is possible that KVM will not show such symmetry for the limited number of elements of the series.

Advantages: Base functions with respect to one variable do not depend on the form of boundary conditions.

6.2.2 Vindiner method (VM)

A solution of equation (6.12) is searched in the form of

$$U_N(x, y) = \sum_{i=1}^{N} \varphi_i(x)\psi_i(y) + U_i(x)V_i(y) , \tag{6.15}$$

where $\psi_i(y)$ and $U_i(x)$ are given, whereas $\varphi_i(x)$ i $V_i(x)$ are the searched functions, determined by the following systems of projection equations

$$(AU_N - q, \psi_j(y))_{H(c,d)} = 0 , \quad (AU_N - q, U_j(x))_{H(a,b)} = 0 , \quad (j = 1, 2, \ldots, N) , \tag{6.16}$$

where $H((a,b) \times (c,d)) = H(a,b) \times H(c,d)$ is the space in which there is the accurate solution U_0 of the formulated problem.

Disadvantages: necessity of choosing the complete set of functions; dependence of the accuracy of the obtained solution on the amount of equations in the system (the amount of such equations is $2N$).

Advantages: the same as in the case of KVM. Moreover, there is a possibility of obtaining the symmetric solution with respect to all the variables, if the expected solution shows such symmetry.

6.2.3 Method of variational iterations (MVI)

This method is a modification of Kantorovich-Vlasov method.

According to MVI, we search for solutions of the system (6.12) in the form of (6.13), where functions φ_i i $\psi_i(y)$ are determined on the basis of the system of equations

$$(AU_N - q, \varphi_j(x))_{H(a,b)} = 0 \ , \tag{6.17}$$

$$(AU_N - q, \psi_j(y))_{H(c,d)} = 0 \ , \quad (j = 1, 2, \ldots, N) \ , \tag{6.18}$$

as follows. First, we establish a certain system of N-functions with respect to one variable, eg. $\varphi_j^0(x)$ $(j = 1, 2, \ldots, N)$. Next, from the system (6.17) we determine the system of functions $\psi_j^{(1)}(y)$, $j = 1, 2, \ldots, N$. The obtained functions are substituted into system (6.18) and a new choice of functions is determined with respect to the variable $x - \varphi_j^{(2)}(x)$. With the use of this choice, new functions with respect to variables (6.17), etc. are determined on the basis of the system $y - \psi_j^{(3)}(y)$, etc. Interrupting the process of determining the functions $\varphi_j(x)$ i $\psi_j(y)$ on the k-th step, we will determine the function

$$U_N = \sum_{i=1}^{N} \varphi_i^{(k-1)}(x) \psi_j^{(k)}(y) \ , \tag{6.19}$$

which is treated as the approximate solution of equation (6.12).

Disadvantages: Accuracy of the approximate solution of equation (6.12) depends on the amount of the elements of the series in formula (6.13).

Advantages: while realizing the procedure in accordance with MVI, there is no need to construct the initial approximation satisfying e.g. boundary conditions of the given problem; symmetry of the approximate solution is maintained only when it corresponds with the accurate solution of the studied problem; even for the limited amount of elements of the series approximating the approximate solution, it can be of high accuracy.

6.2.4 Agranowsky-Baglay-Smirnov method (ABSM)

Method of variational iterations (MVI), which for the first approximation corresponds with obtaining the first element of the series has been proposed by Agranowsky, Baglay and Smirnov. The formal scheme of this method follows.

We search for the solution of equation (6.12) in a similar way as in MVI discussed earlier, and for the first approximation ($N = 1$) we have

$$U_1 = \varphi_1^{(k-1)}(x)\psi_1^{(k)}(y) \ . \tag{6.20}$$

Let us construct a new equation of the form

$$AU(x,y) = q(x,y) - AU_1(x,y) \ , \tag{6.21}$$

i.e., in equation (6.12) the right-hand side has been changed. We solve equation (6.21) again with the use of MVI for the first approximation, and we get as a result

$$U_2(x,y) = \varphi_2^{(k-1)}(x)\psi_2^{(k)}(y) \ . \tag{6.22}$$

Next, we again construct a new equation of the form

$$AU(x,y) = q(x,y) - AU_1(x,y) - AU_2(x,y)$$

and we apply MVI for the first approximation.

Eventually, we take the following series into consideration as an initial solution

$$U(x,y) = \sum_{n=1}^{N} U_n(x,y). \tag{6.23}$$

Disadvantages: This method cannot be applied in solving a wide class of problems (e.g., it cannot be applied in the analyzed plate, where boundary conditions undergo a change along a certain border).

Advantages: similar to those enumerated earlier for MVI. Moreover, for each formulated right-hand side of equation (6.21), it is enough to apply MVI only for the first approximation, which results only in solving the algebraic problems of a low order.

6.2.5 Combined method (CM)

Combined method (CM) is the combination of earlier discussed methods (VM+ABSM+MVI).

According to this method, the approximate solution is searched in the form of

$$U^{(k)}(x,y) = \varphi^{(k)}(x)\psi^{(k-1)}(y) + U^{(k-1)}(x)V^{(k)}(y) \ , \tag{6.24}$$

where $\psi^o(y)$ and $U^o(x)$ are assumed as demanded functions, $\varphi^{(1)}(x)$ and $V^{(1)}(y)$ are the searched functions determined on the basis of the systems of projection equations (6.16). However, whereas in VM, the found functions $\varphi^{(1)}(x)$ and $V^{(1)}(y)$ were taken as the final ones, in this method they are treated as known, and then new functions are defined $\psi^{(2)}(y)$ and $U^{(2)}(x)$, etc., so that the error between $U^{(k)}(x,y)$ and $U^{(k-1)}(x,y)$ is smaller that the demanded quantity. We search for the solution of the following equation

$$AU_2(x,y) = q(x,y) - AU_1(x,y) \ . \tag{6.25}$$

Using the procedure discussed above, we get

$$U_2(x,y) = \varphi_2^{(k)}(x)\psi_2^{(k-1)}(y) + U_2^{(k-1)}(x)V^{(k)}(y) \; .$$

If the error between U_1 and $(U_1 + U_2)$ is smaller than the demanded quantity, the process is interrupted. If not, the next equation is constructed:

$$AU_3(x,y) = q(x,y) - AU_1(x,y) - AU_2(x,y) \; ,$$

from which $U_3(x,y)$, etc is determined.

$$U_0(x,y) = \sum_{j=1}^{N} U_j(x,y) \; . \tag{6.26}$$

Adavantages: This method has the advantages of VM and ABSM mentioned above. Moreover, it turns out that it is possible to apply this method for solving a wider class of problems, e.g., for the analysis of rectangular plates, for which boundary conditions change along certain boundary.

6.2.6 Kantorovich-Vlasov method with the amendment

The core of this method is applying Agranovski-Baglay-Smirnov method to Kantorovich-Vlasov method, but without using the procedure of variational iterations.

6.2.7 Vindiner method with the amendment

The idea of this method is that Agranovski-Baglay-Smirnov procedure is applied in the method of Vindiner method, but without using the procedure of variational iterations (MVI).

6.2.8 Vindiner method and variational iterations

The idea of this method is applying the method of variational iterations to Vindiner method.

Taking the example of bending of a plate (i.e., Germain-Lagrange equation), we will study the accuracy of each of the mentioned methods in relation to the accurate solution.

Let us have the following equation given

$$U(x,y) = D\nabla^2\nabla^2 W(x,y) \; , \tag{6.27}$$

together with the following boundary conditions:

– *simple support*

$$W\,|_{\partial\Omega} = \left.\frac{\partial^2 w}{\partial n^2}\right|_{\partial\Omega} = 0 \; , \tag{6.28}$$

– *clamping*

$$W\,|_{\partial\Omega} = \left.\frac{\partial w}{\partial n}\right|_{\partial\Omega} = 0\,. \tag{6.29}$$

Space $\Omega = (0,a)\times(0,b)$, $\partial\Omega$ is the boundary of area Ω. The differential operator (6.27) and boundary conditions are reduced to the dimensionless form

$$x = \bar{x}a\,, \quad y = \bar{y}b\,, \quad w = \bar{w}h\,, \quad \lambda = a/b\,.$$

Dimensionless parameters are marked with a line above, and they will be omitted in further solutions.

Equations, which are obtained on applying Kantorovich-Vlasov procedure, are solved with the use of the method of finite differences, with the application of Gauss method in the stage of solving the algebraic equations. The solution is obtained with the use of single and double accuracy. The range $[0,1]$ is divided into 15 do 30 nodes. The relative error in calculation is

$$\varepsilon = \left|\frac{W^k(x,y) - W^{k-1}(x,y)}{W^k(x,y)}\right| < 10^{-3}\,.$$

Results of computations for $q(x,y) = const$ of the case of local load are:

$$q(x,y) = \frac{n}{2\pi}\exp\left(-\frac{n(x-0,5)}{2}\right)\exp\left(-\frac{n(y-0,5)}{2}\right)\,, \quad A = \frac{\mathrm{h}}{2\pi}\,,$$

where $A = 50$ are given in the work [28]. The range $0 \le x,\ y \le 1$ is divided into 10, 100, 500 parts.

Methods discussed above and the results of the numerical experiment allow to draw a conclusion that KV is the most precise one and the results obtained thanks to it are very close to exact results. It has to be stressed, though, that increasing the dimension of an system of differential equations and the occurrence of the additional iterational procedure generally demands applying the method of variational iterations (MVI).

6.3 Transverse vibrations of rectangular plates

Let us consider the transverse vibrations of an isotropic rectangular plate supported regularly (supporting ribs are uniform with respect to their geometrical, rigidity, and inertial characteristics and they are distributed at equal distance (see Figure 6.4).

In further considerations, it is assumed that they are mono-dimensional elastic elements (Kirchhoff-Klebsch bars), which are stiffly fixed to the plate and they are distributed symmetrically in relation to its average surface.

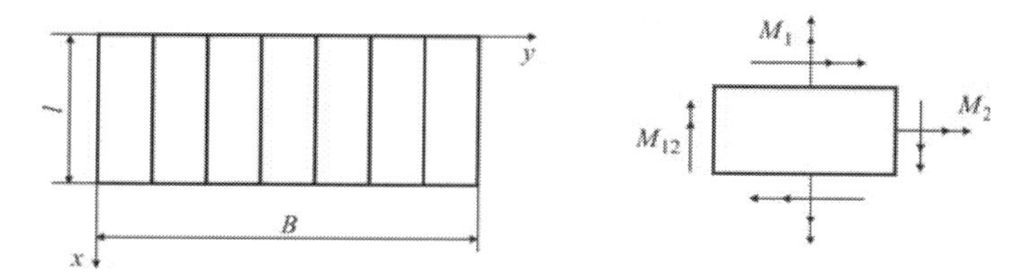

FIGURE 6.4: Computational scheme and the support of the analyzed plate.

If we hypothetically isolate ribs from the sheating, and change the influence if the i-th rib on the plate by using the contact force q_i, *the system of the equations of the plate movements* (without the external load) will have the form of

$$D\nabla^4 w + \rho_0 h \frac{\partial^2 w}{\partial t^2} = \sum_{i=0}^{N-1} q_i \delta\left(y - \frac{B}{N} i\right) , \tag{6.30}$$

$$E_0 I \frac{\partial^4 w_{ci}}{\partial x^4} + \rho_c F \frac{\partial^2 w_{ci}}{\partial t^2} = -q_i ,$$

$$\nabla^4 = \frac{\partial^4}{\partial x^4} + 2\frac{\partial^4}{\partial x^2 \partial y^2} + \frac{\partial^4}{\partial y^4} ; \tag{6.31}$$

where w, w_{ci} are ordinary dislocations of the plate and the i-th stringer; $D = Eh^3/[12(1-\nu^2)]$; E, ν, ρ_0, h are, respectively, Young modulus, Poisson coefficient, the density of plate material and its thickness; E_c, ρ_c, F, I stand for, respectively, Young module and the density of stringer material, intersection and the moment of inertia (in relation to the mean plate surface) of its cross-section, N is the number of ribs, and t is time.

Dislocations of ribs and a plate are associated through the following relations

$$w_{ci} = w\,|_{y=ib} \qquad i = 0, \ldots, N-1 , \tag{6.32}$$

where $b = B/N$ is the distance between the ribs.

Taking the relations (6.30)–(6.32) into account, we get the following *equation of the motion of the supported plate*

$$D\nabla^4 w + E_c I \frac{\partial^4 w}{\partial x^4} \Phi(y) + [\rho_0 h + \rho_c F \Phi(y)] \frac{\partial^2 w}{\partial t^2} = 0 , \tag{6.33}$$

where $\Phi(y) = \sum\limits_{i=0}^{N-1} \delta(y - ib)$.

Physical and geometrical relations can be presented as

$$M_1 = D(k_1 + \nu k_2) + E_c I k_1 \Phi(y) , \quad M_2 = D(k_2 + \nu k_1) , \tag{6.34}$$

$$M_{12} = D(1-\nu)k_{12} \quad Q_1 = \frac{\partial M_1}{\partial x} + \frac{\partial M_{12}}{\partial y} , \quad Q_2 = \frac{\partial M_2}{\partial y} + \frac{\partial M_{21}}{\partial x} ;$$

$$k_1 = -\frac{\partial^2 w}{\partial x^2}, \quad k_2 = -\frac{\partial^2 w}{\partial y^2}, \quad k_{12} = -\frac{\partial^2 w}{\partial x \partial y}, \tag{6.35}$$

where M_1, M_2 are the moments bending, respectively, toward the axes x and y, M_{12} is the torsional moment, Q_1, Q_2 are transverse forces; k_1, k_2 are the changes of curvatures towards the axes x and y, and k_{12} is the change of a curvature caused by torsions.

On the boundaries $x = 0, L$, the following *boundary conditions* are demanded

$$w = 0, \quad M_1 = 0, \tag{6.36}$$

or

$$w = 0, \quad \frac{\partial w}{\partial x} = 0, \tag{6.37}$$

or

$$Q_1 + \frac{\partial M_{12}}{\partial y} = 0, \quad M_1 = 0, \tag{6.38}$$

or

$$Q_2 + \frac{\partial M_{12}}{\partial y} = 0, \quad \frac{\partial w}{\partial x} = 0. \tag{6.39}$$

The analogous conditions may be formulated also for the boundaries $y = 0, B$, i.e.:

$$w = 0, \quad M_2 = 0, \tag{6.40}$$

or

$$w = 0, \quad \frac{\partial w}{\partial y} = 0, \tag{6.41}$$

or

$$Q_2 + \frac{\partial M_{21}}{\partial x} = 0, \quad M_2 = 0, \tag{6.42}$$

or

$$Q_2 + \frac{\partial M_{21}}{\partial x} = 0, \quad \frac{\partial w}{\partial y} = 0. \tag{6.43}$$

Conditions (6.36), (6.40) correspond to the classical simple support; (6.37), (6.41) – to clamping; (6.38), (6.39), (6.42), (6.43) – to free boundary (in the case of (6.39), (6.43) with the limitation for the angle of rotation).

Let us consider the following boundary conditions for $y = 0, B$:

$$\left\{ w; \frac{\partial w}{\partial y}; \frac{\partial^2 w}{\partial y^2}; \frac{\partial^3 w}{\partial y^3} \right\}\bigg|_{y=0} = \left\{ w; \frac{\partial w}{\partial y}; \frac{\partial^2 w}{\partial y^2}; \frac{\partial^3 w}{\partial y^3} \right\}\bigg|_{y=B}. \tag{6.44}$$

The boundary problem includes in this case the vibrations antisymmetric in relation to axis $y = B/2$ for the condition (6.40), and the symmetric vibrations for conditions (6.43). It also allows to move to the problem of the vibrations of a closed cylindrical shell.

In the case of different boundary conditions for $y = 0, B$, solutions can be also found with the use of the method discussed above.

Let us notice that with respect to the assumed hypotheses concerning the type of support, angles of rotation, bending moments and the transverse force Q_1 are continuous functions, whereas the force Q_2 has the discontinuity of the first order on the ribs.

The mentioned conditions can be expressed in the following form

$$\left\{w;\ \frac{\partial w}{\partial y};\ \frac{\partial^2 w}{\partial y^2}\right\}\bigg|_{y=ib+0} = \left\{w;\ \frac{\partial w}{\partial y};\ \frac{\partial^2 w}{\partial y^2}\right\}\bigg|_{y=ib-0},$$

$$Q_2\big|_{y=ib+0} - Q_2\big|_{y=ib-0} = q_i\ .$$

The last condition is equivalent to the following:

$$D\left(\frac{\partial^3 w}{\partial y^3}\bigg|_{y=ib+0} - \frac{\partial^3 w}{\partial y^3}\bigg|_{y=ib-0}\right) = \left(E_c I \frac{\partial^4 w}{\partial x^4} + \rho_c F \frac{\partial^2 w}{\partial t^2}\right)\bigg|_{y=ib}.$$

Let us consider *free vibrations*, assuming that

$$w = w(x,y)e^{i\sqrt{\lambda}t}\ , \tag{6.45}$$

where $\sqrt{\lambda}w(x,y)$ is the frequency of vibrations, corresponding to the assumed vibrations mode.

Let the oscillatory wave directed at axis *oy* be much longer than the distance between ribs (in engineering, *low-frequency vibrations* are among the most dangerous).

Let us move to the averaged system by means of "blurring" the rigidity and density of ribs along their sheating borderline. As a result, we will get *the equation of the ortothropic plate movements* in the form of

$$D_1\frac{\partial^4 w}{\partial x^4} + 2D\frac{\partial^4 w_0}{\partial x^2 \partial y^2} + D\frac{\partial^2 w_0}{\partial y^4} - \rho_1\lambda_0 w_0 = 0\ , \tag{6.46}$$

for the corresponding boundary conditions on boudaries for $x = 0, L$ in the form of

$$w_0 = 0\ , \quad M_1^{(0)} = -(D_1 k_1^{(0)} + D\nu k_2^{(0)}) = 0\ , \tag{6.47}$$

or

$$w_0 = 0\ , \quad \frac{\partial\, w_0}{\partial x} = 0\ , \tag{6.48}$$

or

$$Q_1^{(0)} + \frac{\partial M_{12}^{(0)}}{\partial y} = -D_1\frac{\partial^3 w_0}{\partial x^3} - D(2-\nu)\frac{\partial^3 w_0}{\partial x \partial y^2} = 0\ , \quad M_1^{(0)} = 0 \tag{6.49}$$

or

$$Q_1^{(0)} + \frac{\partial M_{12}^{(0)}}{\partial y} = 0\ , \quad \frac{\partial w_0}{\partial x} = 0\ , \tag{6.50}$$

where

$$D_1 = D + E_c I/b\,, \quad \rho_1 = \rho_0 h + \rho_c F/b\ .$$

On boudaries $y = 0, B$, conditions (6.44) are demanded.

The forms of *proper vibrations of the structurally ortothropic plate* can be presented as follows

$$w_0 = \psi(x)\varphi_0(y)\ ,$$

where:

$$\varphi_0(y) = \begin{cases} \sin(2\pi n y/B) & \text{(for conditions (6.40))}, \\ \cos(2\pi n y/B) & \text{(for conditions (6.43))}, \end{cases}$$

and functions $\psi(x)$ are determined in relation to boundary conditions for $x = 0, L$.

In the case of a simple support (6.36), we have

$$w_0 = A \sin\frac{2\pi m x}{L}\varphi_0(y)\,, \quad m = 1, 2, \ldots,$$

$$\lambda_0 = \frac{16\pi^4 D}{\rho_1}\left[\frac{D_1}{D}\left(\frac{m}{L}\right)^4 + 2\frac{m^2 n^2}{L^2 B^2} + \left(\frac{n}{B}\right)^4\right]\,,$$

where A is an amplitude.

We will present initial dislocation w as the sum of an averaged (determined on the basis of the structural-ortothropic theory) dislocation w_0 and a fast-changing correction w_1 along the coordinate y in the form of

$$w = w_0 + w_1\ . \tag{6.51}$$

The square of the vibration frequency will be also searched in the form of the sum of two components

$$\lambda = \lambda_0 + \lambda_1\ . \tag{6.52}$$

The occurrence of element λ_1 is associated with the change of frequency that results from taking the discreteness of ribs into account. Let us notice that the solution of initial equation (6.33) is not periodic with the period $2\pi/\sqrt{\lambda_0}$, and that is why the occurrence of component w_1 should lead to the change of frequency ($\sqrt{\lambda} \neq \sqrt{\lambda_0}$).

Substituting expressions (6.51) and (6.52) to initial equation (6.33), and to boundary conditions (6.36)–(6.39), and taking relations of the structural-ortothropic theory (6.46)–(6.50) into account, we get the following relations leading to determination of, w_1, λ_1.

$$D\nabla^4 w_1 + E_c I \Phi(y)\frac{\partial^4 w_1}{\partial x^4} - \left[\rho_0 h + \rho_c F \Phi(y)\right](\lambda_0 + \lambda_1) w_1 =$$

$$= -E_c I\left[\Phi(y) - \frac{N}{B}\right]\frac{\partial^4 w_0}{\partial x^4} + \rho_c F(\lambda_0 + \lambda_1)\left[\Phi(y) - \frac{N}{B}\right] w_0 + \lambda_1 w_0\ . \tag{6.53}$$

Boundary conditions on boundaries $x = 0, L$ have the form of

$$w_1 = 0\,, \quad M_1^{(1)} = -D\left(\frac{\partial^2 w_1}{\partial x^2} + \nu\frac{\partial^2 w_1}{\partial y^2}\right) - E_c I \Phi(y)\frac{\partial^2 w_1}{\partial x^2} = 0\,, \tag{6.54}$$

or

$$w_1 = 0\,, \quad \frac{\partial w_1}{\partial x} = 0\,, \tag{6.55}$$

or

$$\tilde{Q}_1^{(1)} = \frac{\partial M_1^{(1)}}{\partial x} + 2\frac{\partial M_{12}^{(1)}}{\partial y} = E_c I\left[\Phi(y) - \frac{N}{B}\right]\frac{\partial^3 w_0}{\partial x^3}\,,$$

$$M_1^{(1)} = E_c I\left[\Phi(y) - \frac{N}{B}\right]\frac{\partial^2 w_2}{\partial x^2}\,, \tag{6.56}$$

or

$$\tilde{Q}_1^{(1)} = 0\,, \quad \frac{\partial w_1}{\partial x} = 0\,. \tag{6.57}$$

Conditions for boundaries $y = 0, B$ have the form of (6.44).

The right-hand side of equation (6.53) can be treated as certain fictional surface load, and the left-hand sides occurring in boundary conditions (6.54), (6.56) can be regarded as certain fictional boundary conditions.

For further considerations, it is convenient to exchange the coordinates as follows

$$\xi = \frac{x}{R}\,, \quad \eta = \frac{y}{R}\,, \quad \tau = \frac{t}{R}\sqrt{\frac{E}{\rho_0(1-\nu^2)}} \quad R = \frac{B}{2\pi}$$

and to introduce the following new parameters

$$a^2 = \frac{h^2}{12R^2}\,, \qquad \rho = \frac{\rho_c F}{\rho_0 h B}\,,$$

$$\alpha = \frac{E_c I}{Db}\,, \qquad l = \frac{2\pi L}{B}\,.$$

Equations leading to determination of *the corrections* w_1 and λ_1, caused by the discretization of ribs, will have the following form

$$\left(a^2\nabla^4 \alpha N \Delta_N \frac{\partial^4}{\partial \xi^4}\right) w_1 - \lambda(1+\rho\Delta_N)w_1 =$$

$$= (\Delta_N - 1)\left(-a^2\alpha N\frac{\partial^4}{\partial \xi^4} + \lambda\rho\right)w_0 + \lambda_1(1+\rho N)w_0\,, \tag{6.58}$$

where:

$$\nabla^4 = \frac{\partial^4}{\partial \xi^4} + 2\frac{\partial^4}{\partial \xi^2 \partial \eta^2} + \frac{\partial^4}{\partial \eta^4}\,,$$

$$\Delta_N = \frac{2\pi}{N}\sum_{i=0}^{N-1}\delta\left(\eta - \frac{2\pi\, i}{N}\right)\,.$$

In solutions, we will limit our considerations to boundary conditions (6.54) and (6.55), which for the new variables will take the form of (for $\xi = 0, l$)

$$w_1 = 0\,, \quad \frac{\partial^2 w_1}{\partial \xi^2} = 0\,, \tag{6.59}$$

or

$$w_1 = 0\,, \quad \frac{\partial w_1}{\partial \xi} = 0\,. \tag{6.60}$$

In the case of other boundary conditions, the solution is constructed in the analogical way.

We will introduce parameter ε, characterizing the following relation of the distance between ribs to the length of the oscillatory wave in the transverse direction

$$\varepsilon = \frac{B/N}{B/n} = \frac{n}{N}\,. \tag{6.61}$$

The above parameter is small, according to the earlier assumption (the oscillatory wave is much longer than the distance between ribs).

Let us write the right-hand side of the equation (6.58) in the form of

$$\Phi_1 + \Phi_0\,,$$

where:

$$\Phi_1 = (\Delta_N - 1)\left(-a^2\alpha N \frac{\partial^4}{\partial \xi^4} + \lambda\rho\right) w_0\,,$$

$$\Phi_0 = \lambda_1(1 + \rho N) w_0\,.$$

The quickness of changes of function Φ_0 along coordinates ξ and η, and of function Φ_1 along ξ are small and correspond to the quickness of change of the structural-isothropic solution. In that time, Φ_1 changes quickly along η, which is easy to notice if we present Φ_1 as follows

$$\Phi_1 = \sum_{j=1}^{\infty} \cos(j\varepsilon^{-1} n\eta)\left(-a^2\alpha N \frac{\partial^4}{\partial \xi^4} + \lambda\rho\right) w_0\,.$$

In accordance with the structure of the right-hand side of equation (6.58), we will search for the particular solution w_{14} in the form of

$$w_{14} = w_{10} + w_{11}\,,$$

where w_{10} is slow-changing, and w_{11} represents quick-changing corrections along coordinate η made in the averaged solution.

The occurrence of the slow-changing component in the solution of equation (6.58) is associated with the occurrence of the slow-changing (with respect to η) function while multiplying quick-changing coefficients and functions.

Therefore, the aim of introducing the function w_{10} is compensation of slow-changing corrections that occurred in equation (6.58) with respect to w_{11}.

It may happen that the found solution w_{11} will not fulfill the demanded boundary conditions, and, what is more, the corresponding corrections will be quickly changing with respect to η. That is why the solution of the boundary problem w_1 should include the component w_{12} of *the boundary layer type*, quick-changing with respect to η, and ξ.

Our next step will be to separate the boundary problems for the quick-changing components w_{11}, w_{12} using the operator D_η in the form of

$$D_\eta w = w - M_\eta w \quad M_\eta w = \frac{\varphi_0(\eta)}{2\pi} \int_0^{2\pi} w\varphi_0(\eta) d\eta \ .$$

As a result, we get

$$a^2 \left(\nabla^4 + \alpha \frac{\partial^4}{\partial \xi^4} \right) w_{11} - \lambda(1+\rho) w_{11} +$$

$$+ D_\eta \left[(\Delta_N - 1) \left(a^2 \alpha \frac{\partial^4}{\partial \xi^4} - \lambda \rho \right) (w_{10} + w_{11}) \right] =$$

$$= -(\Delta_N - 1)(a^2 \alpha + \lambda \rho) w_0 \ , \tag{6.62}$$

$$\nabla^4 w_{12} - \lambda(1+\rho) w_{12} = 0 \ , \tag{6.63}$$

and for $\xi = 0, l$ we have

$$w_{12} = -w_{11} \ , \quad \frac{\partial^2 w_{12}}{\partial \xi^2} = -\frac{\partial^2 w_{11}}{\partial \xi^2} \ , \tag{6.64}$$

$$w_{12} = -w_{11} \ , \quad \frac{\partial w_{12}}{\partial \xi} = -\frac{\partial w_{11}}{\partial \xi} \ . \tag{6.65}$$

In order to determine the function w_{10} one has to separate in equation (6.58) and in boundary conditions (6.59) and (6.60) the slow-changing components, i.e., again apply the procedure of averaging.

As a result, we get

$$a^2 \left(\nabla^4 + \alpha N \frac{\partial^4}{\partial \xi^4} \right) w_{10} - \lambda(1+\rho) w_{10} +$$

$$+ M_\eta \left[(\Delta_N - 1) \left(a^2 \alpha N \frac{\partial^4}{\partial \xi^4} - \lambda \rho \right) (w_{10} + w_{11}) \right] = \lambda_1 (1 + \rho N) w_0 \ , \tag{6.66}$$

and for $\xi = 0, l$, we have

$$w_{10} = 0 \ , \quad \frac{\partial^2 w_{10}}{\partial \xi^2} = 0 \ , \tag{6.67}$$

$$w_{10} = 0 , \quad \frac{\partial^2 w_{10}}{\partial \xi^2} = 0 . \tag{6.68}$$

Equation (6.66) describes vibrations of the structural-ortothropic plate under the influence of a slow-changing fictional load.

Before starting the asymptotic analysis of obtained boundary problems, one has to derivate relations between parameters a^2, α, ε (the parameter ρ does not essentially influence the character of the asymptotic process, and we assume that $N\rho \sim 1$, i.e., the reduced density of stringers has the order of the density of a plate).

As the basic small parameter, with respect to which we determine the orders of magnitude of other parameters, characterizing the frequency of the decomposition of ribs (ε) and their total rigidity (αN), we will choose the parameter characterizing thin-walledness a, so that

$$\varepsilon \sim a^p , \quad \alpha N \sim a^{p_1} . \tag{6.69}$$

With such formulation of the problem, the classification given below allows to generalize the solutions for the case of a shell.

Assessing the parameters of asymptotic integration is by no means trivial [75, 76]. It is usually realised on the basis of the formal choice of adequate values of *index exponents of the small parameter*, which is described in a more detailed way in the work [153]. In the given case, on the basis of such "free game with the parameters", we get the following values p_1 and p,yielding the following, qualitatively different, systems

$$p_1 > -1 , \tag{6.70}$$

$$-1 \geq p_1 > -2 , \tag{6.71}$$

$$-2 \geq p_1 , \tag{6.72}$$

$$p \leq 1/2 , \tag{6.73}$$

$$1/2 < p < 1 , \tag{6.74}$$

$$1 \leq p . \tag{6.75}$$

The obtained assessments allow for relatively simple considerations.

We will be considering the ribs of average rigidity, distributed with "average frequency" (assessment 6.71, 6.74). The obtained results can be applied in the case of the analysis of densely distributed ribs of average rigidity, or for the ribs of small rigidity.

In the case of rigid ribs, one may use a simple scheme [1, 2, 79, 92], in calculations, i.e., a plate fixed on ribs. Let us notice that searching for the solution with the small number of ribs should not begin with the averaging of coefficients.

Let us proceed to "constructing" a quick (with respect to η) solution w_{11}. The analysis, analogous to the one conducted in p. 6.1, shows that equation

(6.65) can be solved in the range $0 \leq \eta \leq 2\pi/N$, with the following conditions demanded for $\eta = 0,\ 2\pi/N$

$$w_{11} = 0 \ , \tag{6.76}$$

$$\frac{\partial w_{11}}{\partial \eta} = 0 \ . \tag{6.77}$$

The quickness of the change of function w_{11} with respect to η is higher than with respect to ξ, because

$$\frac{\partial w_{11}}{\partial \eta} \sim \varepsilon^{-1} w_{11} \ , \quad \frac{\partial w_{11}}{\partial \xi} \sim w_{11} \ , \quad \left(0 < \eta < \frac{2\pi}{N}\right) \ . \tag{6.78}$$

That is why on the left-hand side of equation (6.65) one may leave only the derivative η of the form

$$a^2 \frac{\partial^4 w_{11}}{\partial \eta^4} = N \left(\lambda\rho - a^2 \alpha \frac{\partial^4}{\partial \xi^4}\right) w_0 \ . \tag{6.79}$$

During the integration of equation (6.79), function w_0 can be treated as constant, which is justified by the small quickness of its changes with respect to $\eta(\partial w_0/\partial\eta \sim \varepsilon^0 w_0)$. Taking boundary conditions (6.76), (6.77) into account, we get the following from (6.79)

$$w_{11} = \frac{N}{24}\left(-\lambda a^2 \rho + \alpha \frac{\partial^4}{\partial\, \xi^4}\right) w_0 F_4(\eta) \ ,$$

where $F_4(\eta)$ is the periodic function, that in the area of changes $0 \leq \eta \leq 2\pi/N$ has the following form

$$F_4(\eta) = \eta^2 \left(\eta - \frac{2\pi}{N}\right)^2 \ .$$

After the application of initial variables, we get

$$w_{11} = \frac{1}{24}\left(\frac{E_c I}{Db}\frac{\partial^4}{\partial x^4} - \frac{\rho_c F \lambda}{\rho_0 h b}\right) w_0 y^2 (y-b)^2 \ . \tag{6.80}$$

Let us now proceed to the solution w_{12} of a boundary layer type. The function w_{12} is quick-changing with respect to both coordinates

$$\frac{\partial w_{12}}{\partial \xi} \sim \frac{\partial w_{12}}{\partial \eta} \sim \varepsilon^{-1} w_{12} \ , \quad \left(0 \leq \eta \leq \frac{2\pi}{N}\right) \ , \tag{6.81}$$

which allows to omit the dynamic element in equation (6.63), and eventually we get

$$\nabla^4 w_{12} = 0 \ . \tag{6.82}$$

Taking the symmetry of the problem into account with respect to the ribs, the following condition should be fulfilled for $\eta = 0.2\pi/N$

$$\frac{\partial w_{12}}{\partial \eta} = 0 . \qquad (6.83)$$

From the assessment (6.81) we obtain $w_{12}|_{\eta=0.2\pi/N} = 0$. Let us consider more carefully the physical interpretation of this condition. It does not stem from the conditions of periodicity and symmetry that w_{12} should be equal to zero on the ribs, because the boundary corrections with respect to w_{11} do not counterbalance. On the other hand, it is obvious that dislocations of the ribs of a plate are related to each other in the same way as their reduced rigidities are related to one another.

Since the rigidity of ribs is relatively high (see assessment 6.71), their deflections are small in relation to the deflection of the boundary sheating, and therefore we can assume that for $\eta = 0.2\pi/N$, we get

$$w_{12} = 0 . \qquad (6.84)$$

Boundary conditions for $\xi = 0, l$ have the following form

$$w_{12} = -w_{11}\,, \qquad \frac{\partial^2 w_{12}}{\partial \xi^2} = -\frac{\partial^2 w_{11}}{\partial \xi^2}\,, \qquad (6.85)$$

$$w_{12} = -w_{11}\,, \qquad \frac{\partial w_{12}}{\partial \xi} = -\frac{\partial\, w_{11}}{\partial \xi}\,. \qquad (6.86)$$

Moreover, for the boundary $\xi = 0$, additional conditions should be formulated

$$w_{12} \to 0\,, \quad \frac{\partial w_{12}}{\partial \xi} \to 0 \quad \text{for} \quad \xi \to \infty\,, \qquad (6.87)$$

and for the condition $\xi = l$,

$$w_{12} \to 0\,, \quad \frac{\partial w_{12}}{\partial \xi} \to 0 \quad \text{for} \quad \xi \to -\infty\,. \qquad (6.88)$$

During the formulation of conditions (6.87), (6.88) it was assumed that the reciprocal influence of boundary layers in the vicinity of boundaries $\xi = 0$, $\xi = l$ may be neglected. It is justified only in the case $L \gg b$.

In order to solve the problem concerning the deflection of a rib-supported plate, Kantorovich-Vlasov method is applied by presenting function w_{12} in such a form, so as the following boundary conditions for $\eta = 0.2\pi/N$ can be fulfilled

$$w_{12} = \psi(\xi)\eta^2(\eta - 2\pi/N)^2\,. \qquad (6.89)$$

The above approximation leads to satisfying results during calculations.

Applying the standard Kantorovich-Vlasov procedure, we get the ordinary differential equation with respect to coordinate ξ in the form of

$$\frac{d\psi}{d\varepsilon} - 2\left(\frac{N}{\pi}\right)^2 \frac{d^2\psi}{d\xi^2} + 31.5\left(\frac{N}{\pi}\right)^4 \psi = 0 \ . \tag{6.90}$$

The general solution for equation (6.90) has the following form

$$\psi = \exp(\delta_1\xi)\left[C_1\cos(\delta_2\xi) + C_2\sin(\delta_2\xi)\right] +$$

$$+ \exp\left[\delta_1(\xi - l)\right]\left[C_3\cos(\delta_2(\xi - l)) + C_4\sin(\delta_2(\xi - l))\right] \ ,$$

where $\delta_1 = 0.661N$, $\delta_2 = 0.364N$, C_1-C_4 are the constants enabling the satisfaction of boundary conditions.

For initial variables, the solution of a boundary layer can be presented as follows:

$$w_2 = \{\exp(-0.661x/b)\left[C_1\cos(0.364x/b) + C_2\sin(0.364x/b)\right] +$$

$$+ \exp\left(0.661(x - L)/b\right)\left[C_3\cos(0.364(x - L)/b) +\right.$$

$$\left. + C_4\sin(0.364(x - L)/b)\right]\}\, y^2(y - b)^2 \ .$$

The above computations yielded the corrections made in the form of vibrations. The correction made in the periodicity of proper vibrations $\sqrt{\lambda_1}$ can be determined on the basis of conditions of orthogonality that guarantee the solvability of boundary problem (6.58)–(6.60) /or (6.66)–(6.68). Physically, this means that the work of additional forces associated with the discretization on dislocations of structural-ortothropic plate should be equal to zero. Satysfying those conditions guarantees the lack of occurrence of the secular (infinitely increasing for $t \to \infty$) [14], elements in the solution, i.e., we have

$$\int_0^{2\pi}\int_0^{l}\left[-a^2\alpha(\Delta_N - N)\frac{\partial^4 w_{11}}{\partial\xi^4} + \rho(\Delta_N - N)\lambda_0 w_{11} +\right.$$

$$\left. + \lambda_1(1 + \rho N)w_0\right] w_0 d\xi d\eta = 0 \ . \tag{6.91}$$

The above condition yields

$$\lambda_1 = \int_0^{2\pi}\int_0^{l}\left[a^2\alpha(\Delta_N - N)\frac{\partial^4 w_{11}}{\partial\xi^4} -\right.$$

$$\left. - \rho(\Delta_N - N)\lambda_0 w_{11}\right] w_0 d\xi d\eta \left[\int_0^{2\pi}\int_0^{l}(1 + \rho N)w_0^2 w_0 d\xi d\eta\right]^{-1} ,$$

and for initial variables, we have

$$\lambda_1 = \int_0^B \int_0^L \left\{ E_c I \left[\sum_{i=0}^{N-1} \delta(y - ib) - N \right] \frac{\partial^4 w_{11}}{\partial x^4} - \right.$$

$$\left. - \rho_c F \left[\sum_{i=0}^{N-1} \delta(y - ib) - N \right] \lambda_0 w_{11} \right\} w_0 dx dy \times$$

$$\times \left[\int_0^B \int_0^L \left(1 + \frac{\rho_- F}{\rho_0 b} \right) w_0^2 dx dy \right]^{-1} . \tag{6.92}$$

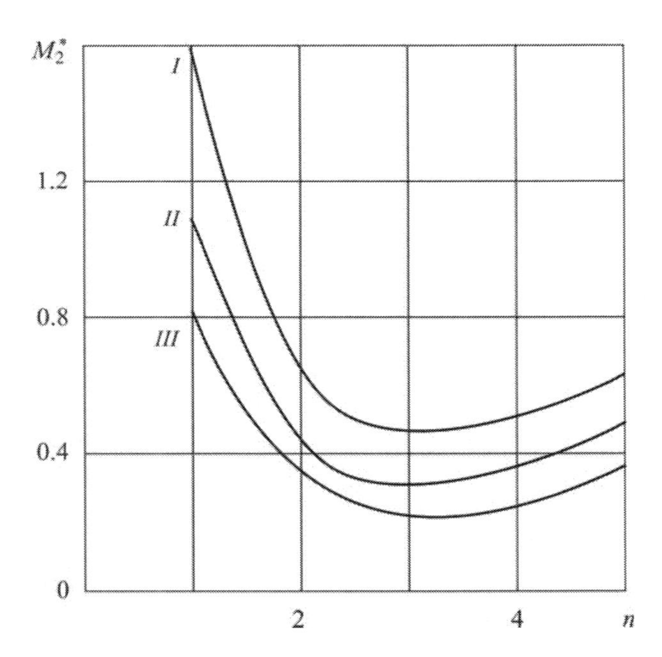

FIGURE 6.5: Results of computations M_2^* with respect to the wave number.

The discussed method allows to determine the decomposition of searched frequencies with the accuracy of any power of ε (determination of the successive recurrent asymptotic process does not pose any difficulties). In practice, however, very often it is enough to limit considerations to the first few elements of particular decompositions.

Finally, we will draw certain conclusions concerning the applicability of the structural-ortothropic theory. For $\varepsilon \ll 1$, we have got the following estimations

$$\lambda_1 \sim \varepsilon^4 \lambda_0 , \quad w_1 \sim \varepsilon^4 w_0 ,$$
$$M_1^{(1)} \sim M_1^{(0)} , \quad M_2^{(1)} \sim M_2^{(0)} , \quad M_{12}^{(1)} \sim \varepsilon^2 M_{12}^{(0)} . \tag{6.93}$$

Hence, the structural-ortothropic scheme allows to satisfactorily estimate lower frequencies and amplitudes of dislocations, whereas moments can be reliably determined only if the discreteness of the decomposition of ribs is taken into consideration. Let us notice that regardless of the importance of the correction made in the form in the area of the boundary (component w_{12} of the boundary layer type is small in comparison to the correction in the basic area w_{11}, $w_{12} \sim \varepsilon^2 w_{11}$), corrections made in bending moments in the vicinity of the boundary are of the same order as in the basic area ($M_1^{(12)} \sim M_1^{(1)}$).

In Figure 6.5, the results of computations M_2^* with respect to the wave number of the form

$$M_2^* = \max |M_2^{(1)}| \Big/ \max |M_2^{(0)}| .$$

$N = 8$ corresponds to curve I, curve II – $N = 12$, curve III – $N = 16$. The values of established parameters are: $L/R = 4.73$, $\rho H = 1$, $\alpha N = 100$.

6.4 Deflections of rectangular plates

Let us proceed to the problem of statics. *The equation of equilibrium* initial for the analysis has the form of

$$D\nabla^4 w + E_c I \sum_{i=0}^{N-1} \delta(y - ib) \frac{\partial^4 w}{\partial x^4} = q_z , \tag{6.94}$$

and *the averaged equation* is as follows

$$D\nabla^4 w_0 + \frac{E_c I}{b} \frac{\partial^4 w_0}{\partial x^4} = q_z , \tag{6.95}$$

where $q_z(x, y)$ is the ordinary surface load.

Boundary conditions for equations (6.94) and (6.95) can be presented in the form of (6.36)–(6.39), (6.44) and (6.47)–(6.50), (6.44). Discretness is taken into account the use of the formulas (6.80), (6.81), for which $\rho_0 = \rho_c = 0$ is assumed.

As an example, we will consider the deflection of the infinite band fixed along the boundary $x = 0, L$ and subjected to the ordinary surface load

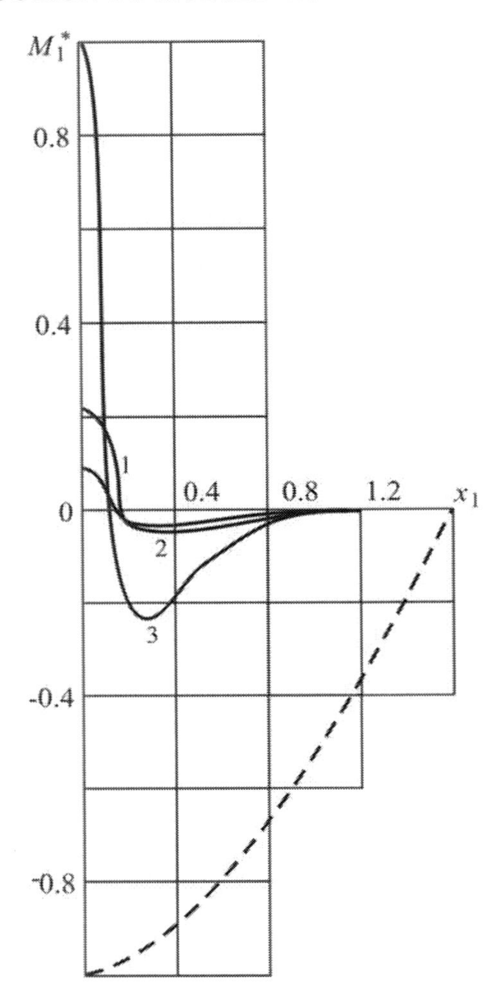

FIGURE 6.6: Numerical computations of relation $M_1^*(x_1)$.

$q_z = q_0 \cos x_1$, where $q_0 = const$, $x_1 = x/L_1$, $L_1 = L/(2\pi)$. The structural-ortothropic solution has the form of

$$w_0 = q_1 L_1^4 (\cos x_1 - 1) \ ,$$

where $q_1 = q_0/D_1$. This solution is constant along axis y, and the quickness of its changes along axis x is determined by the length of the band and L.

The condition of the smooth change of the ortothropic solution can be presented as follows: $\varepsilon = b/L \ll 1$.

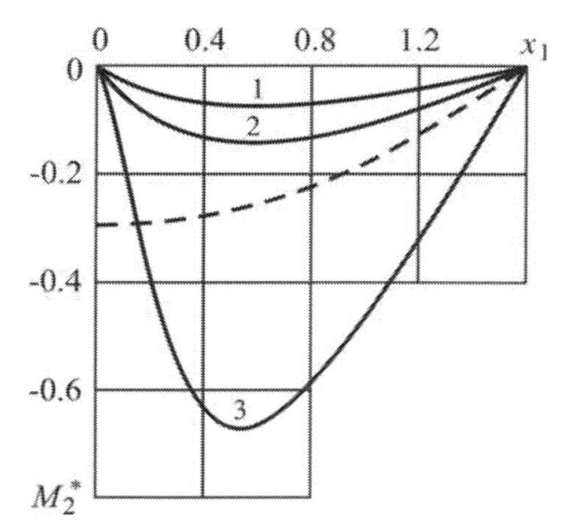

FIGURE 6.7: Numerical computations of relation $M_2^*(x_1)$.

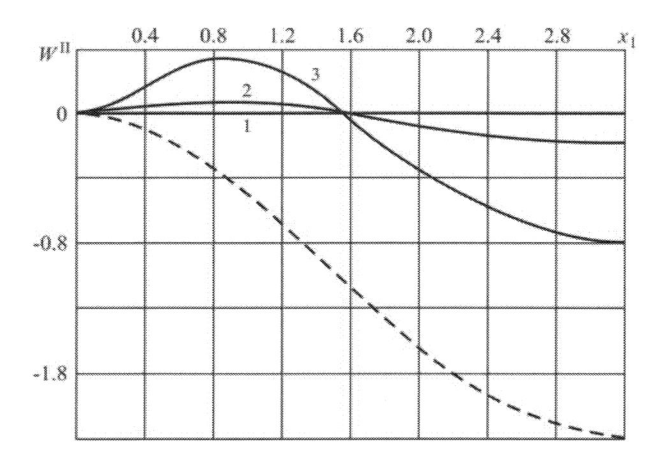

FIGURE 6.8: Numerical computations of relation $W^{II}(x_1)$.

The additional solution on the segment $0 \leq y \leq b$ (further, it is periodically prolonged) has the form

$$w_1 = q_1 \alpha y^2 (y-b)^2 \left[\cos x_1 - \frac{1}{24} e^{-\delta x_1} \times \left(\cos(\delta_2 x_1) + 1.816 \sin(\delta_2 x_1) \right) \right] ,$$

where:

$$\delta_1 = 0.661 \varepsilon^{-1} , \quad \delta_2 = 0.364 \varepsilon^{-1} .$$

Results of numerical computations for different values of ε, α (for $\nu = 0.3$) are shown in a graphic form in Figures 6.6–6.8, where the following notations are assumed:

$$w^* = \frac{w}{q_1 L_1^4}, \quad M_1^* = \frac{M_1}{Dq_1 L_1^2}, \quad M_2^* = \frac{M_2}{Dq_1 L_1^2},$$

and in Figures 6.6, 6.7 $\varepsilon = 0.1$, $y = 0.05b$, whereas in Figure 6.8 $\varepsilon = 0.25$, $y = 0.125b$.

Discontinuous curves represent the structural-orthotropic solution, and continuous curves stem from the amendment resulting from discretization. Values corresponding to curves 1 - 3 are αN equal to 5, 10, 50.

It is visible in Figures 6.6, 6.7 that corrections made in bending moments with high enough relative rigidities of ribs are essential (in our case $\alpha N = 10$). Apart from that, bending moment M_2 is few times smaller than M_1, in the assumed structural-orthotropic scheme of computations.

Dislocations w for $\varepsilon = 0.1$ with the sufficient accuracy are determined according to the structural-orthotropic scheme of computations (the correction stemming from taking discreteness into account is not higher than 2-4%). On increasing the parameter ε however, the influence of taking the process of discretization into consideration becomes essential also in this case (Figure 6.8).

Chapter 7

Chaos foresight

There exists an extensive research devoted to the analysis of low- and high-dimensional systems with friction. Some fundamental problems of nonsmooth dynamical systems with friction are addressed for example, in references [38, 39, 69, 93, 96, 137, 158]. However we are not going to cite many of them, but a reader may go through over 400 bibliography items devoted to nonsmooth regular and chaotic dynamics included in a monograph by Awrejcewicz and Lamarque [40]. Beginning from the pioneering work of Melnikov [115], the Melnikov-like approaches spread into different branches of science. We briefly address the Melnikov-like techniques to predict the onset of chaos in systems governed by ODEs or maps. For example, in reference [49], Melnikov function (integral) is successfully applied in fluid particle kinematics analysis in weakly perturbed integrable dynamical systems. The method proposed by Melnikov allows also to predict the onset of chaos via homoclinic (or heteroclinic) tangencies in periodically perturbed $2D$ flows (see also [80, 177]).

An existence of transversal homoclinic orbits of systems of singularly perturbed two first-order differential equations using exponential dichotomies is illustrated and discussed in [175]. The exponential dichotomy and a unified geometrical approach to calculate Melnikov vector function assuming the existence of transversal homoclinic points for high-dimensional maps with a saddle connection are studied in [159].

Splitting of separatrices for high-frequency perturbations of a planar Hamiltonian system using Melnikov technique is also examined (see [74]). In reference [155], it is shown that although the original Melnikov's approach correctly estimates parameter values for bifurcation and transverse intersections of separatrices and manifolds, it does not correctly approximate solutions in a neighbourhood of the associated fixed point of the homoclinic orbit. In the latter paper, a multiple scales technique, in which inner solutions are matched with a regular outer solution, is proposed. Finally, we finish our brief review of recent modifications and for extensions of Melnikov's original work addressing results obtained by Salam [67]. In the mentioned reference, an extension of Melnikov approach to a class of highly dissipative systems is proposed, and obtained results are illustrated with the use of numerical simulations. However, mainly Gruendler's work [81] served for us as the basic reference to start with a construction of a homoclinic orbit in our $4D$ mechanical system perturbed by friction and harmonic excitation, and then to derive the associated

Melnikov's function. It is worth noticing that an important opened problem of Melnikov approach relies on its extension into analysis of higher order dynamical systems. This problem seems to be unsolved since it is difficult to establish a priori a homoclinic orbit associated with a highly dimensional system considered.

It is needless to say that a prediction of chaos in an analytical way in nonsmooth objects modeled as systems in R^4 plays a crucial role for both theoretical and applicable reasons. A key point in the research carried out in this direction is the paper by Awrejcewicz and Holicke [37], where a chaotic threshold for both smooth and stick-slip chaotic behaviour in one degree-of-freedom system with friction is obtained using directly Melnikov technique. On the other hand, it is impossible to extend directly the original Melnikov method devoted to the analysis of an analytic system in R^4. Therefore, the authors apply Gruendler extension of Melnikov method to R^4, which is further referred to as Melnikov-Gruendler approach. However, in the cited Gruendler's work [81] again an emphasis if put on C^2 systems. On the contrary, in our research we extend the results obtained earlier (see [37]) to R^4. Although we do not give rigorous definitions and proofs of a C^n vector field on R^n, we show the computations of related integrals yielding a sought chaotic threshold defined by the appropriate Melnikov function (see [30], [31]).Furthermore, a reduction of obtained Melnikov integrals to those associated with previously considered one-degree-of-freedom mechanical system and illustrated numerical examples indicate a validity of our approach.

7.1 The analyzed system

The analyzed mechanical object consists of two stiff bodies with the masses m coupled via linear and nonlinear springs in the way shown in Figure 7.1. Note that when the system is autonomous, i.e., $\Gamma = 0$, self-excited oscillations appear, which are generated by frictional characteristics. The latter ones have a decreasing part versus relative velocity between both bodies and the tape moving with a constant velocity w. Although this problem belongs to classical ones and has been studied by vast number of researchers, an attempt to formulate threshold for chaos occurence in the analytical way failed. In what follows we show how to solve this problem using Melnikov technique applied to our discontinuous system. It is also recommended to be familiar with the reference [37], where a similar approach is applied to predict chaos in a similar system, but with one degree-of-freedom.

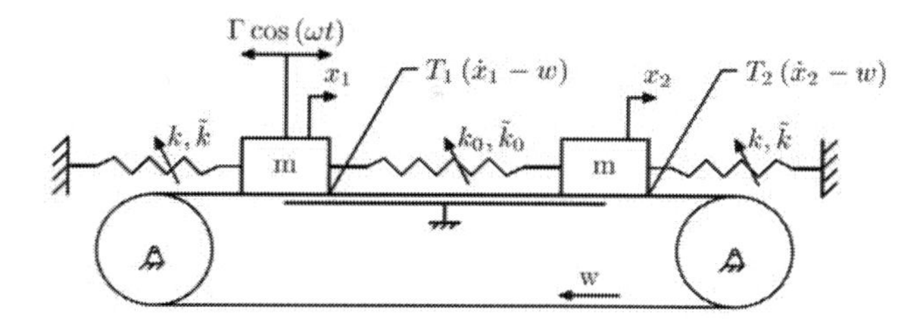

FIGURE 7.1: The analyzed system.

Dynamics of our system is governed by the following ODEs:

$$\begin{cases} \dot{x}_1 & = p_1/m \\ \dot{p}_1 & = kx_1 - \tilde{k}x_1^3 + k_0\left(x_1 - x_2\right) - \tilde{k}_0\left(x_1 - x_2\right)^3 + \\ & +\varepsilon_1 \Gamma \cos\left(\omega t\right) - \varepsilon_2 T_1\left(p_1/m - w\right) \\ \dot{x}_2 & = p_2/m \\ \dot{p}_2 & = kx_2 - \tilde{k}x_2^3 - k_0\left(x_1 - x_2\right) + \tilde{k}_0\left(x_1 - x_2\right)^3 - \varepsilon_3 T_2\left(p_2/m - w\right) \end{cases} \tag{7.1}$$

where:

$$T_i\left(p_i/m - w\right) = T_{i0}\mathrm{sgn}\left(p_i/m - w\right) - B_{i1}\left(p_i/m - w\right) + B_{i2}\left(p_i/m - w\right)^3 \tag{7.2}$$

and w is the tape velocity, whereas $B_{11}, B_{12}, B_{21}, B_{22}, T_{10}, T_{20}$ are friction coefficients. Introducing the following scaling

$$t \to t\sqrt{\frac{k}{m}}, \quad x = x_1\sqrt{\frac{\tilde{k}}{k}}, \quad u = p_1\sqrt{\frac{\tilde{k}}{mk^2}}, \quad y = x_2\sqrt{\frac{\tilde{k}}{k}}, \quad v = p_2\sqrt{\frac{\tilde{k}}{mk^2}} \tag{7.3}$$

and the following relations

$$k_0 = \xi k, \quad \tilde{k}_0 = \xi\tilde{k} \quad \text{where}\, \xi \geq 0, \tag{7.4}$$

the analyzed ODEs are cast in the nondimensional form

$$\begin{pmatrix} \dot{x} \\ \dot{u} \\ \dot{y} \\ \dot{v} \end{pmatrix} = \begin{pmatrix} u \\ x - x^3 + f_\xi\left(x, y\right) \\ v \\ y - y^3 - f_\xi\left(x, y\right) \end{pmatrix} + \begin{pmatrix} 0 \\ \varepsilon_1 \Gamma' \cos\left(\omega' t\right) - \varepsilon_2 T_1'\left(u - w'\right) \\ 0 \\ -\varepsilon_3 T_2'\left(v - w'\right) \end{pmatrix}, \tag{7.5}$$

where:

$$T_1'\left(u - w'\right) = T_{10}'\mathrm{sgn}\left(u - w'\right) - B_{11}'\left(u - w'\right) + B_{12}'\left(u - w'\right)^3, \tag{7.6}$$

$$T_2'\left(u - w'\right) = T_{20}'\mathrm{sgn}\left(v - w'\right) - B_{21}'\left(v - w'\right) + B_{22}'\left(v - w'\right)^3, \tag{7.7}$$

$$\Gamma' = \Gamma\sqrt{\frac{\tilde{k}}{k^3}}, \; \omega' = \omega\sqrt{\frac{m}{k}}, \; T'_{i0} = T_{i0}\sqrt{\frac{\tilde{k}}{k^3}},$$
$$B'_{i1} = \frac{B_{i1}}{\sqrt{mk}}, \; B'_{i2} = \frac{k^2 B_{i2}}{\sqrt{m^3\tilde{k}^3}}, \; w' = w\sqrt{\frac{\tilde{k}m}{k^2}}, \tag{7.8}$$
$$f_\xi(x,y) = \xi(x-y) - \xi(x-y)^3 .$$

7.2 Melnikov-Gruendler's approach

The method applied in the paper is due to Gruendler [81]. Although the theory is a generalization to a non-Hamiltonian case we apply it to a Hamiltonian one. Here we consider a mechanical system governed by the equation:

$$\dot{x}(t) = f(x(t)) + h(x(t), t, \varepsilon), \tag{7.9}$$

where $f : R^4 \to R^4$ is a Hamiltonian vector field and $h : R^4 \times R \times B \subset R^4 \to R^4$ is periodic in t with frequency ω and satisfies $h(x(t), t, 0) = 0$. For $\varepsilon = 0$, we obtain the unperturbed system. Let the unperturbed system possess a homoclinic orbit $\gamma(t)$ to a hyperbolic point at the origin. The variational equation along $\gamma(t)$ is the following:

$$\dot{y}(t) = Df(\gamma(t))\, y(t). \tag{7.10}$$

We seek a fundamental solution $\{\psi^{(1)}(t), \psi^{(2)}(t), \psi^{(3)}(t), \psi^{(4)}(t)\}$ to the equation (7.10) possessing some special properties. The properties are the following:

1. $\psi^{(4)}(t) = \dot{\gamma}(t)$[1]
2. Initial vectors $\psi^{(i)}(0)$ span a vector space;
3. Each $\psi^{(i)}(t)$ has the exponential behavior as $t \to \pm\infty$. Namely:
   $\psi^{(i)}(t) \sim t^{k_i} e^{\lambda_i t} v^{(i)}$ as $t \to +\infty$. $k_i \in N$
   $\psi^{(i)}(t) \sim t^{k_{\sigma(i)}} e^{\lambda_{\sigma(i)} t} \bar{v}^{(i)}$ as $t \to -\infty$. $k_{\sigma(i)} \in N$
   where σ is a permutation on four symbols and $\{\lambda_1, \lambda_2, \lambda_3, \lambda_4\}$ are the eigenvalues of $Df(0)$;
4. Signs of $\Re(\lambda_i)$ and $\Re(\lambda_{\sigma(i)})$ in the exponential behavior have to be such that:

$$\psi^{(1)}(t) = \begin{cases} \Re(\lambda_i) > 0 \\ \Re(\lambda_{\sigma(i)}) > 0 \end{cases} \tag{7.11}$$

[1] It is easy to show that $\dot{\gamma}(t)$ satisfies the equation (7.10)

$$\psi^{(2)}(t) = \begin{cases} \Re(\lambda_i) > 0 \\ \Re(\lambda_{\sigma(i)}) < 0 \end{cases} \tag{7.12}$$

$$\psi^{(3)}(t) = \begin{cases} \Re(\lambda_i) < 0 \\ \Re(\lambda_{\sigma(i)}) < 0 \end{cases} \tag{7.13}$$

$$\psi^{(4)}(t) = \begin{cases} \Re(\lambda_i) < 0 \\ \Re(\lambda_{\sigma(i)}) > 0 \end{cases} \tag{7.14}$$

Next, we define an index set I by $i \in I$ if and only if $\psi^{(i)}(t) \xrightarrow{t\to\pm\infty} \infty$. Moreover we form the functions:

$$D(t) = \det\left\{\psi^{(1)}(t), \psi^{(2)}(t), \psi^{(3)}(t), \psi^{(4)}(t)\right\} \mathrm{e}^{-\int_0^t \nabla f(\gamma(s))ds}. \tag{7.15}$$

Since f is a Hamiltonian vector field we obtain $\nabla f = 0$. Thus, the function $D(t)$ reduces to simpler form:

$$D(t) = \det\left\{\psi^{(1)}(t), \psi^{(2)}(t), \psi^{(3)}(t), \psi^{(4)}(t)\right\}. \tag{7.16}$$

Let $K_{ij}(t, t_0)$[2] denote the result of replacing $\psi^{(i)}(t)$ in $D(t)$ by $\frac{\partial h(\gamma(t), t+t_0, 0)}{\partial \varepsilon_j}$. We define the function:

$$M_{ij}(t_0) = -\int_{-\infty}^{\infty} K_{ij}(t, t_0)\, dt, \quad i \in I \tag{7.17}$$

The function above measures the separation of stable and unstable manifolds. Melnikov function is defined as follows:

$$M(t_0) = \sum_{j=1}^{4} M_{ij}(t_0)\, \varepsilon_j, \quad i \in I \tag{7.18}$$

7.3 Melnikov-Gruendler function

Let us denote by $\gamma(t)$ the homoclinic orbit of the point {0,0,0,0}. It has (in our case) the following form

$$\gamma(t) = \begin{pmatrix} q(t) \\ \dot{q}(t) \\ -q(t) \\ -\dot{q}(t) \end{pmatrix}, \quad \text{where} \quad q(t) = \sqrt{\frac{2(1+2\xi)}{1+8\xi}} \operatorname{sech}\left(t\sqrt{1+2\xi}\right). \tag{7.19}$$

[2]This function represents the projection onto the direction of $\psi^{(i)}(t)$ of the ε_j of the h evaluated along $\gamma(t)$.

The linearized system of the unperturbed equations (7.5) in the vicinity of the homoclinic orbit $\gamma(t)$ reads

$$\dot{\psi} = F(t)\,\psi, \tag{7.20}$$

where

$$F(t) = \begin{pmatrix} 0 & 1 & 0 & 0 \\ 1+\xi-3(1+4\xi)\,q^2(t) & 0 & -\xi+12\xi q^2(t) & 0 \\ 0 & 0 & 0 & 1 \\ -\xi+12\xi q^2(t) & 0 & 1+\xi-3(1+4\xi)\,q^2(t) & 0 \end{pmatrix}.$$

Next, we obtain the following equations

$$\begin{cases} \ddot{\psi}_1 = \left(1+\xi-3(1+4\xi)\,q^2(t)\right)\psi_1 + \xi\left(12q^2(t)-1\right)\psi_3 \\ \ddot{\psi}_3 = \left(1+\xi-3(1+4\xi)\,q^2(t)\right)\psi_3 + \xi\left(12q^2(t)-1\right)\psi_1 \end{cases}. \tag{7.21}$$

A combination of equations (7.21) yields

$$\ddot{\phi}_1 = (1+2\xi)\left(1-6\mathrm{sech}^2\left(t\sqrt{1+2\xi}\right)\right)\phi_1, \quad \phi_1 \equiv \psi_1 - \psi_3\,. \tag{7.22}$$

It is easy to see that $\psi^{(4)}(t) = \dot{\gamma}(t)$ satisfies the above equation. In order to find another solution, the following substitution is applied: $\dot{q}(t) \to r(t)\,\dot{q}(t)$. Since $\dot{q}(t)$ is a solution to (7.22) one gets

$$\ddot{r}\dot{q} + 2\dot{r}\ddot{q} = 0\,. \tag{7.23}$$

Integrating of (7.23), owing to obtained results, the solution reads

$$\phi_1(t) = r(t)\,\dot{q}(t) = \left(\frac{3}{4}C_1 t - \frac{1}{2}C_1\mathrm{ctgh}(t) + \frac{1}{8}C_1\sinh(2t) + C_2\right)\dot{q}(t)\,. \tag{7.24}$$

The above solution possesses the following asymptotics $\phi_1(t) \xrightarrow{t\to\pm\infty} \mathrm{e}^{\pm t\sqrt{1+2\xi}}$, so according to (7.12), we obtain the next solution $\psi^{(2)}(t)$. Next, summing up equations (7.21), we obtain

$$\ddot{\phi}_2 = g(t,\xi)\,\phi_2, \quad g(t,\xi) = 1 - 6\frac{1+2\xi}{1+8\xi}\mathrm{sech}^2\left(t\sqrt{1+2\xi}\right), \quad \phi_2 \equiv \psi_1 + \psi_3\,. \tag{7.25}$$

Suppose that $y_1(t)$ is a solution of the above equation. Then, $y_2(t) = y_1(-t)$ is also the solution because $g(t,\xi)$ is an even function with respect to t.

In our case, a perturbation term associated with (7.5) reads

$$h(t,\varepsilon) = \{0, \varepsilon_1\Gamma'\cos(\omega' t) - \varepsilon_2 T_1'(u-w'), 0, -\varepsilon_3 T_2'(v-w')\}^T. \tag{7.26}$$

Therefore, one gets

$$\frac{\partial h(\gamma(t), t+t_0, 0)}{\partial\varepsilon_1} = \begin{pmatrix} 0 \\ \Gamma'\cos(\omega'(t+t_0)) \\ 0 \\ 0 \end{pmatrix}, \tag{7.27}$$

$$\frac{\partial h(\gamma(t), t+t_0, 0)}{\partial \varepsilon_2} = \begin{pmatrix} 0 \\ -T_1'(\dot{q}(t) - w') \\ 0 \\ 0 \end{pmatrix},$$

$$\frac{\partial h(\gamma(t), t+t_0, 0)}{\partial \varepsilon_3} = \begin{pmatrix} 0 \\ 0 \\ 0 \\ -T_2'(\dot{q}(t) - w') \end{pmatrix}, \frac{\partial h(\gamma(t), t+t_0, 0)}{\partial \varepsilon_4} = \begin{pmatrix} 0 \\ 0 \\ 0 \\ 0 \end{pmatrix}. \tag{7.28}$$

Observe that only $K_{2j}(t, t_0)$ should be found, since $\psi^{(2)}(t) \xrightarrow{t \to \pm\infty} \infty$. First, K_{21} is found

$$\begin{aligned} K_{21}(t, t_0) &= \det \begin{pmatrix} y_1 & 0 & y_2 & \dot{q} \\ \dot{y}_1 & \Gamma' \cos(\omega'(t+t_0)) & \dot{y}_2 & \ddot{q} \\ y_1 & 0 & y_2 & -\dot{q} \\ \dot{y}_1 & 0 & \dot{y}_2 & -\ddot{q} \end{pmatrix}. \\ &= 2\Gamma' \dot{q} \cos(\omega'(t+t_0))(y_1\dot{y}_2 - \dot{y}_1 y_2) \\ &= 2(y_1\dot{y}_2 - \dot{y}_1 y_2)\Gamma' \dot{q} \cos(\omega'(t+t_0)) \end{aligned} \tag{7.29}$$

Secondly, K_{22} and K_{23} are found

$$K_{22}(t, t_0) = -2(y_1\dot{y}_2 - \dot{y}_1 y_2)\dot{q}T_1'(\dot{q} - w'),$$

$$K_{23}(t, t_0) = 2(y_1\dot{y}_2 - \dot{y}_1 y_2)\dot{q}T_2'(\dot{q} - w'). \tag{7.30}$$

Note that in each K_{2i} we have the same term $(y_1\dot{y}_2 - \dot{y}_1 y_2)$. It can be shown that $\Omega(\xi) = (y_1\dot{y}_2 - \dot{y}_1 y_2)$, i.e., this function is time-independent. Hence we obtain

$$K_{21}(t, t_0) = 2\Omega(\xi)\Gamma' \dot{q}(t)\cos(\omega'(t+t_0)), \tag{7.31}$$

$$K_{22}(t, t_0) = -2\Omega(\xi)\dot{q}(t)T_1'(\dot{q} - w'), \tag{7.32}$$

$$K_{23}(t, t_0) = 2\Omega(\xi)\dot{q}(t)T_2'(\dot{q} - w'). \tag{7.33}$$

According to (7.17) we get

$$M_{21}(t_0) = -2\sqrt{2}\Gamma'\Omega(\xi)\pi\omega'\sqrt{\frac{1+2\xi}{1+8\xi}}\operatorname{sech}\left(\frac{\pi\omega'}{2\sqrt{1+2\xi}}\right)\sin(\omega' t_0), \tag{7.34}$$

$$M_{22}(t_0) = 2\Omega(\xi)\int_{-\infty}^{\infty} \dot{q}T_1'(\dot{q}-w')\,dt = 2\Omega(\xi)\,T_{10}'\int_{-\infty}^{\infty} \dot{q}\,\mathrm{sgn}(\dot{q}-w')\,dt$$
$$-2\Omega(\xi)\,B_{11}'\int_{-\infty}^{\infty} \dot{q}(\dot{q}-w')\,dt + 2\Omega(\xi)\,B_{12}'\int_{-\infty}^{\infty} \dot{q}(\dot{q}-w')^3\,dt$$
$$= -\frac{8}{3}\Omega(\xi)\,B_{11}'\frac{1+2\xi}{1+8\xi}\sqrt{1+2\xi} + \frac{32}{35}\Omega(\xi)\,B_{12}'\frac{(1+2\xi)^3}{(1+8\xi)^2}\sqrt{1+2\xi}$$
$$+8\Omega(\xi)\,B_{12}'w'^2\frac{1+2\xi}{1+8\xi}\sqrt{1+2\xi} + 2\Omega(\xi)\,T_{10}'\int_{-\infty}^{\infty} \dot{q}\,\mathrm{sgn}(\dot{q}-w')\,dt \tag{7.35}$$

Consider the last integral in the above term:

$$\int_{-\infty}^{\infty} \dot{q}(t)\,\mathrm{sgn}(\dot{q}(t)-w')\,dt = \frac{1+2\xi}{1+8\xi}\sqrt{1+2\xi}\int_{-\infty}^{\infty} \dot{\tilde{q}}(t)\,\mathrm{sgn}\left(\dot{\tilde{q}}(t)-\tilde{w}'\right)dt, \tag{7.36}$$

where $\dot{\tilde{q}}(t) = -\sqrt{2}\mathrm{sech}(t)\,\mathrm{tgh}(t)$ and $\tilde{w}' = w'\frac{\sqrt{1+8\xi}}{1+2\xi}$.
Assume first that $\tilde{w}' > 1/\sqrt{2}$, then

$$\int_{-\infty}^{\infty} \dot{\tilde{q}}(t)\,\mathrm{sgn}\left(\dot{\tilde{q}}(t)-\tilde{w}'\right)dt = \mathrm{sgn}\left(-\tilde{w}'\right)\int_{-\infty}^{\infty} \dot{\tilde{q}}(t)\,dt = 0\,. \tag{7.37}$$

Assume now that $\tilde{w}' < 1/\sqrt{2}$, then

$$\int_{-\infty}^{\infty} \dot{\tilde{q}}\,\mathrm{sgn}\left(\dot{\tilde{q}}-\tilde{w}'\right)dt =$$
$$-\int_{-\infty}^{t_1} \dot{\tilde{q}}dt + \int_{t_1}^{t_2} \dot{\tilde{q}}dt - \int_{t_2}^{\infty} \dot{\tilde{q}}dt = 2\sqrt{2}\left(\mathrm{sech}(t_2) - \mathrm{sech}(t_1)\right)\,. \tag{7.38}$$

where:

$$t_1 = \ln\left(\frac{1}{\tilde{w}'}\sqrt{1+\sqrt{1-2\tilde{w}'^2}}\left(1-\sqrt{\frac{1}{2}+\frac{1}{2}\sqrt{1-2\tilde{w}'^2}}\right)\right),$$

$$t_2 = \ln\left(\frac{1}{\tilde{w}'}\sqrt{1-\sqrt{1-2\tilde{w}'^2}}\left(1-\sqrt{\frac{1}{2}-\frac{1}{2}\sqrt{1-2\tilde{w}'^2}}\right)\right).$$

Substituting the obtained result we find

$$\frac{1+8\xi}{4(1+2\xi)^{3/2}}M_{22}(t_0) = -\frac{2}{3}\Omega(\xi)\,B_{11}' + 2\Omega(\xi)\,B_{12}'\left(w'^2 + \frac{4(1+2\xi)^2}{35(1+8\xi)}\right)$$
$$+\Omega(\xi)\,T_{10}'\sqrt{2}\theta\left(\frac{1}{\sqrt{2}}-\tilde{w}'\right)\left(\mathrm{sech}(t_2)-\mathrm{sech}(t_1)\right), \tag{7.39}$$

where $\theta(x)$ is Heaviside's function. Similarly, we obtain function $M_{23}(t_0)$. Finally, we find Melnikov-Gruendler function

$$
\begin{aligned}
M(t_0) = & -\sqrt{2}\Gamma'\pi\omega'\operatorname{sech}\left(\frac{\pi\omega'}{2\sqrt{1+2\xi}}\right)\sin(\omega' t_0) - \frac{4(1+2\xi)}{3\sqrt{1+8\xi}}\left(B'_{11}-B'_{21}\right) \\
& + 4\sqrt{\frac{1+2\xi}{1+8\xi}}\left(B'_{12}-B'_{22}\right)\left(w'^2 + \frac{4(1+2\xi)^{5/2}}{35(1+8\xi)}\right) \\
& + 2\sqrt{2}\left(T'_{10}-T'_{20}\right)\frac{1+2\xi}{\sqrt{1+8\xi}}\left(\operatorname{sech}(t_2)-\operatorname{sech}(t_1)\right).
\end{aligned}
\tag{7.40}
$$

7.4 Numerical results

It is clear that having analytical form of Melnikov function, various control parameters can be taken to show regular and/or chaotic dynamics. Let us take, following the paper [37], two of them i.e., $\{\Gamma', w'\}$ (see Figure 7.2).

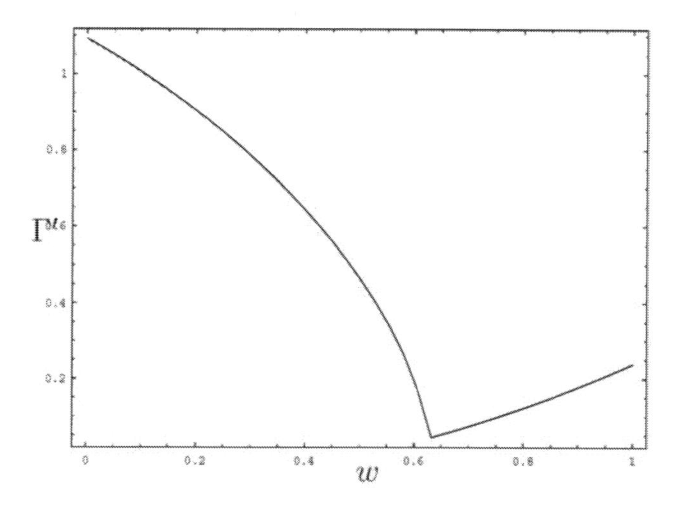

FIGURE 7.2: The threshold curve.

The obtained curves define a chaotic threshold. Namely, above the mentioned curves chaos is expected, whereas below a regular behaviour is expected. The cusp corresponds to a switch between smooth and stick-slip dynamics. Note that the switch takes place exactly for the tape velocity value

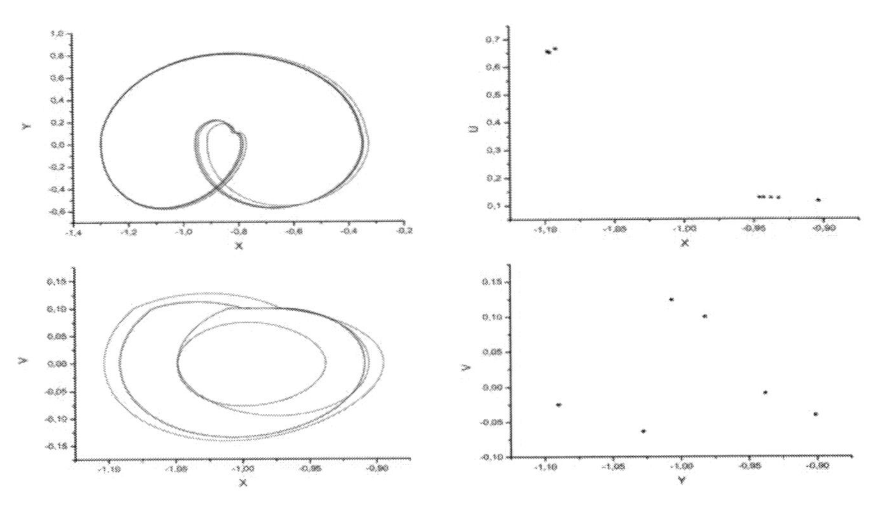

FIGURE 7.3: Phase portraits and Poincaré maps ($\Gamma' = 0.98, w' = 0.1$).

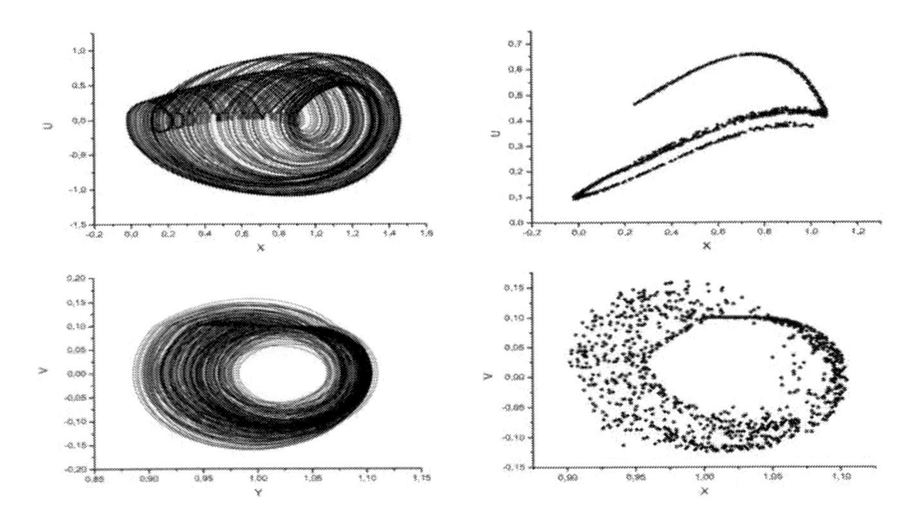

FIGURE 7.4: Phase portraits and Poincaré maps ($\Gamma' = 1.02, w' = 0.1$).

$w' = (1+2\xi) \, / \, \left(\sqrt{2}\sqrt{1+8\xi}\right)$. One may state the following question. Why extra numerical examples are added having the analytical construction of the Melnikov's function? Some of the reasons are given below:

1. It may happen that the obtained chaotic set is unstable, and hence it is impossible to show it applying a standard initial value problem.

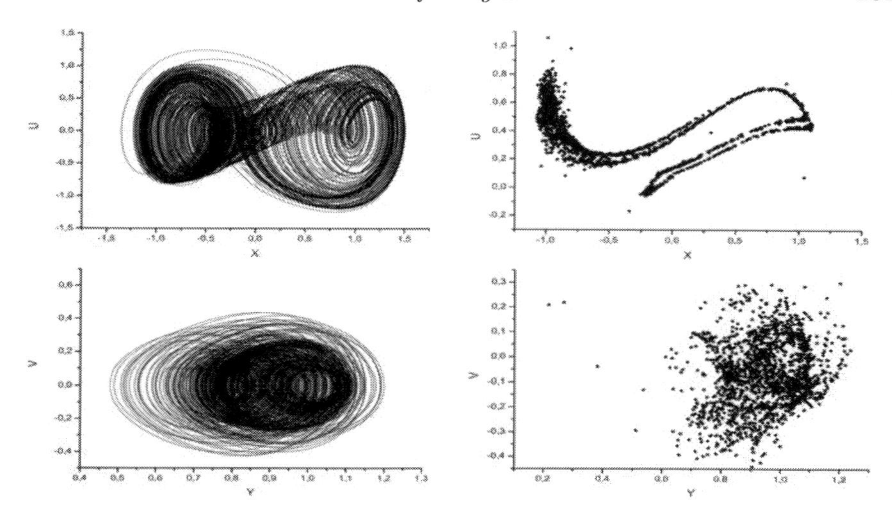

FIGURE 7.5: Phase portraits and Poincaré maps ($\Gamma' = 1.1, w' = 0.1$).

2. Numerical tests allow for estimation of validity of our perturbational approach.

3. Numerical simulations can verify smooth and stick-slip chaotic dynamics. Note that in general approach given in reference [81], the introduced main theorem works only for C^2 systems.

In our numerical simulations we have taken $T_{10} = 0.45, T_{20} = 0.05, B_{11} = 0.25, B_{21} = 0.15, B_{12} = 0.2, B_{22} = 0.1, \xi = 0.1$. The results are presented in the form of phase portraits $(x, y), (y, v)$ and Poincaré maps which correspond to the first block and the second one, respectively. For $\{\Gamma' = 0.98, w' = 0.1\}$, we obtain periodic orbits (see Figure 7.3). Observe that in these figures there are cusps which correspond to a sign change of the relative velocity. Moreover, there are horizontal parts corresponding to the stick phases during the motion. While we cross the threshold curve we arrive at point $\{\Gamma' = 1.02, w' = 0.1\}$, where qualitatively different behavior is observed (see Figure 7.4). We can still observe stick phases during the motion (especially in (y, v) Poincaré section) and many cusps. Increasing Γ' to 1.1 chaotic behaviour is observed (see Figure 7.5).

To conclude, Melnikov integrals are computed for qualitatively different cases i.e., for regular and discontinous onset of chaos, and the analytical prediction of chaotic threshold is verified by numerical computations.

Chapter 8

Continuous approximation of discontinuous systems

There are many physical systems where mathematical modeling leads to discontinuous dynamical systems that switch between different states, and dynamics of each state is given by a different set of differential equations [38, 39, 40, 55, 102]. From the mathematical point of view, several ways exist to handle such discontinuous differential equations. One way is to use the theory of differential inclusions [64, 70]. Another way is a continuous approximation of discontinuities to get smooth differential equations [41].

In this chapter we follow the second way. We consider differential equations with discontinuous nonlinearities. Then, we continuously approximate those nonlinearities by using mono-parametric families of continuous functions. The parameter is $\varepsilon > 0$, and ε tends to 0. To study the dynamics of the approximated equation, we split it into variables near and far from the discontinuities. We scale the variables near discontinuities to get singular differential equations. Then, we use results from the theory of singularly perturbed differential equations, like Tichonov theorem [170]. Finally, we combine dynamics of singularly perturbed and normal differential equations to get the dynamics of the original approximated differential equations. We use this method in the study of persistence and stability of a periodic solution of the discontinuous systems under the continuous approximation. Summarizing, the method is based on construction of Poincaré maps along periodic solutions of discontinuous systems and their continuous approximations. Then Tichonov theorem is applied to study the relationship between those Poincaré maps. Some transversal assumptions are needed to derive those Poincaré maps. This chapter follows results obtained in reference [42].

8.1 An illustrative example

Let us consider a one-degree-of-freedom mechanical system (see Figure 8.1) consisting of a mass m oscillating on a belt which moves with constant velocity v_b, and which is connected to a nonlinear oscillator with the elastic support characterized by constants k_1 and k_2.

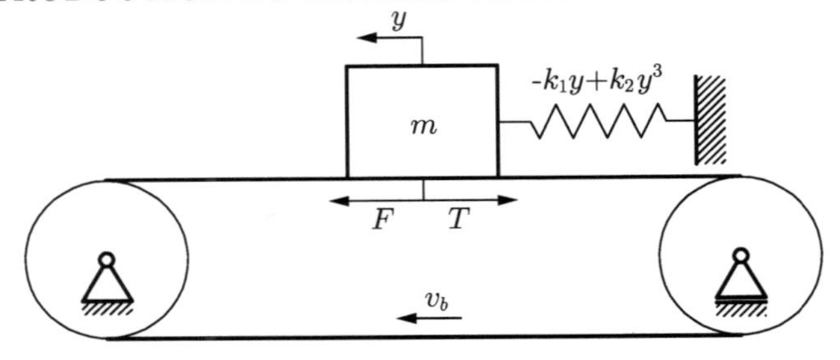

FIGURE 8.1: The considered one-degree-of-freedom mechanical system with friction.

Such a model is governed by the following equation of motion

$$m\ddot{y} - k_1 y + k_2 y^3 - T = 0\,, \tag{8.1}$$

where friction force T is applied in the following form

$$T = -\frac{\mu_0}{1 + |\dot{y} - v_b|}\operatorname{sgn}(\dot{y} - v_b)\,.$$

Considering $m = k_1 = k_2 = v_b = 1$ and static friction coefficient $\mu_0 = 0.2$, the following simplified equation governs dynamics of our dynamical system:

$$\ddot{y} - y + y^3 - \frac{0.2}{1 + |\dot{y} - 1|}\operatorname{sgn}(1 - \dot{y}) = 0\,. \tag{8.2}$$

By putting $\dot{y} = -w + 1$, we get the system

$$\begin{aligned} \dot{y} &= -w + 1\,, \\ \dot{w} &= y^3 - y - \frac{0.2}{1 + |w|}\operatorname{sgn} w\,. \end{aligned} \tag{8.3}$$

Now we approximate (8.3) by the system

$$\begin{aligned} \dot{y} &= -w + 1\,, \\ \dot{w} &= y^3 - y - f_\varepsilon(w) \end{aligned} \tag{8.4}$$

for $\varepsilon > 0$ small and a function $f_\varepsilon : \mathbb{R} \to \mathbb{R}$ defined as follows

$$f_\varepsilon(w) := \begin{cases} \dfrac{0.2}{1 + |w|}\operatorname{sgn} w & \text{for} \quad |w| \geq \varepsilon\,, \\[2ex] \dfrac{0.2w}{(1 + \varepsilon)\varepsilon} & \text{for} \quad |w| \leq \varepsilon\,. \end{cases}$$

Then (8.4) for $w \geq \varepsilon$ has the form

$$\begin{aligned} \dot{y} &= -w + 1\,, \\ \dot{w} &= y^3 - y - \frac{0.2}{1 + w}\,, \end{aligned} \tag{8.5}$$

which is (8.3) for $w > 0$. For $|w| \leq \varepsilon$, we take $w = \varepsilon v$, $|v| \leq 1$ and (8.4) has the form

$$\begin{aligned} \dot{y} &= -\varepsilon v + 1\,, \\ \varepsilon \dot{v} &= y^3 - y - \frac{0.2}{1+\varepsilon} v\,. \end{aligned} \tag{8.6}$$

If we check the vector field of (8.5) (see Fig. 8.2) near the line $w = 0$ for $w > 0$, we see that for $y < y_0$ the line $w = 0$ is attracting, and for $y > y_0$ the line $w = 0$ is repelling. Of course, the variable y is increasing. Here $y_0^3 - y_0 = 0.2$, $y_0 = 1.08803$.

Now we can check by the program *Mathematica* that the solution of (8.5) with the initial conditions $y(0) = y_0$, $w(0) = 0$ hits the line $w = 0$ at the time $t_0 = 7.117155465$ in $y(t_0) := \bar{y}_0 = 0.3218918837 \in (-y_0, y_0)$. Of course, for the discontinuous system (8.3), we get a periodic solution $p_0(t)$ starting from the point $(y_0, 0)$, which is infinitely stable, i.e., all solutions starting near periodic solution $p_0(t)$ collapse after a finite time to $p_0(t)$. We expect that its approximation (8.4) would also possess a unique periodic solution near $p_0(t)$ with a rapid attractivity. This phenomenon is numerically demonstrated in [41] for a two-degrees-of-freedom autonomous system with friction. To show analytically this property for our simple system (8.4), we consider the dynamics of a Poincaré map of (8.4) near periodic orbit $p_0(t)$ of (8.3). For the construction of this Poincaré map, we take the interval

$$I := [y_0 - \delta, y_0 + \delta]$$

for a fixed small $\delta > 0$. For any $\bar{y} \in I$, we consider the solution $(y(t), w(t))$ of (8.5) with the initial value conditions $y(0) = \bar{y}$, $w(0) = \varepsilon$. Then for a small $\delta > 0$, there is $\bar{t} \sim t_0$ such that $w(\bar{t}) = \varepsilon$. We put

$$\Phi_\varepsilon(\bar{y}) := y(\bar{t})\,.$$

We get a mapping $\Phi_\varepsilon : I \to \bar{I}$ for $\bar{I} = [\bar{y}_0 - \bar{\delta}, \bar{y}_0 + \bar{\delta}]$ and $\bar{\delta} = \bar{\delta}(\delta)$ is small.

Concerning (8.6), we put $\varepsilon = 0$ and we get

$$V(y) := 5(y^3 - y)$$

as a solution of the equation

$$f(V, y) := y^3 - y - 0.2V = 0\,.$$

Moreover, we have

$$\frac{\partial f}{\partial V}(V(y), y) < -0.2 < 0\,.$$

Let us consider the rectangle

$$Q := [-y_0 - 1, y_0 + 1] \times [-1, 1]\,.$$

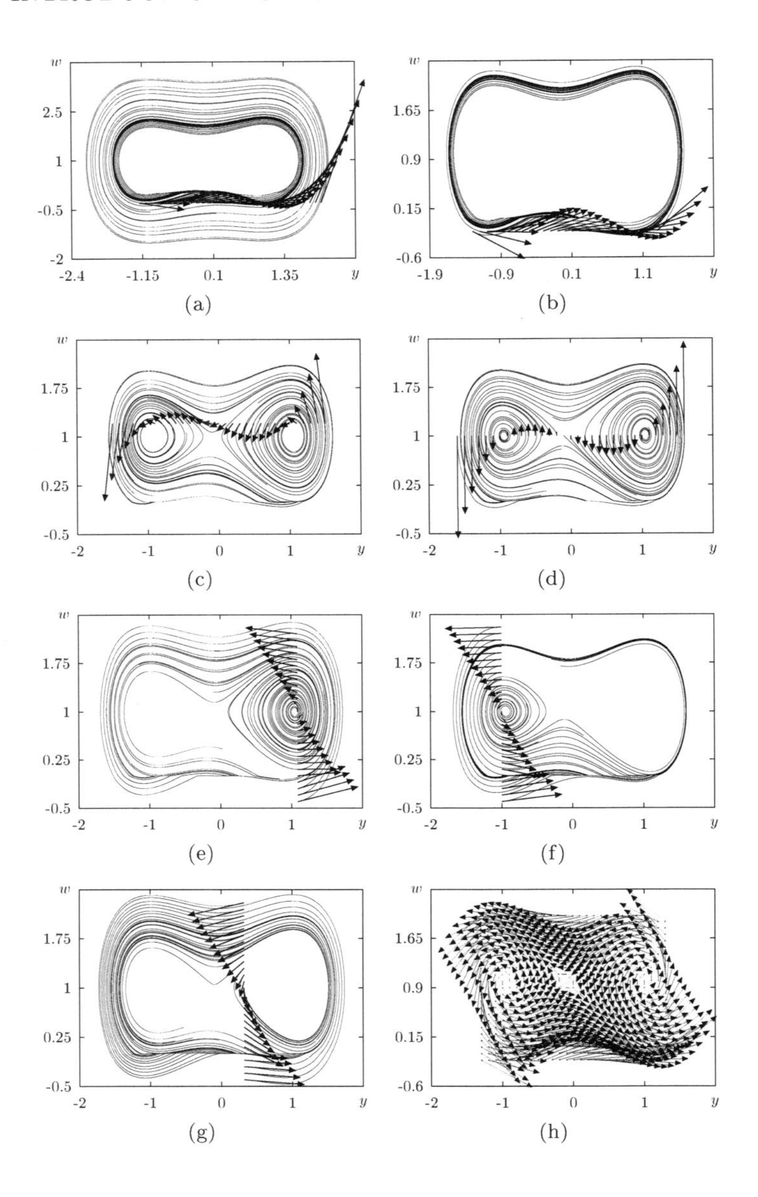

FIGURE 8.2: Vector field of (2.3) for (a) $w_0 = -0.3$, $y_{0i} \in (-1.2, 2)$ for $i = 0 \ldots 32$, (b) $w_0 = -0.2$, $y_{0i} \in (-1.3, 1.3)$ for $i = 0 \ldots 26$, (c) $w_0 = 1.2$, $y_{0i} \in (-1.5, 1.5)$ for $i = 0 \ldots 30$, (d) $w_0 = 1.0$, $y_{0i} \in (-1.6, 1.6)$ for $i = 0 \ldots 32$, (e) $y_0 = 1.08803$, $w_{0i} \in (-0.4, 2.2)$ for $i = 0 \ldots 26$, (f) $y_0 = -1.0$, $w_{0i} \in (-0.4, 2.3)$ for $i = 0 \ldots 27$, (g) $y_0 = 0.321889$, $w_{0i} \in (-0.4, 2.3)$ for $i = 0 \ldots 27$, (h) $w_{0k} \in (-0.2, 2.1)$ for $k = 0 \ldots 23$, $y_{0i} \in (-1.3, 1.3)$ for $i = 0 \ldots 26$.

The graph of $V(y)$ leaves Q only in points $(-y_0, -1)$ and $(y_0, 1)$. Therefore, we can apply Tichonov theorem [170] to the singularly perturbed system (8.6). We take $y(0) = \bar{y} \in \bar{I}$, $v(0) = 1$, and the corresponding solution $(y(t), v(t))$ of (8.6) leaves Q near y_0 at the time $\tilde{t}$. We put

$$\Psi_\varepsilon(\bar{y}) := y(\tilde{t}) .$$

We get a map $\Psi_\varepsilon : \bar{I} \to I$. Finally, we put

$$P_\varepsilon(y) := \Psi_\varepsilon(\Phi_\varepsilon(y))$$

for $y \in I$. Clearly $P_\varepsilon : I \to I$ and this is the desired Poincaré map of (8.4) near periodic solution $p_0(t)$ of (8.3). The map Φ_ε depends smoothly on ε small and $y \in I$. Similarly, the map Ψ_ε depends smoothly on $\varepsilon > 0$ small and $y \in I$. We need to find the limit of $\Psi_\varepsilon(y)$ as $\varepsilon \to 0_+$. To do this, we apply Tichonov theorem [170]. For the system (8.6), we have already verified all assumptions of this theorem except V.-th on page 31 of [170]: Namely, we must show that for any $(y, v^0) \in Q$, it holds

$$\begin{aligned} (v(\tau), y) \in Q \quad &\text{for} \quad \tau \geq 0 , \\ v(\tau) \to V(y) \quad &\text{as} \quad \tau \to \infty , \end{aligned} \tag{8.7}$$

where $v(\tau)$ is the solution of the equation

$$\begin{aligned} \dot{v}(\tau) &= y^3 - y - 0.2v(\tau) , \\ v(0) &= v^0 . \end{aligned}$$

Clearly, we have

$$v(\tau) = \mathrm{e}^{-0.2\tau}\Big(v^0 - 5(y^3 - y)\Big) + 5(y^3 - y) .$$

If $v^0 \geq 5(y^3 - y)$ then $5(y^3 - y) \leq v(\tau) \leq v^0$. Since $|v^0| \leq 1$ and $|5(y^3 - y)| \leq 1$ then $|v(\tau)| \leq 1$, and condition (8.7) holds. Similarly, if $v^0 \leq 5(y^3 - y)$ then $5(y^3 - y) \geq v(\tau) \geq v^0$ and again $|v(\tau)| \leq 1$, and condition (8.7) still holds. Summarizing, we can apply Tichonov theorem to (8.6). Consequently, the solution $(y(t), v(t))$ of (8.6) with the initial value conditions $y(0) = \bar{y} \in \bar{I}$ and $v(0) = 1$ has the asymptotic expansion

$$\begin{aligned} y(t) &= \bar{y} + t + O(\varepsilon) , \\ v(t) &= 5\big((\bar{y} + t)^3 - \bar{y} - t\big) + \mathrm{e}^{-0.2t/\varepsilon}\Big(1 - 5(y^3 - y)\Big) + O(\varepsilon) . \end{aligned}$$

The time $\tilde{t}$ is determined from equation $v(\tilde{t}) = 0$, and we get

$$\bar{y} + \tilde{t} \sim y_0 + O(\varepsilon) .$$

Hence

$$\Psi_\varepsilon(\bar{y}) = y(\tilde{t}) = \bar{y} + \tilde{t} + O(\varepsilon) = y_0 + O(\varepsilon) .$$

This gives

$$P_\varepsilon(y) = \Psi_\varepsilon(\Phi_\varepsilon(y)) = y_0 + O(\varepsilon)\,. \tag{8.8}$$

We note that the limit map $P_0(y) = y_0$ in (8.8) is just the Poincaré map along the periodic solution $p_0(t)$ of (8.3). Furthermore, the identity (8.8) holds also in the C^1-topology, i.e., it holds

$$P'_\varepsilon(y) = O(\varepsilon)\,. \tag{8.9}$$

Hence, the map $P_\varepsilon : I \to I$ has a unique fixed point $y_\varepsilon \in I$ of the form $y_\varepsilon = y_0 + O(\varepsilon)$, which is, according to (8.9), also rapidly attractive. Summarizing, we get the next theorem.

Theorem 1. *The discontinuous system* (8.3) *has the periodic solution* $p_0(t)$ *starting from the point* $(y_0, 0)$, *which is infinitely stable, i.e., all solutions starting near* $p_0(t)$ *collapse after a finite time to* $p_0(t)$. *Its approximation* (8.4) *has also a unique periodic solution* p_ε *starting from the point* $(y_\varepsilon, \varepsilon)$ *which approximates* $p_0(t)$ *and which is rapidly attracting. This coincides with the infinite stability of* $p_0(t)$.

Theorem 1 analytically explains numerical results of [41] concerning stable periodic solutions taken into account.

Finally, we note that function f_ε is an approximation of the multivalued mapping

$$\operatorname{Sgn} w := \begin{cases} \operatorname{sgn} w & \text{for} \quad w \neq 0\,, \\ [-1, 1] & \text{for} \quad w = 0\,. \end{cases}$$

Hence, (8.4) is the approximation of the discontinuous differential inclusion

$$\begin{aligned} \dot{y} &= -w + 1\,, \\ \dot{w} - y^3 + y &\in -\frac{0.2}{1 + |w|}\operatorname{Sgn} w\,, \end{aligned}$$

which is a differential inclusion version of (8.3). But of course, these arguments fit into the general theory of differential inclusions [64, 70].

8.2 Higher dimensional systems

In this section, we consider a general discontinuous system in $\mathbb{R}^n$. For the sake of simplicity, we assume that a system has a discontinuity at level $x = 0$ for $\dim x = 1$. Hence $z = (x, y) \in \mathbb{R}^n$ and $\dim y = n - 1$. The system is given by the equation

$$\dot{z} = \tilde{f}(z)\,, \tag{8.10}$$

where

$$\tilde{f}(z) = \begin{cases} f_+(x, y) & \text{for} \quad x > 0\,, \\ f_-(x, y) & \text{for} \quad x < 0\,. \end{cases}$$

The functions $f_\pm : \mathbb{R}^n \to \mathbb{R}^n$ are smooth and $f_+(0,y) \neq f_-(0,y)$ in general. We put

$$f_\pm = (h_\pm(x,y), g_\pm(x,y)),$$

where $h_\pm : \mathbb{R}^n \to \mathbb{R}$ and $g_\pm : \mathbb{R}^n \to \mathbb{R}^{n-1}$. Now we take $\varepsilon > 0$ small and consider a continuous approximation of (8.10) given by

$$\dot{z} = \tilde{f}(z) \quad \text{for} \quad |x| \geq \varepsilon, \tag{8.11}$$

and for $|x| \leq \varepsilon$:

$$\begin{aligned} \dot{x} &= \frac{h_+(\varepsilon,y) - h_-(-\varepsilon,y)}{2\varepsilon}x + \frac{h_+(\varepsilon,y) + h_-(-\varepsilon,y)}{2}, \\ \dot{y} &= \frac{g_+(\varepsilon,y) - g_-(-\varepsilon,y)}{2\varepsilon}x + \frac{g_+(\varepsilon,y) + g_-(-\varepsilon,y)}{2}. \end{aligned} \tag{8.12}$$

We put $x = \varepsilon w$, $|w| \leq 1$ in (8.12) to get the system

$$\begin{aligned} \varepsilon\dot{w} &= \frac{h_+(\varepsilon,y) - h_-(-\varepsilon,y)}{2}w + \frac{h_+(\varepsilon,y) + h_-(-\varepsilon,y)}{2}, \\ \dot{y} &= \frac{g_+(\varepsilon,y) - g_-(-\varepsilon,y)}{2}w + \frac{g_+(\varepsilon,y) + g_-(-\varepsilon,y)}{2}. \end{aligned} \tag{8.13}$$

In order to apply Tichonov theorem, we consider the assumption

$$h_+(0,y) - h_-(0,y) < 0. \tag{8.14}$$

If (8.14) fails, then we use (8.11)-(8.12), since the right-hand side of (8.11)–(8.12) belongs to the set

$$\begin{gathered} \tilde{F}(z) := \operatorname{conv}[f_-(x,y), f_+(x,y)] := \\ \Big\{\lambda f_-(x,y) + (1-\lambda) f_+(x,y) \mid \lambda \in [0,1]\Big\}, \end{gathered}$$

and

$$\dot{z} \in \tilde{F}(z) \tag{8.15}$$

is the differential inclusion corresponding to (8.10).

Now we take condition (8.14) along with the assumption

(A1) There is a solution $p(t)$ of (8.10) defined on $[0,T]$ such that the x-coordinate $p_1(t)$ of $p(t)$ satisfies $p_1(t) > 0$ for $t \in (0,T)$ and $p_1(0) = p_1(T) = 0$. Moreover, $h_+(p(0)) = 0$, $h_+(p(T)) < 0$, and the gradient $\nabla_y h_+(p(0)) \neq 0$.

The condition $\nabla_y h_+(p(0)) \neq 0$ implies that the set $h_+^{-1}(0)$ is a manifold M on R^{n-1} near the point $p(0)$. Here we consider the restriction $h_+(0,\cdot) : \mathbb{R}^{n-1} \to \mathbb{R}$.

The reduced system of (8.13) for $\varepsilon = 0$ has the form

$$\dot{y} = H(y) := \frac{g_+(0,y) - g_-(0,y)}{2} V(y) + \frac{g_+(0,y) + g_-(0,y)}{2} \tag{8.16}$$

for

$$V(y) = \frac{h_+(0,y) + h_-(0,y)}{h_-(0,y) - h_+(0,y)}.$$

We assume

(A2) Let $p_2(t)$ be the y-coordinate of $p(t)$. Then the solution $y_0(t)$ of (8.16) with the initial condition $y_0(0) = p_2(T)$ passes through the point $p_2(0)$ at the time T_0, and $h_+(0, y_0(t)) < 0$ for $t \in [0, T_0)$, $h_-(0, y_0(t)) > 0$ for $t \in [0, T_0]$. Moreover, $\dot{y}_0(T_0)$ is transversal to M, i.e., $g_+(p(0))$ is not orthogonal to $\nabla_y h_+(p(0))$.

We note that equation (8.16) is related to the formula (2.12) of [102]. The condition (A2) implies condition (8.14) along $y = y_0(t)$, $t \in [0, T_0]$, and it also gives a sliding solution $(0, y_0(t))$, $t \in [0, T_0]$ of (8.15). Moreover, we get a periodic solution $p_0(t)$ of (8.15) given by

$$p_0(t) := \begin{cases} p(t) & \text{for} \quad t \in [0, T], \\ (0, y_0(t - T)) & \text{for} \quad t \in [T, T + T_0]. \end{cases}$$

Now we construct a Poincaré map P_ε of (8.11)-(8.12) along $p_0(t)$ as follows. Let $B_\delta(p_2(0))$ be the small ball in $\mathbb{R}^{n-1}$ centered in $p_2(0)$ with the small radius $\delta > 0$. We take a solution $z(t)$ of (8.11) starting from the point (ε, y), $y \in B_\delta(p_2(0))$. This hits the surface $x = \varepsilon$ near $p(T)$ in the point $(\varepsilon, y(\bar{t}))$. We consider the map

$$\Phi_\varepsilon(y) := y(\bar{t}),$$
$$\Phi_\varepsilon : B_\delta(p_2(0)) \to B_{\delta_1}(p_2(T))$$

for a small $\delta_1 > 0$. Now we consider the solution $(w(t), y(t))$ of (8.13) starting from the point $(1, y)$, $y \in B_{\delta_1}(p_2(T))$. It will hit the surface $w = 1$ at a time $\tilde{t}$ near the point $(1, p_2(0))$. Like in Section 2, Tichonov theorem implies that

$$\begin{aligned} 1 = w(\tilde{t}) &= V(y(\tilde{t})) + O(\varepsilon), \\ y(\tilde{t}) &= \bar{y}(\tilde{t}) + O(\varepsilon), \end{aligned} \tag{8.17}$$

where $\bar{y}(t)$ is the solution of (8.16) with the initial condition $\bar{y}(0) = y$. Since y is near to $p_2(T)$, condition (A2) and (8.17) imply that $w(t)$ transversally crosses the line $w = 1$ at $\tilde{t}$. Hence, $\tilde{t}$ is locally uniquely defined. We also note that $V(\bar{y}(\tilde{t})) + O(\varepsilon) = 1$. Hence, $\bar{y}(\tilde{t})$ is $O(\varepsilon)$-near to the point $\Theta(y) \in M$, which is the crossing point of M with $\bar{y}(t)$. Thus, we can put

$$\Psi_\varepsilon(y) := y(\tilde{t}),$$
$$\Phi_\varepsilon : B_{\delta_1}(p_2(0)) \to \mathbb{R}^{n-1}.$$

Finally, we consider the Poincaré map

$$P_\varepsilon(y) := \Psi_\varepsilon(\Phi_\varepsilon(y)), \quad y \in B_\delta(p_2(0)).$$

We have $\Psi_\varepsilon(y) = \Theta(y) + O(\varepsilon)$ and

$$\begin{aligned} P_\varepsilon(y) &= \Theta(\Phi_0(y)) + O(\varepsilon), \\ DP_\varepsilon(y) &= D\Theta(\Phi_0(y)) + O(\varepsilon). \end{aligned} \tag{8.18}$$

In order to find a periodic solution of (8.11)-(8.12) near $p_0(t)$, we must solve equation

$$y = P_\varepsilon(y). \tag{8.19}$$

We have $p_0(0) = P_0(p_0(0))$. Now we take a tubular/normal neighborhood $M \times W$ of M in $\mathbb{R}^{n-1}$ near the point $p_0(0)$, where $W \subset \mathbb{R}$ is an open neighborhood of $0 \in \mathbb{R}$. The corresponding projections are as follows: $\Gamma_1 : M \times W \to W$ and $\Gamma_2 : M \times W \to M$. Then equation (8.19) is splited to the system

$$\begin{aligned} y_1 &= \Gamma_1 P_\varepsilon(y_1, y_2), \quad y_1 \in W, \\ y_2 &= \Gamma_2 P_\varepsilon(y_1, y_2), \quad y_2 \in M. \end{aligned} \tag{8.20}$$

Since $\Theta(y) \in M$, (8.18) gives

$$\begin{aligned} y_1 &= \Gamma_1 P_\varepsilon(y_1, y_2) = O(\varepsilon), \\ y_2 &= \Gamma_2 P_\varepsilon(y_1, y_2) = \Omega(y_2) + O(\varepsilon), \end{aligned}$$

where we get a map $\Omega : M \to M$ defined near $p_0(0)$ by

$$\Omega(y_2) := \Theta(\Phi_0(0, y_2)).$$

The map Ω is a Poincaré map of $p_0(t)$ for (8.15) when the dynamics on the level $x = 0$ is given by (8.16) like on p. 111 of [102]. Clearly $\Omega(p_0(0)) = p_0(0)$. Hence if the linearization $I - D\Omega(p_0(0))$ is nonsingular, then we can solve (8.20) near $p_0(0)$ by using the implicit function theorem. Moreover, if $D\Omega(p_0(0))$ is stable, i.e., all eigenvalues of $D\Omega(p_0(0))$ are inside the unit circle of $\mathbb{C}$, then also the corresponding periodic solution $p_\varepsilon(t)$ of (8.11)–(8.12) is stable. Of course, the stability of $D\Omega(p_0(0))$ gives the stability of periodic solution $p_0(t)$ for (8.15) when again the dynamics on the level $x = 0$ is given by (8.16). Summarizing we arrive at the following result.

Theorem 2. *Let the assumptions* (A1) *and* (A2) *hold. If the linearization* $I - D\Omega(p_0(0))$ *is nonsingular, then the approximate system* (8.11)–(8.12) *possesses a periodic solution* $p_\varepsilon(t)$ *near* $p_0(0)$. *If* $D\Omega(p_0(0))$ *is stable, then also the corresponding periodic solution* $p_\varepsilon(t)$ *of* (8.11)–(8.12) *is stable.*

Remark 1. The conditions (A1), (A2) and Theorem 2 contain some transversal/generic assumptions, namely that $\nabla_y h_+(p(0)) \neq 0$, $g_+(p(0))$ is transversal to M and that the linearization $I - D\Omega(p_0(0))$ is nonsingular. If one of them fails, then the construction of the Poincaré map P_ε as well as the solvability of equation (8.19) become problematic. Some bifurcations of periodic solutions are expected.

Remark 2. Theorem 2 states about the persistence of a generic periodic solution of discontinuous systems crossing a discontinuity level under a continuous approximation. This persistence could be proved by using the Leray-Schauder degree theory [64], but since we use the implicit function theorem, we get uniqueness and stability of periodic solutions as well. Also this approach is constructive. Finally, we assume that the discontinuity has a codimension 1. Higher codimension problems could be also interesting to study.

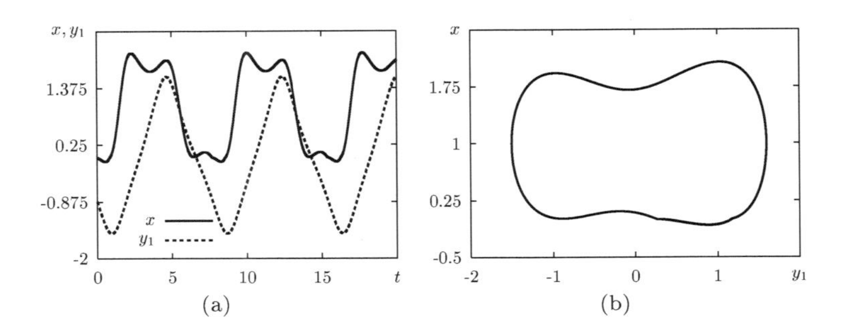

FIGURE 8.3: Stable solution to the (3.16) ($x_0 = 0$): (a) $x(t)$, $y_1(t)$, and (b) $x(y_1)$ for $\bar{y}_0 = [0.321889, 0, 0]$, $\delta = 0.1$.

We note that $p_0(0) = p(0)$. For application of Theorem 2, we need to find $D\Omega(p(0))$ as a map from $T_{p(0)}M$ to $T_{p(0)}M$, where $T_{p(0)}M$ is the tangent space of manifold M at point $p(0)$ given by

$$T_{p(0)}M := \left\{ \eta \in \mathbb{R}^{n-1} \mid \eta \perp \nabla_y h_+(p(0)) \right\}.$$

From the definition of maps Θ and Φ_0, we can derive after some algebra that for any $\eta \in T_{p(0)}M$, the vector $D\Omega(p(0))\eta \in T_{p(0)}M$ is given by

$$D\Omega(p(0))\eta = w(T_0) - \frac{\langle \nabla_y h_+(p(0)), w(T_0) \rangle}{\langle \nabla_y h_+(p(0)), \dot{y}_0(T_0) \rangle} \dot{y}_0(T_0), \tag{8.21}$$

where $\langle \cdot, \cdot \rangle$ is the scalar product and the function $w(t)$ (depending on η) is the solution of the initial value problem

$$\begin{aligned} \dot{w}(t) &= DH(y_0(t))w(t), \\ w(0) &= -\frac{\dot{p}_2(T)}{\dot{p}_1(T)} x(T) + y(T), \end{aligned} \tag{8.22}$$

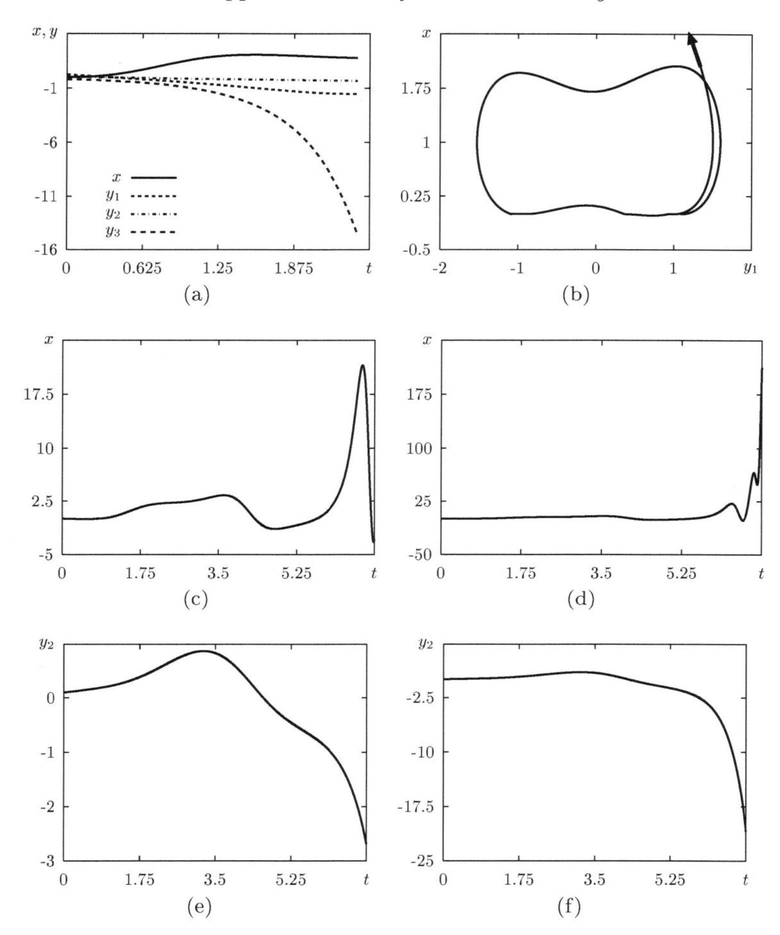

FIGURE 8.4: Unstable solution to the (3.16) ($x_0 = 0$): (a) $x(t)$, $\bar{y}(t)$, and (b) $x(y_1)$ for $\bar{y}_0 = [1.08803, 0.01, 0.01], \delta = 0.1$; $x(t)$ for $\bar{y}_0 = [0.321889, 0.1, 0.1]$ if (c) $\delta = 0.2$ and (d) $\delta = 0.07$; $y_2(t)$ for $\bar{y}_0 = [0.321889, 0.1, 0.1]$ if (e) $\delta = 0.5$ and (f) $\delta = 0.2$.

where the functions $x(t)$, $y(t)$ (depending also on η) are the solutions of the initial value problem

$$\begin{gathered}\dot{x}(t) = h_{+x}(p(t))x(t) + h_{+y}(p(t))y(t)\,,\\ \dot{y}(t) = g_{+x}(p(t))x(t) + g_{+y}(p(t))y(t)\,,\\ x(0) = 0, \quad y(0) = \eta\,.\end{gathered} \tag{8.23}$$

Formulas (8.21)-(8.23) can be used to compute $D\Omega(p(0))$. For instance, let us consider an extension of system (8.2) to higher dimension given by

$$\begin{gathered}\ddot{v} - v + v^3 - \frac{0.2}{1+|\dot{v}-1|}\mathrm{sgn}(1-\dot{v}) + zf(z,v) = 0\,,\\ \ddot{z} + \delta\dot{z} + zg(z,v) = 0\,,\end{gathered} \tag{8.24}$$

where $z \in \mathbb{R}$, $\delta > 0$ and f, g are smooth functions.(8.24) has the form

$$\begin{aligned}\dot{x} &= y_1^3 - y_1 - \frac{0.2}{1+|x|}\mathrm{sgn}x + y_2 f(y_2, y_1)\,,\\ \dot{y}_1 &= 1 - x\,,\\ \dot{y}_2 &= y_3\,,\\ \dot{y}_3 &= -\delta y_3 - y_2 g(y_2, y_1)\,.\end{aligned} \tag{8.25}$$

Since now

$$h_\pm(x,y) = y_1^3 - y_1 \mp \frac{0.2}{1\pm x} + y_2 f(y_2, y_1)$$

for $y = (y_1, y_2, y_3)$, the assumption (A1) holds with $T = t_0 = 7.117155465$ and with $p(t) = (w(t), y(t), 0, 0)$, where $w(t)$, $y(t)$ solve (8.5) with the initial conditions $y(0) = y_0 = 1.08803$ and $w(0) = 0$. We note that $p(T) = (0, \bar{y}_0, 0, 0)$ for $\bar{y}_0 = 0.3218918837 \in (-y_0, y_0)$. Furthermore, the reduced system (8.16) has now the form

$$\begin{aligned}\dot{y}_1 &= 1\,,\\ \dot{y}_2 &= y_3\,,\\ \dot{y}_3 &= -\delta y_3 - y_2 g(y_2, y_1)\,.\end{aligned}$$

Hence we have $y_0(t) = (\bar{y}_0 + t, 0, 0)$ with $T_0 = y_0 - \bar{y}_0$ for assumption (A2). Next, since

$$\dot{y}_0(T_0) = (1,0,0), \quad \nabla_y h_+(p(0)) = (3y_0^2 - 1, f(0, y_0), 0),$$

and $3y_0^2 - 1 = 2.551428 \neq 0$, we see that $\dot{y}_0(T_0)$ is not orthogonal to $\nabla_y h_+(p(0))$. Summarizing, we get that also assumption (A2) holds. So in order to apply Theorem 2 for system (8.25), we need to find the spectrum of $D\Omega(p(0))$. We note that now

$$\begin{aligned}T_{p(0)}M &:= \Big\{\eta \in \mathbb{R}^3 \mid \eta \perp (3y_0^2 - 1, f(0, y_0), 0)\Big\}\\ &= \Big\{\Big(-\frac{f(0,y_0)}{3y_0^2-1}\eta_2, \eta_2, \eta_3\Big) \mid \eta_2, \eta_3 \in \mathbb{R}\Big\}\,.\end{aligned}$$

According to (8.21), (8.22) and (8.23), we now have

$$\Omega(p(0))\eta = \Big(-\frac{f(0,y_0)}{3y_0^2-1}w_2(T_0), w_2(T_0), w_3(T_0)\Big)\,, \tag{8.26}$$

where $w_1(t), w_2(t), w_3(t)$ solve the following ordinary differential equations

$$\begin{aligned}\dot{w}_1 &= 0\,,\\ \dot{w}_2 &= w_3\,,\\ \dot{w}_3 &= -\delta y_3 - g(0,\bar{y}_0+t)w_2\end{aligned}\tag{8.27}$$

with the initial value conditions

$$\begin{aligned}w_1(0) &= y_1(T) - \frac{\dot{y}(T)}{\dot{w}(T)}x_1(T)\,,\\ w_2(0) = y_2(T), &\quad w_3(0) = y_3(T)\,,\end{aligned}$$

when $x_1(t), y_1(t), y_2(t), y_3(t)$ solve the following ordinary differential equations

$$\begin{aligned}\dot{x}_1 &= (3y(t)^2 - y(t))y_1 + \frac{0.2}{(1+w(t))^2}x_1 + f(0,y(t))y_2\,,\\ \dot{y}_1 &= -x_1\,,\\ \dot{y}_2 &= y_3\,,\\ \dot{y}_3 &= -\delta y_3 - g(0,y(t))y_2\end{aligned}\tag{8.28}$$

with the initial value conditions

$$\begin{aligned}x_1(0) = 0, &\quad y_1(0) = -\frac{f(0,y_0)}{3y_0^2-1}\eta_2\,,\\ y_2(0) = \eta_2, &\quad w_3(0) = \eta_3\end{aligned}$$

for $\eta_2, \eta_3 \in \mathbb{R}$. We can easily see from (8.26) and from the initial value problems (8.27), (8.28) that the spectrum $\sigma(D\Omega(p(0)))$ is the spectrum of the fundamental matrix solution of the ordinary differential equation

$$\begin{aligned}\dot{y}_2 &= y_3\,,\\ \dot{y}_3 &= -\delta y_3 - q(t)y_2\,,\end{aligned}$$

where

$$q(t) = \begin{cases} g(0,y(t)) & \text{for} \quad t \in [0,T]\,,\\ g(0,\bar{y}_0+t-T) & \text{for} \quad t \in [T, T+T_0]\,.\end{cases}$$

We note that $q(t)$ is $(T+T_0)$-periodic and $T+T_0 = 7.883297537$. By using results of Section 2.5 from [68] we obtain that if

$$\begin{aligned} q(t) &> \frac{\delta^2}{4}\,,\\ \int_0^{T+T_0} q(t)\,dt &\leq \frac{\delta^2}{4}(T+T_0) + \frac{4}{T+T_0}\,,\end{aligned}\tag{8.29}$$

then the spectrum $\sigma(D\Omega(p(0)))$ is inside the unit disc. Summarizing, condition (8.29) implies the stability of periodic solution $p_0(t)$ as well as the existence and stability of periodic solutions $p_\varepsilon(t)$ from Theorem 2 applied to system (8.25).

More general criteria for system (8.25) than condition (8.29) can be derived by using results from [58]. Of course, condition (8.29) trivially holds for a smooth function $g(z,v)$ when $g(0,v) = \frac{\delta^2}{4} + \frac{2}{(T+T_0)^2}$. To find a nontrivial example, we use numerical method. We take $f(y_1,y_2) = y_2$ and $g(y_2,y_1) = y_2 - y_1$. So $g(0,v) = -v$. Since $y(t)$ changes the sign over the interval $[0,T]$, so $q(t)$ defined above also changes the sign over the interval $[0,T+T_0]$. Hence we cannot use criterion (8.29). Instead, we use a criterion by R. Einaudi from [58] of the form

$$\int_0^{T+T_0} q(t)\,dt \tan\frac{\delta(T+T_0)}{2} \le 2\delta$$

since now $\int_0^{T+T_0} q(t)\,dt = 0.0594259 > 0$. Consequently, we get the following inequality

$$\tan(3.94165\delta) \le 33.6554\delta$$

for the parameter δ when our theory is applied to get the existence and stability of periodic solutions $p_\varepsilon(t)$ from Theorem 2 to this concrete system (8.25). Finally, we present several numerical solutions to (8.25). Taking into account various initial conditions as well as δ parameter, an occurrence of stable (Figure 8.3) and unstable (Figure 8.4) periodic orbits is confirmed.

Chapter 9

Nonlinear dynamics of a swinging oscillator

In order to investigate nonlinear vibrations of a swinging lumped oscillator, the method of Poincaré-Birkhoff normal form is applied. As it is known, owing to this method [133, 52] the Hamiltonian of a system is outlined in a form of both a square part referred to as a nonperturbed one, and a sum of terms of powers higher than two.

It is worth noticing that through canonical Hamilton transformations, a system under investigation is essentially simplified. Namely, the Hamiltonian up to the terms of fourth order and higher is integrated. Therefore, an asymptotic solution to nonlinear problem is obtained. Traditional methods of normalization of a system with two-degrees-of-freedom are rather complicated and unsuitable for direct investigations [52, 54, 20, 21, 22]. A change of the variables is either sought using the guiding functions, or applying a guiding Hamiltonian.

A method of construction of canonical variables transformation in the parametric form is proposed, which differs from the existing approaches in Hamiltonian mechanics [128, 129]. A criterion of existence of parametric variables transformation is first formulated, and then a rule of Hamiltonian transformation is given. The proposed method can be used to find normal forms of Hamiltonians.

It should be emphasized that the introduced definition of a normal form [179, 180], does not require any partition to either autonomous - nonautonomous, or resonance - nonresonance cases [163], but it is treated in the frame of one approach. In order to find the corresponding normal form asympotics, the system of equations is derived analogous to the equations chain obtained earlier [179, 180]. Instead of the generator method and the guiding Hamilonian, the parametrized guiding function is used [128]. It allows for direct (without the transformation to an autonomous system) [179, 180] computation of the equations chain for nonautonomous Hamiltonians [129]. For autonomous systems the methods of computation of normal forms coincide in the first and second approximations.

In the first part of this chapter the parametric form of canonical transformation and the method of normalization are introduced and illustrated. Note that latter method is sometimes called the method of invariant normalization (see [179, 180]).

In the second part, dynamics of a swinging oscillator is analyzed. In what follows, the Hamiltonian of the studied system is defined, and also its normal form is constructed. After integration of the system, the asymptotic solution of the investigated nonlinear system is obtained (see also [43]).

9.1 Parametrical form of canonical transformations

The general result of parametrisation of the Hamiltonian systems of canonical variables transformations is formulated in the frame of the following theorem [128].

Theorem 1. Let a transformation of the variables $\mathbf{q}, \mathbf{p} \to \mathbf{Q}, \mathbf{P}$ be given in the following parametric form

$$\begin{array}{ll} \mathbf{q} = \mathbf{x} - \frac{1}{2}\Psi_{\mathbf{y}} & \mathbf{Q} = \mathbf{x} + \frac{1}{2}\Psi_{\mathbf{y}} \\ \mathbf{p} = \mathbf{y} + \frac{1}{2}\Psi_{\mathbf{x}} , & \mathbf{P} = \mathbf{y} - \frac{1}{2}\Psi_{\mathbf{x}} . \end{array} \tag{9.1}$$

Then, for any arbitrary function $\Psi(t, \mathbf{x}, \mathbf{y})$, the following properties hold:

1. Jacobians of two transformations $\mathbf{q} = \mathbf{q}(t, \mathbf{x}, \mathbf{y}), \mathbf{p} = \mathbf{p}(t, \mathbf{x}, \mathbf{y})$ and $\mathbf{Q} = \mathbf{Q}(t, \mathbf{x}, \mathbf{y}), \mathbf{P} = \mathbf{P}(t, \mathbf{x}, \mathbf{y})$ are identity ones:

$$\frac{\partial(\mathbf{q}, \mathbf{p})}{\partial(\mathbf{x}, \mathbf{y})} = \frac{\partial(\mathbf{Q}, \mathbf{P})}{\partial(\mathbf{x}, \mathbf{y})} = J(t, \mathbf{x}, \mathbf{y}); \tag{9.2}$$

2. In the space $J > 0$ relation (9.1) with respect to the variables $\mathbf{q}, \mathbf{p} \to \mathbf{Q}, \mathbf{P}$ transforms the Hamiltonian system $H = H(t, \mathbf{q}, \mathbf{p})$ owing to the following rule:

$$\Psi_t(t, \mathbf{x}, \mathbf{y}) + H(t, \mathbf{q}, \mathbf{p}) = \tilde{H}(t, \mathbf{Q}, \mathbf{P}), \tag{9.3}$$

where the arguments $\mathbf{q}, \mathbf{p}$ and $\mathbf{Q}, \mathbf{P}$ in the Hamiltonians H and $\tilde{H}$ are expressed through the parameters $\mathbf{x}, \mathbf{y}$ in formulas (9.1).

In the next step we investigate for which canonical transformations the parametrization exists.

9.2 Function derivative

A canonical transformation can be represented through the guiding functions $S_1(t, \mathbf{q}, \mathbf{P})$ and $S_2(t, \mathbf{p}, \mathbf{Q})$ in the following way

$$\begin{aligned} dS_1 &= \mathbf{p}d\mathbf{q} + \mathbf{Q}d\mathbf{P} + (\tilde{H} - H)dt, \quad \det S_{1\mathbf{qP}} \neq 0, \\ dS_2 &= -\mathbf{q}d\mathbf{p} - \mathbf{P}d\mathbf{Q} + (\tilde{H} - H)dt, \quad \det S_{2\mathbf{pQ}} \neq 0. \end{aligned}$$

The following new guiding function is introduced

$$\Phi = \frac{1}{2}\left[S_1(t,\mathbf{q},\mathbf{P}) - \mathbf{qP} + S_2(t,\mathbf{Q},\mathbf{p}) + \mathbf{Qp}\right] . \tag{9.4}$$

Its differential form follows

$$d\Phi = \frac{1}{2}\sum_{i=1}^{n}\begin{vmatrix} Q_i - q_i & P_i - p_i \\ dQ_i + dq_i & dP_i + dp_i \end{vmatrix} + (\tilde{H} - H)dt . \tag{9.5}$$

For $dt = 0$ the differential form $d\Phi$ has been introduced by Poincaré (see [133], page 141, and [20], page 337). He showed that if $\mathbf{Q}(\mathbf{q},\mathbf{p}), \mathbf{P}(\mathbf{q},\mathbf{p})$ is the canonical transformation, then $d\Phi$ is the full differential and the function $\Phi(\mathbf{q},\mathbf{p})$ does exist.

Solving (9.1) with respect to $\mathbf{x},\mathbf{y}$ and $\Psi_{\mathbf{y}}, \Psi_{\mathbf{x}}$ one gets

$$\mathbf{x} = \frac{1}{2}(\mathbf{q}+\mathbf{Q}), \quad \mathbf{y} = \frac{1}{2}(\mathbf{p}+\mathbf{P}), \tag{9.6}$$

$$\Psi_{\mathbf{y}} = \mathbf{Q} - \mathbf{q}, \quad \Psi_{\mathbf{x}} = -\mathbf{P} + \mathbf{p}.$$

Assuming that the Jacobian of transformation (9.6) differs from zero ($\partial(\mathbf{x},\mathbf{y}) / \partial(\mathbf{q},\mathbf{p}) \neq 0$), one gets $d\Phi = d\Psi$, and hence the identity of the function Φ and Ψ follows

$$\Psi(\mathbf{x},\mathbf{y}) = \Psi\left(\frac{\mathbf{q}+\mathbf{Q}(\mathbf{q},\mathbf{p})}{2}, \frac{\mathbf{p}+\mathbf{P}(\mathbf{q},\mathbf{p})}{2}\right) = \Phi(\mathbf{q},\mathbf{p}).$$

The relations (9.2) and (9.6) yield

$$\frac{\partial(\mathbf{x},\mathbf{y})}{\partial(\mathbf{q},\mathbf{p})} = 1/J = 2^{-2n}\det(E+A), \quad A = \frac{\partial(\mathbf{Q},\mathbf{P})}{\partial(\mathbf{q},\mathbf{p})},$$

and the condition of nonsingularity of transformation (9.6) is cast in the form: $\det(E+A) \neq 0$, where: A - Jacobi matrix; E - unit matrix.

The obtained result is formulated in the form of the following theorem.

Theorem 2. If in the space $(\mathbf{q},\mathbf{p}) \in \Omega$ the transformation $\mathbf{Q}(\mathbf{q},\mathbf{p}), \mathbf{P}(\mathbf{q},\mathbf{p})$ is canonical one and none of the eigenvalues of the Jacobi matrix A is not equal to -1, then the parametrization (9.1) exists in the space Ω.

In monograph [20] an emphasis is put on noninvariance of the guiding function with respect to a choice of a basis of the canonical system coordinates and invariance of the differential Poincaré's formula (9.5) are outlined. It follows that the parametric function $\Psi(\mathbf{x},\mathbf{y})$ also has an invariant character. If function $\Psi(\mathbf{x},\mathbf{y})$ exists for some arbitrary parameters, then it will exist also for any arbitrary canonical variables transformation.

The condition of existence of parametrization $J \neq 0$ is invariant with respect to the canonical variables choice, since the equation $\det S_{1\mathbf{qP}} \neq 0$ depends on a choice of canonical variables. The condition on $\det S_{1\mathbf{qP}} \neq 0$ can be violated during a change of canonical variables. Besides, the class of parametrized canonical transformations is essentially wider than the class of canonical transformation through the use of a guiding function. For instance, the rotation on amount of 90° : $q = -P, p = Q$ can not be achieved through the guiding function $S(q, P)$, but it can be realized through the parametric function of the form: $\Psi = x^2 + y^2$. These and other advantages of the parametrization in comparison to the method of guiding function have been already mentioned [128].

In what follows we, illustrate how an application of equation (9.3) yields the earlier developed [179, 180] method of invariant normalization of Hamiltonians.

9.3 Invariant normalization of Hamiltonians

The normal form of a Hamilton system is called the normal Birkhoff form [52]. The general compact definition of this form is given in reference [54]. In all cases the generated Hamiltonian is chosen in the form of the simplest quadratic function associated with a linear vibrational system. On the other hand the definition of the normal form is associated with a choice of the generated Hamiltonian and has noninvariant character [21, 22, 179, 180, 128].

There exist mainly two widely used methods of construction of canonical transfomations leading to the normal form. One of them is based on application of the so-called guiding functions. This way has been chosen by Birkhoff [52]. In the second approach, instead of the guiding functions, Li generators are applied. It seems that the latter approach is more suitable, since it does not require an inverse of the power series required in the case of the guiding functions use.

It is worth noticing that Zhuravlev [179, 180] has proposed a general criterion of normal Birkhoff form for the perturbed Hamiltonian of the form

$$\bar{H}(t, \mathbf{q}, \mathbf{p}, \varepsilon) = H_0(t, \mathbf{q}, \mathbf{p}) + \bar{F}(t, \mathbf{q}, \mathbf{p}, \varepsilon),$$

$$\bar{F}(t, \mathbf{q}, \mathbf{p}, \varepsilon) = \varepsilon \bar{F}_1(t, \mathbf{q}, \mathbf{p}) + \varepsilon^2 \bar{F}_2(t, \mathbf{q}, \mathbf{p}) + \dots$$

Definition. A perturbed Hamiltonian possesses the normal form if and only if the associated perturbation is the first integral of the nonperturbed form $\frac{\partial F}{\partial t} + \{H_0, F\} = 0$, when $\{f, g\} = f_{\mathbf{p}} g_{\mathbf{q}} - f_{\mathbf{q}} g_{\mathbf{p}}$ are the Poisson brackets.

There are at least three main advantages of this approach with respect to known ones [21, 22, 179, 180, 128].

1°. A solution to full system of differential Hamilton equations with the normal form Hamiltonians is obtained through the superposition of the solutions of nonperturbed system and solution of the system with autonomous Hamiltonian equal to $\bar{F}(0, \mathbf{q}, \mathbf{p}, \varepsilon)$.

The result has been formulated as the theorem (see [180]).

Theorem (Zhuravlev). If a system with the Hamiltonian $\bar{H}$ satisfies the normal form condition, then the following steps are required to construct a general solution of the corresponding Hamilton equations:

(a) find a general solution of the system with Hamiltonian $H_0(t, p, q)$;

(b) find a general solution of the system defined through the perturbation $\bar{F}(0, p, q, \varepsilon)$ under the condition that time occurred in this system should be equal to zero.

Then the general solution of the input nonautonomous system can be presented as matching (in an arbitrary manner) of the obtained solutions (instead of arbitrary constants in solution of the second system, those of first system are substituted and vice versa).

2°. An invariant character of this criterion allows for normalization without an initial simplification of the nonperturbed part and without splitting into cases of autonomous - nonautonomous, and resonance - nonresonance ones.

3°. Asymptotics of normal form and variables transformations associated with Hamiltonian normalization are found through succesive quadratures of known (on each step) functions.

9.4 Algorithm of invariant normalization with the help of parame-
tric transformations

In what follows we illustrate how equation (9.3) of Theorem 1 can be transformed to an equivalent one of Zhuravlev normalization method.

Let the normalized Hamiltonian has the following form

$$H(t, \mathbf{q}, \mathbf{p}) = H_0(t, \mathbf{q}, \mathbf{p}) + F(t, \mathbf{q}, \mathbf{p}, \varepsilon),$$

$$F(t, \mathbf{q}, \mathbf{p}, \varepsilon) = \varepsilon F_1(t, \mathbf{q}, \mathbf{p}) + \varepsilon^2 F_2(t, \mathbf{q}, \mathbf{p}) + \dots$$

and let

$\bar{H}^{(k)}(t, \mathbf{Q}, \mathbf{P}, \varepsilon) = H_0(t, \mathbf{Q}, \mathbf{P}) + \bar{F}^{(k)}(t, \mathbf{Q}, \mathbf{P}, \varepsilon)$

is the k-th order asymptotics of the normal form

$\bar{F}^{(k)}(t, \mathbf{q}, \mathbf{p}, \varepsilon) = \varepsilon \bar{F}_1(t, \mathbf{Q}, \mathbf{P}) + \dots + \varepsilon^k \bar{F}_k(t, \mathbf{Q}, \mathbf{P})$

with respect to canonical transformation (9.1), and let

$\Psi^{(k)}(t, \mathbf{x}, \mathbf{y}, \varepsilon) = \varepsilon \Psi_1(t, \mathbf{x}, \mathbf{y}) + \dots + \varepsilon^k \Psi_k(t, \mathbf{x}, \mathbf{y})$

is the k-th order asymptotics of the function $\Psi(t, \mathbf{x}, \mathbf{y}, \varepsilon)$ in relations (9.1).

Then, it follows from Theorem 1, that the asymptotics $\Psi^{(k)}$ satisfies equation (9.3), which can be written in the following form

$$\begin{aligned} &\tfrac{\partial \Psi^{(k)}}{\partial t} + H_0(t, \mathbf{x} - \tfrac{1}{2}\Psi^{(k)}_{\mathbf{y}}, \mathbf{y} + \tfrac{1}{2}\Psi^{(k)}_{\mathbf{x}}) - H_0(t, \mathbf{x} + \tfrac{1}{2}\Psi^{(k)}_{\mathbf{y}}, \mathbf{y} - \tfrac{1}{2}\Psi^{(k)}_{\mathbf{x}}) + \\ &+ F^{(k)}(t, \mathbf{x} - \tfrac{1}{2}\Psi^{(k)}_{\mathbf{y}}, \mathbf{y} + \tfrac{1}{2}\Psi^{(k)}_{\mathbf{x}}) = \bar{F}^{(k)}(t, \mathbf{x} + \tfrac{1}{2}\Psi_{\mathbf{y}}, \mathbf{y} - \tfrac{1}{2}\Psi_{\mathbf{x}}). \end{aligned} \tag{9.7}$$

The latter result yields a chain of the coefficients of the canonical transformation Ψ_i and the normalized Hamiltonians $\bar{F}_i$ of the form

$$\frac{\partial \Psi_i}{\partial t} + \{H_0, \Psi_i\} + R_i = \bar{F}_i, \quad \frac{\partial \bar{F}_i}{\partial t} + \{H_0, \bar{F}_i\} = 0; \quad i = 1, 2, \ldots \tag{9.8}$$

The functions R_i are computed successively through formulas

$$R_1 = F_1, \quad R_2 = F_2 + \frac{1}{2}\{F_1 + \bar{F}_1, \Psi_1\}, \ldots \tag{9.9}$$

Observe that if H_0 is the polynom of powers not higher than second order with respect to $\mathbf{q}$ and $\mathbf{p}$, then R_i, $i \leq k$ are the series coefficients of the function

$$F(t, \mathbf{x} - \tfrac{1}{2}\Psi_{\mathbf{y}}, \mathbf{y} + \tfrac{1}{2}\Psi_{\mathbf{x}}) - \bar{F}^{(k)}(t, \mathbf{x} + \tfrac{1}{2}\Psi_{\mathbf{y}}, \mathbf{y} - \tfrac{1}{2}\Psi_{\mathbf{x}}) + \bar{F}^{(k)}(t, \mathbf{x}, \mathbf{y}) = \\ = \varepsilon R_1 + \varepsilon^2 R_2 + \varepsilon^2 R_3 + \ldots . \tag{9.10}$$

The obtained chain of equations (9.8) has been earlier obtained in [179, 180]. The obtained equations are called homoclinic ones and they are presented in the following form

$$R_i = \bar{F}_i - \frac{d\Psi_i}{dt}, \quad \frac{d\bar{F}_i}{dt} = 0; \quad i = 1, 2, \ldots \tag{9.11}$$

Here, full derivatives d/dt are computed using the rule of complex functions $\Psi_i(t, \mathbf{x}, \mathbf{y}), F_i(t, \mathbf{x}, \mathbf{y})$ differentiation, where $\mathbf{x}(t)$, $\mathbf{y}(t)$ (the functions of time) are defined through a solution of the nonperturbed system of the form

$$\dot{\mathbf{x}} = H_{0\mathbf{y}}, \quad \dot{\mathbf{y}} = -H_{0\mathbf{x}}, \quad \mathbf{x}(t_0) = \mathbf{x}_0, \quad \mathbf{y}(t_0) = \mathbf{y}_0. \tag{9.12}$$

Instead of $\mathbf{x}$ and $\mathbf{y}$, the solution of (9.12) into relations (9.11) is substituted, then from the second equation of (9.11) one may conclude that the function $\bar{F}_i$ does not depend on time. Therefore, the integral of the first equation reads

$$\int_{t_0}^{t} R_i(t)dt = (t - t_0)\bar{F}_i(t_0, \mathbf{x}_0, \mathbf{y}_0) + \Psi_i(t_0, \mathbf{x}_0, \mathbf{y}_0) - \Psi_i(t, \mathbf{x}, \mathbf{y}). \tag{9.13}$$

In fact, this is a key result, since the quadrature (9.13) defines both normal form and functions Ψ_i through variable transformation (9.1). However, the proposed integral representation (9.13) is not always achieved uniquely. The uniqueness is realized, if the function R_i with the substituted solution (9.12) is quasi-periodic, i.e., it is the sum of periodeic functions with respect to t. In the mentioned case, the integral of R_i is expressed through linear and quasi-periodic functions $f(t)$. One may compute then the averaged part $f(t)$ (independent on time), and then match it with the second part of the right hand side of (9.13). In what follows, the representation (9.13) defines uniquely $\bar{F}_i(t_0, \mathbf{x}_0, \mathbf{y}_0)$, and the function $\Psi_i(t_0, \mathbf{x}_0, \mathbf{y}_0)$ with zero time averaged value: $\overline{\Psi_i(t, \mathbf{x}(t), \mathbf{y}(t))} = 0$. The quasi-periodicity condition of R_i yields constraints on the parameters, for which the normal form exists. Below, the obtained result is formally formulated.

Main Theorem. Asymptotics of the k-th solution of a normal form and the associated variables transformation exist and they are unique, if after a substitution of solution (9.12) into the functions R_i, $(i = 1, 2, \ldots, k)$, they are quasi-periodic functions with respect to time. Then, in the right hand side of integral (9.13), $\bar{F}_i(t_0, \mathbf{x}_0, \mathbf{y}_0)$ is the coefficient of a linear term with respect to t, and $\bar{\Psi}_i(t_0, \mathbf{x}_0, \mathbf{y}_0)$ are terms independent on time.

The fundamental peculiarities of the discussed algorithm outlining differencies in comparison to that proposed in references [179, 180] are reported in Table 9.1.

TABLE 9.1: Properties of algorithms

Proposed algorithm	Algorithm [179, 180]
1°. Input system $H(\mathbf{q}, \mathbf{p})$ – nonautonomous	System $H(\mathbf{q}, \mathbf{p})$ – autonomous; initial system is reduced to autonomous one with increase of its order
2°. Function $\Psi(t, \varepsilon, \mathbf{x}, \mathbf{y})$	Guiding Hamiltonian $G(\varepsilon, \mathbf{Q}, \mathbf{P})$
3°. Canonical transformation $(\mathbf{q}, \mathbf{p}) \Rightarrow (\mathbf{Q}, \mathbf{P})$ in the parametrical form (9.1)	Canonical transformation $d\mathbf{Q}/d\varepsilon = \partial G/\partial \mathbf{P}$, $d\mathbf{P}/d\varepsilon = -\partial G/\partial \mathbf{Q}$, $\mathbf{Q}(0) = \mathbf{q}$, $\mathbf{P}(0) = \mathbf{p}$

The link between the guiding Hamiltonian G and function Ψ. In the method proposed in [179, 180], the transformation $\mathbf{q}, \mathbf{p} \to \mathbf{Q}, \mathbf{P}$ is sought on the phase flow of a Hamiltonian system, i.e.:

$$
\begin{gathered}
d\mathbf{X}/d\tau = G_Y, \quad d\mathbf{Y}/d\tau = -G_X, \\
\mathbf{X}(0) = \mathbf{q}, \ \mathbf{Y}(0) = \mathbf{p}; \quad \mathbf{X}(\varepsilon) = \mathbf{Q}, \ \mathbf{Y}(\varepsilon) = \mathbf{P},
\end{gathered} \tag{9.14}
$$

where: $\tau \in (0 \le \tau \le \varepsilon)$ is the helping parameter playing the role similar to that of time t.

The mentioned transformation is realized in our algorithm using the parametrization. The function governing transformation on the phase flow of the Hamiltonian (9.14) is defined through the equation

$$
\Psi_\tau(\tau, \mathbf{x}, \mathbf{y}) = G(\mathbf{x} + \frac{1}{2}\Psi_\mathbf{y}, \mathbf{y} - \frac{1}{2}\Psi_\mathbf{x}), \quad \Psi(0, \mathbf{x}, \mathbf{y}) = 0.
$$

Notice that $\tau^3 = \varepsilon^3$ is computed with the accuracy of $\Psi = \varepsilon G$. Therefore, asymptotics $\Psi_1 = G_0$, $\Psi_2 = G_1$, of two first methods overlap, and consequently R_1 and R_2 are identical in both approaches. However, one finds differences in next approximations for $R_3, R_4, \ldots$, in spite of that the normal form does not depend on the method choice.

9.5 Algorithm of invariant normalization for asymptotical determination of the Poincaré series

The proposed algorithm allows to find asymptotics of a general k-th order solution. It is worth noticing that it can be even more simplified for Hamiltonians periodical in time. In the latter case, instead of tracking trajectories $\mathbf{q}(t), \mathbf{p}(t)$ of motion, it is more convinient to choose the series of points $\mathbf{q}_m = \mathbf{q}(Tm), \mathbf{p}_m = \mathbf{p}(Tm)$ in time instants $t = t_m = Tm$, $(m = 0, 1, 2, \dots)$. Such set of points on a trajectory is further referred to as Poincaré map PM.

Asymptotical solution of PM is constructed in the following way. From quadratures (9.13) for $i = 1, \dots k$, the functions $\bar{F}_i(0, \mathbf{Q}, \mathbf{P})$ and $\Psi_i(0, \mathbf{x}, \mathbf{y})$ are found, and the last quadrature can be simplified taking $t_0 = 0$ in (9.13). Therefore, asymptotics of the k-th approximation are defined in the following way

$$\bar{F}^{(k)}(0, \mathbf{Q}, \mathbf{P}, \varepsilon) = \varepsilon \bar{F}_1(0, \mathbf{Q}, \mathbf{P}) + \dots + \varepsilon^k F_k(0, \mathbf{Q}, \mathbf{P}),$$
$$\bar{\Psi}^{(k)}(0, \mathbf{x}, \mathbf{y}, \varepsilon) = \varepsilon \bar{\Psi}_1(0, \mathbf{x}, \mathbf{y}) + \dots + \varepsilon^k \Psi_k(0, \mathbf{x}, \mathbf{y}),$$

and in the next step the Zhuravlev's theorem is applied.

Let $\mathbf{Q}(Tm, \mathbf{a}, \mathbf{b})$, $\mathbf{P}(Tm, \mathbf{a}, \mathbf{b})$ be PM of an unperturbed system, and let $\mathbf{X}(Tm, \mathbf{Q}_0, \mathbf{P}_0)$, $\mathbf{Y}(Tm, \mathbf{Q}_0, \mathbf{P}_0)$ be PM found from solution of the following equations

$$\dot{\mathbf{X}} = \frac{\partial}{\partial \mathbf{Y}} \bar{F}^{(k)}(0, \mathbf{X}, \mathbf{Y}, \varepsilon),$$

$$\dot{\mathbf{Y}} = -\frac{\partial}{\partial \mathbf{X}} \bar{F}^{(k)}(0, \mathbf{X}, \mathbf{Y}, \varepsilon),$$

$$\mathbf{X}(0) = \mathbf{Q}_0, \ \mathbf{Y}(0) = \mathbf{P}_0.$$

Then PM defined as $\mathbf{Q}_m, \mathbf{P}_m$ is the full Hamiltonian in new variables $\mathbf{Q}, \mathbf{P}$ obtained through substitution of $\mathbf{a} = \mathbf{X}(Tm, \mathbf{Q}_0, \mathbf{P}_0)$, $\mathbf{b} = \mathbf{Y}(Tm, \mathbf{Q}_0, \mathbf{P}_0)$ into the functions $\mathbf{Q}(Tm, \mathbf{a}, \mathbf{b})$, $\mathbf{P}(Tm, \mathbf{a}, \mathbf{b})$, $m = 0, 1, \dots$.

Observe that PM is expressed through the parametric exchange with the function $\bar{\Psi}^{(k)}(0, \mathbf{x}, \mathbf{y}, \varepsilon)$. Furthermore, the parameters $\mathbf{x}, \mathbf{y}$ can be excluded expressing them by $\mathbf{q}, \mathbf{p}$ in the following form

$$\mathbf{x}(\mathbf{q}, \mathbf{p}) = \mathbf{q} + \tfrac{1}{2}\Psi_{\mathbf{p}}(\mathbf{q}, \mathbf{p}) + \tfrac{1}{4}\{\Psi, \Psi_{\mathbf{p}}\} + \dots,$$
$$\mathbf{y}(\mathbf{q}, \mathbf{p}) = \mathbf{p} - \tfrac{1}{2}\Psi_{\mathbf{q}}(\mathbf{q}, \mathbf{p}) - \tfrac{1}{4}\{\Psi, \Psi_{\mathbf{q}}\} + \dots.$$

New variables are expressed through old ones in the following way: $\mathbf{Q} = 2\mathbf{x}(\mathbf{q}, \mathbf{p}) - \mathbf{q}$, $\mathbf{P} = 2\mathbf{y}(\mathbf{q}, \mathbf{p}) - \mathbf{p}$.

As a result, the following link is obtained between old and new parameters

$$\begin{aligned} \mathbf{Q}(\mathbf{q}, \mathbf{p}) &= \mathbf{q} + \Psi_{\mathbf{p}}(\mathbf{q}, \mathbf{p}) + \tfrac{1}{2}\{\Psi, \Psi_{\mathbf{p}}\} + \dots, \\ \mathbf{P}(\mathbf{q}, \mathbf{p}) &= \mathbf{p} - \Psi_{\mathbf{q}}(\mathbf{q}, \mathbf{p}) - \tfrac{1}{2}\{\Psi, \Psi_{\mathbf{q}}\} + \dots. \end{aligned} \tag{9.15}$$

In order to express old variables through new ones, it is sufficient to exchange them in formulas (9.15), and then to change the sign before Ψ:

$$\begin{aligned} \mathbf{q}(\mathbf{Q},\mathbf{P}) &= \mathbf{Q} - \Psi_{\mathbf{P}}(\mathbf{Q},\mathbf{P}) + \tfrac{1}{2}\{\Psi,\Psi_{\mathbf{P}}\} + \ldots, \\ \mathbf{p}(\mathbf{Q},\mathbf{P}) &= \mathbf{P} + \Psi_{\mathbf{Q}}(\mathbf{Q},\mathbf{P}) - \tfrac{1}{2}\{\Psi,\Psi_{\mathbf{Q}}\} + \ldots. \end{aligned} \tag{9.16}$$

It should be emphasized that in the method of invariant normalization [179, 180], the mentioned approach can be carried out applying Cambell-Hausdorf formulas, which (up to second approximation) overlap with (9.15) and (9.16) with accuracy of transformation of Ψ into the guiding Hamiltonian G.

9.6 Examples of asymptotical solutions

Highly didactic examples given in references [179, 180] demonstrate essential simplicity of the proposed method in comparison to all existing classical ones. Our method, from a point of view of its simplification, is equivalent to that proposed in [179, 180]. However, it has one more advantage. Namely, the obtained chain of equations for the asymptotics is completely independent of the input Hamiltonian form, i.e., there is no need to distinguish between autonomous and nonautonomous cases. Recall that in the method proposed in reference [179, 180], the nonautonomous system can be reduced to autonomous one by an increase of the system order, and then the chain of asymptotics equations must be derived.

We illustrate the introduced method using two examples of solutions to the problems of excited oscillators in resonance case. In order to solve these problems using classical approaches, one has to introduce another normal form definition [54]. This method does not require this additional operation. The normal form is computed directly from the quadrature and then a solution is constructed.

Example 1. Find a solution of excited oscillators of the linear oscillator in resonance.

Example 2. Let the excited oscillations of the nonlinear Duffing oscillator read: $\ddot{q} + q = \varepsilon(\sin t - q^3 + 2\lambda q)$. Find λ, for which a solution is periodic with the period 2π, and then analyze its stability.

In both examples, the invesigated equations are of Hamiltonian type, and they possess the same unperturbed Hamiltonian $H_0 = \frac{1}{2}(q^2 + p^2)$. The associated solution follows

$$q = q_0 \cos(t - t_0) + p_0 \sin(t - t_0), \quad p = -q_0 \sin(t - t_0) + p_0 \cos(t - t_0). \tag{9.17}$$

Solution to Example 1: $R_1 = F_1 = -q \sin t$. Substituting (9.17), the time periodic function is obtained with the associated integral $\int_{t_0}^{t} R_1(t)dt =$

$-\frac{1}{2}(q_0 \sin t_0 + p_0 \cos t_0)(t - t_0) - \frac{1}{4}(q_0 \cos t_0 + p_0 \sin t_0) + f(t)$. Therefore, one may easy derive normal form $\bar{F}_1$ of coefficients and the function Ψ_1. A solution to the first approximation is obtained in the following way. Since the Hamiltonian $\bar{F}(0, Q, P) = -\frac{\varepsilon}{2}P$ is associated with the system $\dot{Q} = -\frac{\varepsilon}{2}$, $\dot{P} = 0$, its solution reads: $Q = Q_0 - \frac{1}{2}\varepsilon t$, $P = P_0$. Substituting Q and P (instead of q_0 and p_0) into (9.17), the following solution is obtained $Q = (Q_0 - \varepsilon t/2)\cos t + P_0 \sin t$, $\quad P = -(Q_0 - \varepsilon t/2)\sin t + P_0 \cos t$ (Zhuravlev's theorem [180]). Then the transformation $Q = q - \frac{1}{2}\varepsilon \sin t$, $P = p + \frac{1}{2}\varepsilon \cos t$ is carried out with respect to the function Ψ, and the exact solution expressed through input variables reads: $q = (q_0 - \varepsilon t/2)\cos t + (p_0 + \frac{\varepsilon}{2})\sin t$. Since the solution of Example 2 is found with the use of the averaging procedure, and it is reported in references [179, 53], the normal form method (in order to compare both approaches) will be further applied.

It is sufficient to find PM using the algorithm on the basis of the quadrature (9.13). One finds $R_1 = F_1 = -q \sin t - \lambda p^2 + q^4/4$. Substituting the solution (9.17) into R_1, one finds the coefficient of the normal form series: $\bar{F}_1(0, Q, P) = -\frac{1}{2}P - \frac{\lambda}{2}(Q^2 + P^2) + \frac{3}{32}(Q^2 + P^2)^2$. A periodic solution is represented by a fixed point corresponding to the system

$$\partial \bar{F}_1/\partial Q = Q(-\lambda + \frac{3}{8}A^2) = 0, \quad \partial \bar{F}_1/\partial P = -\frac{1}{2} + P(-\lambda + \frac{3}{8}A^2) = 0.$$

One finds $Q = 0$, $P = \pm A$ for $\lambda = \frac{3}{8}A^2 \pm \frac{1}{2A}$, where $A = \sqrt{Q^2 + P^2}$ is the amplitude. The dependence $\omega = 1 - \varepsilon\lambda$ vs A is called the amplitude-frequency characteristics.

A fixed point is stable, if the corresponding function F_1 achieves its extremum. This yields stability condition for the periodic solution, i.e.:

$$(\lambda - \frac{3}{8}A^2)(\lambda - \frac{9}{8}A^2) > 0,$$

which fully overlaps with the condition obtained through the averaging procedure.

Let us give one more example. In this case, in order to get a solution, one has to compute higher order approximations, and rather complicated classical procedures are required. Our method yields the solution in much simpler way.

Example 3. Find PM for time instants $t_n = 2\pi n$ for the nonlinear equation $\ddot{q} = \varepsilon^2 \cos t' \cos q$ with accuracy of ε^6.

The equation governs various problem of mechanics and physics. One of them is that of vibrational motion of a spherical particle in a fluid, where a flat standing wave occurs [73, 124]. Consider a vertical tube with stiff horizontal roof. In the tube the standing wave is generated and its velocity is governed by the formula $v = A\omega \sin \omega t \cos kz$, where: ω is the wave frequency; t is the time; k is the wave number; z is the axis going vertically up; A is the amplitude of fluid particle displacement, which is assumed to be small. The frequency and wave number are dependent through sound velocity in the fluid $\omega = kc$.

Notice that for a particle with radius "a" the following inequality is satisfied $\mu/(\rho kca^2) \ll 1$, then the Stokes friction and the Bassé force are neglegible in comparison to inertial forces. Then, the equation governing particle dynamics reads

$$(\rho+2\rho_0)\ddot{z}_0 = 3\rho w - 2(\rho_0-\rho)g, \quad w = \partial v + v\partial v/\partial z \approx \partial v/\partial t = A\omega^2 \cos\omega t \cos kz,$$

where: ρ and ρ_0 are densities of fluid and solid particle, μ is coefficient of dynamical fluid viscosity.

For the particle of neutral floating $\rho = \rho_0$, the governing equation can be transformed to that of Example 3, in which $q = kz_0$, $t' = \omega t$, $\varepsilon = Ak$. In references [73, 124], in order to solve the defined problem the classical averaging technique is applied [53]. The series is developed with respect to the parameter ε. In order to solve the problem, three approximations are required. The solution is obtained in the form $q = \varepsilon f_1 + \varepsilon^2 f_2 + \varepsilon^3 f_3 + O(\varepsilon^4)$.

In what follows, we illustrate how to obtain the solution to the stated problem using our method. The splitting is carried out using the parameter $\delta = \varepsilon^2$. Therefore, in order to achieve essentially higher accuracy, i.e., of ε^6 order, only two approximations are required. Again, they are obtained in more simpler way in comparison to classical approaches.

Solution. The equation of our example is obtained from the system of Hamilton equations with the Hamiltonian $H = \frac{1}{2}p^2 + \delta F_1(t,q,p)$, $F_1 = -\cos t \sin q$.

A solution of the unperturbed system has the form $q = q_0 + p_0(t-t_0)$, $p = p_0$.

The first integration gives $R_1 = F_1$, and the quadrature $\int_{t_0}^t R_1 dt = -\frac{\cos(t_0+q_0)}{2+2p_0} + \frac{\cos(-t_0+q_0)}{2-2p_0} + f_1(t)$. Therefore, one gets

$$\bar{F}_1 = 0, \quad \bar{\Psi}_1(t,q,p) = -\frac{\cos(t+q)}{2+2p} + \frac{\cos(-t+q)}{2-2p}. \tag{9.18}$$

The second integration gives

$$R_2 = -\frac{1}{2}\frac{\partial F_1}{\partial q}\frac{\partial \bar{\Psi}_1}{\partial p} = \frac{1}{4}\cos t \cos q \left(\frac{\cos(t+q)}{(1+p)^2} + \frac{\cos(t-q)}{(1-p)^2} \right).$$

The integral (9.13) yields the linear part F_2, which is independent on time Ψ_2. The final normal form and the function defining the parametric transformation read

$$\begin{gathered} \bar{H} = \tfrac{1}{2}P^2 + \tfrac{\delta^2}{16}\left[\tfrac{1}{(1+P)^2} + \tfrac{1}{(1-P)^2}\right] + O(\delta^3), \\ \Psi(0,x,y) = \tfrac{\delta y}{1-y^2}\cos x - \tfrac{\delta^2(1-3y^2-2y^4)}{16y(1-y^2)^3}\sin 2x + O(\delta^3). \end{gathered}$$

However, note that the obtained normal form can not be applied in vicinity of the resonance points: $P = 0$, $P = \pm 1$.

9.7 A swinging oscillator

Consider an elastic pendulum with two-degrees-of-freedom, i.e., the heavy point mass swinging in the vertical plane and linked to the massless spring (Figure 9.1). The problem is formulated, for instance, in monographs [119, 157, 50], where also some investigations with respect to partial results are given.

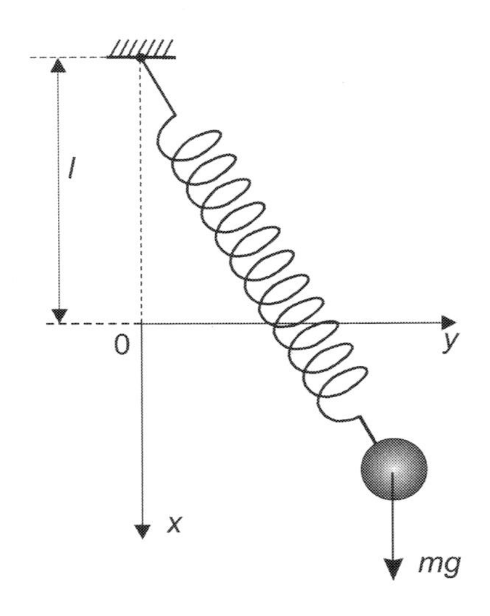

FIGURE 9.1: Scheme of a swinging oscillator.

However, high complexity of the used approaches do not allow to carry out a full analysis. Let us apply our method of invariant normalization with the help of parametric variable transformation [130].

The following notation is used: k is the spring stiffness; l is its length in the lumped body equilibrium position; m is the lumped body mass. In addition, $\omega = \sqrt{g/l}$ (frequency of small vibrations of mathematical pendulum of length l), and

$$\mu = \sqrt{\frac{k}{mg} + 1}.$$

Let us introduce the Cartesian co-ordinates with origin in the lumped system equilibrium position and with the axes along the vertical and horizontal directions. Denote by lx, ly the mass co-ordinates (Figure 9.1). The spring length is denoted as lR, where

$$R = \sqrt{(1+x)^2 + y^2}.$$

The spring tension is $T = k(lR - l_0)/l_0$, where l_0 is the nonstretched spring length. On the other hand, since l is the spring length in the equilibrium position, then $k(l - l_0)/l_0 = mg$. Substituting $l_0 = kl/(k + mg)$ into T, one gets $T = (k + mg)R - k$. Hence, it is clear that $\sqrt{(k + mg)/ml}$ is the vibration mass frequency for vertical spring position, whereas μ is the ratio of this frequency over ω.

The force components acting on the lumped system are: $F_x = mg - T(1 + x)/R$, $F_y = -Ty/R$. Newton's equations follow: $ml\ddot{x} = F_x$, $ml\ddot{y} = F_y$.

Potential E_p and kinetic E_c energies of the system have the form

$$E_p = -mglx + \frac{k}{2l_0}(lR - l_0)^2 - \frac{k}{2l_0}(l - l_0)^2,$$

$$E_c = \frac{m}{2}\left[\left(\frac{ldx}{dt'}\right)^2 + \left(\frac{ldy}{dt'}\right)^2\right] = \frac{mgl}{2}\left[\left(\frac{dx}{dt}\right)^2 + \left(\frac{dy}{dt}\right)^2\right],$$

where t' is dimensional time and $t = \omega t'$ is undimensional time.

Assuming undimensional impulses $u = \dot{x}$ and $v = \dot{y}$, the following Hamiltonian function $H = (E_c + E_p)/(mgl)$ is found

$$H = \frac{1}{2}(u^2 + v^2) + \frac{\mu^2}{2}(R^2 - 1) - (\mu^2 - 1)(R - 1) - x.$$

The associated constant in H is chosen in the way that $H(0, 0, 0) = 0$.

The system of Hamilton's equations has the form

$$\frac{dx}{d\tau} = \frac{\partial H}{\partial u}, \quad \frac{du}{dt} = -\frac{\partial H}{\partial x}, \quad \frac{dy}{dt} = \frac{\partial H}{\partial v}, \quad \frac{dv}{dt} = -\frac{\partial H}{\partial y}.$$

Next, we are going to investigate the lumped system motion in the neighbourhood of equilibrium (or large time durations t).

Let us change x, y, u, v into $\epsilon x, \epsilon y, \epsilon u, \epsilon v$ and H into $\epsilon^2 H$. Then the investigated system remains unchanged, whereas H takes the following form [50]

$$\begin{aligned}
H &= H_0 + \epsilon F_1 + \epsilon^2 F_2, \\
H_0 &= (1/2)(u^2 + v^2 + \mu^2 x^2 + y^2), \\
F_1 &= (1/2)(\mu^2 - 1)xy^2, \\
F_2 &= (1/2)(\mu^2 - 1)(y^4/4 - x^2 y^2).
\end{aligned}$$

9.8 Normal form

Owing to the illustrated algorithm, the following general solution of the unperturbed system with the Hamilton H_0 is found

$$\begin{aligned} x(t) &= X\cos\mu t + \frac{U}{\mu}\sin\mu t,\\ y(t) &= Y\cos t + V\sin t,\\ u(t) &= U\cos\mu t - \mu X\sin\mu t,\\ v(t) &= V\cos t - Y\sin t. \end{aligned} \tag{9.19}$$

First approximation. The function $R_1 = F_1 = \frac{(\mu^2-1)}{2}xy^2$ is obtained and then a solution of the unperturbed system is substituted to it. In the integral only terms independent of time $\left(\mu^4-4\mu^2\right)\int_0^t x(t)y^2(t)dt = \mu^2UY^2 - 2UV^2 - 2UY^2 - 2XYV\mu^2 + ...$ remain. In fact, they define the first approximation of the form

$$\Psi_1 = \frac{(\mu^2-1)}{2\left(\mu^4-4\mu^2\right)}\times \left[\mu^2\left(UY^2-2XYV\right) - 2UV^2 - 2UY^2\right]$$

of the parametric transformation.

Since linear (in time) term is equal to zero, the first approximation of the normal form perturbation is equal to zero, i.e. $\bar{F}_1 = 0$.

To conclude, both the normal form and variable transformation associated with the first approximation are computed.

Second approximation. The function $R_2 = F_2 + \frac{1}{2}\{F_1, \Psi_1\}$ is found, where Poisson's bracket is introduced $\{f,g\} = f_u g_x + f_v g_y - f_x g_u - f_y g_v$. We obtain

$$\frac{R_2(x,y,u,v)}{(\mu^2-1)} = \frac{1}{2}\left(\frac{y^4}{4} - x^2y^2\right) + \frac{(\mu^2-1)}{8\left(\mu^4-4\mu^2\right)}\times \left[-y^4\mu^2 + 2y^2v^2 + 2y^4 + 4x^2y^2\mu^2 + 8xyuv\right].$$

Instead of x, y, u, v the solution of unperturbed system is substituted and integrated again. The multiplier standing by linear (in time) part of this integral defines the normal form of the second approximation

$$\bar{F}_2 = \frac{3\left(\mu^2-1\right)\left(\mu^2X^2+U^2\right)\left(Y^2+V^2\right)}{8\,\mu^2\left(\mu^2-4\right)} - \frac{\left(\mu^2-1\right)\left(8+\mu^2\right)\left(Y^2+V^2\right)^2}{64\,\mu^2\left(\mu^2-4\right)}.$$

Note that function Ψ_2 is the part of integral, which does not depend on time, and it is

$$\Psi_2 = \left(\frac{-1}{64\,\mu^2\,(\mu^2-4)}\right) \times$$
$$(8\,U\,V^2\,X + 16\,U^2\,V\,Y - 8\,V^3\,Y + 40\,U\,X\,Y^2 -$$
$$8\,V\,Y^3 + 16\,U\,V^2\,X\,\mu^2 - 40\,U^2\,V\,Y\,\mu^2 + 7\,V^3\,Y\,\mu^2 +$$
$$+ 32\,V\,X^2\,Y\,\mu^2 - 64\,U\,X\,Y^2\,\mu^2 + V\,Y^3\,\mu^2 +$$
$$V^3\,Y\,\mu^4 - 8\,V\,X^2\,Y\,\mu^4 + 7\,V\,Y^3\,\mu^4).$$

9.9 Normal form integral

Let us investigate the system within new coordinates X, Y, U, V and with the Hamiltonian $H_0(X,Y,U,V)+\epsilon^2\bar{F}_2(X,Y,U,V)$. Owing to Zhuravlev's theorem, it is sufficient to find integral of the system with Hamiltonian $\epsilon^2\bar{F}_2(X,Y,U,V)$. The associated system of equations is

$$\dot{X} = \frac{3\,(\mu^2-1)\,\epsilon^2}{4\mu^2\,(\mu^2-4)}\,(Y^2+V^2)\,U,$$
$$\dot{U} = -\frac{3\,(\mu^2-1)\,\epsilon^2}{4\,(\mu^2-4)}\,(Y^2+V^2)\,X,$$
$$\dot{Y} = -\frac{3\,(\mu^2-1)\,\epsilon^2}{4\mu^2\,(\mu^2-4)}\,(\mu^2X^2+U^2)\,V +$$
$$\frac{(\mu^2-1)(\mu^2+8)\epsilon^2}{16\mu^2(\mu^2-4)}\,(Y^2+V^2)\,V,$$
$$\dot{V} = \frac{3\,(\mu^2-1)\,\epsilon^2}{4\mu^2\,(\mu^2-4)}\,(\mu^2X^2+U^2)\,Y -$$
$$\frac{(\mu^2-1)(\mu^2+8)\epsilon^2}{16\mu^2(\mu^2-4)}\,(Y^2+V^2)\,Y.$$

The system possesses two integrals: $Y^2+V^2=A$, $\mu^2X^2+U^2=B$, and after their account, the linear equations are obtained in the following form

$$\dot{X} = A\frac{3\,(\mu^2-1)\,\epsilon^2}{4\mu^2\,(\mu^2-4)}U,$$
$$\dot{U} = -A\frac{3\,(\mu^2-1)\,\epsilon^2}{4\,(\mu^2-4)}X,$$

$$\dot{Y} = \frac{(\mu^2 - 1)\epsilon^2}{\mu^2(\mu^2 - 4)} \left(-\frac{3}{4}B + \frac{\mu^2 + 8}{16\mu^2}A \right) V,$$

$$\dot{V} = -\frac{(\mu^2 - 1)\epsilon^2}{\mu^2(\mu^2 - 4)} \left(-\frac{3}{4}B + \frac{\mu^2 + 8}{16\mu^2}A \right) V.$$

Finally the following linear system is found

$$\dot{X} = \frac{\omega_1}{\mu}U, \quad \dot{U} = -\mu\omega_1 X,$$

$$\dot{Y} = \omega_2 V, \quad \dot{V} = -\omega_2 Y,$$

$$\omega_1 = \frac{3(\mu^2 - 1)\epsilon^2}{4\mu(\mu^2 - 4)}A,$$

$$\omega_2 = \frac{(\mu^2 - 1)\epsilon^2}{\mu^2(\mu^2 - 4)} \left(\frac{\mu^2 + 8}{16}A - \frac{3}{4}B \right)$$

possessing the following solutions

$$X = x_0 \cos\omega_1 t + (u_0/\omega_1)\sin\omega_1 t,$$

$$U = -\mu x_0 \sin\omega_1 t + \mu u_0 \cos\omega_1 t,$$

$$Y = y_0 \cos\omega_2 t + (v_0/\omega_2)\sin\omega_2 t,$$

$$V = -y_0 \sin\omega_2 t + (v_0/\omega_2)\cos\omega_2 t.$$

One may substitute the functions into (9.19) in order to get the full solution. To sum up, the solution with accuracy of ϵ^4 is successfully constructed.

References

[1] Amiro I.Ya., Zarutskiy V.A., *Theory of Ribbed Shells.* Kiev, Naukova Dumka, 1980, in Russian.

[2] Amiro I.Ya., Zarutskiy V.A., Poliakov P.S., *Ribbed Cylindrical Shells.* Kiev, Naukova Dumka, 1970, in Russian.

[3] Andrianov I.V., Asymptotic solutions for nonlinear systems with high degrees of nonlinearity. PMM, *Journal of Applied Mathematics and Mechanics* 38(2), 1993, 56–57.

[4] Andrianov I.V., Sequential construction of the asymptotic solution in the essentially nonlinear systems. *Doklady Maths* 50(1), 1994, 194–195.

[5] Andrianov I.V., Application of Padé approximants in perturbation methods. *Advances in Mechanics* 14(2), 1991, 3–25.

[6] Andrianov I.V., Awrejcewicz J., New trends in asymptotic approaches: Summation and interpolation methods. *Applied Mechanics Review* 54(1), 2001, 69–92.

[7] Andrianov I.V., Awrejcewicz J., Analysis of jump phenomena using Padé approximations. *Journal of Sound and Vibrations* 260, 2003, 577–588.

[8] Andrianov I.V., Awrejcewicz J., A role of initial conditions choice on the results obtained using different perturbation methods. *Journal of Sound and Vibration* 236(1), 2000, 161–165.

[9] Andrianov I.V., Awrejcewicz J., Asymptotic approaches to strongly non-linear dynamical systems. *Systems Analysis Modeling Simulation* 43(3), 2003, 255–268.

[10] Andrianov I.V., Awrejcewicz J., Matyash M., On application of a perturbation method with a few perturbation parameters. *Machine Dynamics Problems* 24(3), 2000, 5–10.

[11] Andrianov I.V., Awrejcewicz J., Manevitch L., *Asymptotical Mechanics of Thin-Walled Structures. A Handbook.* Berlin, Springer Verlag, 2004.

[12] Andrianov I.V., Awrejcewicz J., Iterative determination of homoclinic orbit parameters and Padé approximants. *Journal of Sound and Vibration* 240(2), 2001, 394–397.

[13] Andrianov I.V., Awrejcewicz J., Methods of small and large δ in the nonlinear dynamics – a comparative analysis. *Nonlinear Dynamics* 23, 2000, 57–66.

[14] Andrianov I.V., Ivanov A.O., New asymptotic method for solving of mixed boundary value problem. *International Series of Numerical Mathematics* 106, 39–45, 1993.

[15] Andrianov I.V., Lesnichnaya V.A., Manevich L.I., *Avaraging in Static and Dynamics of Ribbed Shells.* Moscow, Nauka, 1985, in Russian.
[16] Andrianov I.V., Lesnichaya V.A., Manevich L.I., On application of Padé approximation. *Izviestia AN SSSR, MZhG* 3, 1984, 160–167, in Russian.
[17] Andrianov I.V., Matyash M.V., Methods of small and large δ in the nonlinear dynamics. *Theoretical Foundations of Civil Engineering. VI Polish-Ukrainian Seminar*, Warsaw, 1998, 399–402, in Russian.
[18] Andronov I.V., Vitt A.A., Khaykin S.E., *Theory of Vibrations.* Moscow, Nauka, 1981, in Russian.
[19] Apresyan L.A., Padé approximation. Izviestia VUZ, *Radiophysics* 22(6), 1979, in Russian.
[20] Arnold V.I., *Mathematical Methods of Classical Mechanics.* Moscow, Editorial URSS, 2000, in Russian.
[21] Arnold V. I., *Additional Problems of Theory of Oridinary Differential Equations.* Moscow, Nauka, 1978, in Russian.
[22] Arnold V.I., Kozlov V.V., Neishtadt A.I., *Mathematical Aspect of Classical and Universe Mechanics, Achievements of Science and Technics: Series of Today Problems of Mathematics.* Dynamical Systems - 3 VINITI, 1985, in Russian.
[23] Awrejcewicz J., *Asymptotic Methods - a New Perespective of Knowledge.* Łódź, Łodź TU Press, 1995, in Polish.
[24] Awrejcewicz J., The analytical method to detect Hopf bifurcation solutions in the unstationary nonlinear systems. *Journal of Sound and Vibration* 129(1), 1989, 175–178.
[25] Awrejcewicz J., Determination of periodic oscillations in nonlinear autonomous discrete-continuous systems with delay. *International Journal of Solids and Structures* 27(7), 1991, 825–832.
[26] Awrejcewicz J., Analysis of the biparameter Hopf bifurcation. *Nonlinear Vibration Problems* 24, 1991, 63–76.
[27] Awrejcewicz J., Hopf bifurcations in Mathieu-Duffing oscillator. *Nonlinear Vibration Problems* 24, 1991, 161–172.
[28] Awrejcewicz J., Analysis of double Hopf bifurcation. *Nonlinear Vibration Problems* 24, 1991, 123–140.
[29] Awrejcewicz J., Andrianov I.V., Manevitch L.I., *Asymptotic Approaches in Nonlinear Dynamics: New Trends and Applications.* Berlin, Heidelberg, Springer Verlag, 1998.
[30] Awrejcewicz J. , Sendkowski D., How to predict stick-slip chaos in R^4. *Physics Letters A* 330, 2004, 371–376.
[31] Awrejcewicz J., Sendkowski D., Stick-slip chaos detection in coupled oscillators with friction. *Special Issue of International Journal of Solids and Structures* 42(21–22), 2005, 5669–5682.
[32] Awrejcewicz J., Someya T., Analytical condition for the existence of two-parameter family of periodic orbits in the autonomous system. *Journal of the Physical Society of Japan* 60(3), 1991, 784–787.
[33] Awrejcewicz J., Someya T., Analytical conditions for the existence of a two-parameter family of periodic orbits in nonautonomous dynamical

systems. *Nonlinear Dynamics* 4, 1993, 39–50.

[34] Awrejcewicz J., *Deterministic Vibrations of Lumped Systems.* Warszawa, WNT, 1996, in Polish.

[35] Awrejcewicz J., Andrianov I.V., Oscillations of nonlinear system with restoring force close to sign (x). *Journal of Sound and Vibration* 252(2), 2002, 962–966.

[36] Awrejcewicz J., Krysko V.A., Vakakis A., *Nonlinear Dynamics of Continuous Elastic Systems.* Berlin, Springer Verlag, 2004.

[37] Awrejcewicz J., Holicke M.M., Melnikov's method and stick-slip chaotic oscillations in very weakly forced mechanical systems. *International Journal of Bifurcation and Chaos* 9(3), 1999, 505–518.

[38] Awrejcewicz J., Delfs J., Dynamics of a self-excited stick-slip oscillator with two degrees of freedom, Part I: Investigation of equilibria. *European Journal of Mechanics, A/Solids* 9(4), 1990, 269–282.

[39] Awrejcewicz J., Delfs J., Dynamics of a self-excited stick-slip oscillator with two degrees of freedom, Part II: Slip-stick, slip-slip, stick-slip transitions, periodic and chaotic orbits. *European Journal of Mechanics, A/Solids* 9(5), 1990, 397–418.

[40] Awrejcewicz J., Lamarque C-H., *Bifurcation and Chaos in Nonsmooth Mechanical Systems.* Singapore, World Scientific, 2003.

[41] Awrejcewicz J., Olejnik P., Stick-slip dynamics of a two-degree-of-freedom system. *International Journal of Bifurcation and Chaos* 13(4), 2003, 843–861.

[42] Awrejcewicz J., Fečkan M., Olejnik P., On continuous approximation of discontinuous systems. *Nonlinear Analysis* 62(7), 2005, 1317–1331.

[43] Awrejcewicz J., Petrov A.G., On the method of asymptotical integration of the Hamiltonian systems. *Fifth EUROMECH Nonlinear Dynamics Conference, ENOC2005*, August 7-12, 2005, Eindhoven, the Netherlands, 1496–1504.

[44] Babitsky V.J., *Theory of Vibro-Impact Systems and Applications.* Berlin, Springer-Verlag, 1988.

[45] Babitsky V.J., Krupenin V.L., *Oscillations in Strongly Nonlinear Systems.* Moscow, Nauka, 1985, in Russian.

[46] Baj F., Champneys A.R., Numerical detection and continuation of saddle-node homoclinic bifurcations of codimension one and two. *Dynamics and Stability of Systems* 11(4), 1996, 325–346.

[47] Baker Jr G.A., Graves-Morris P., *Padé Approximants.* Cambridge, Cambridge University Press, 1996.

[48] Baker Jr G.A., Graves-Morris P., *Padé Approximants I and II.* Encyclopedia of Mathematics and Its Applications, Vols. 13 and 14, Addison-Wesley, Reading MA, 1981.

[49] Balasuriya S., Mezic I., Jones C.K.R.T, Weak finite-time Melnikov theory and 3D viscous perturbations of Euler flows. *Physica D* 176, 2003, 82–106.

[50] Bagaevskiy V.N., Povzner A.Ya., *Algebraic Methods in Nonlinear Theory of Perturbation.* Moscow, Nauka, 1987, in Russian.

[51] Bender C.M., Milton K.A., Pinsky S.S., Simmonds L.M., Nonperturbative approach to nonlinear problems. *Journal of Mathematical Physics* 30(7), 1989, 1447–1455.
[52] Birkhoff D.D., *Dynamical Systems.* Moscow-Leningrad, Gostekhizdat, 1941, in Russian.
[53] Bogoliubov N.N., Mitropolskiy Yu.A., *Asymptotical Methods in Theory of Nonlinear Vibrations.* Moscow, Nauka, 1974, in Russian.
[54] Briuno A.D., *Bounded Three Bodies Problem.* Moscow, Nauka, 1990, in Russian.
[55] Brogliato B., *Nonsmooth Mechanics.* London, Springer-Verlag, 1999.
[56] Burgers J.M., A mathematical model illustrating the theory of turbulence. *Advances in Applied Mechanics* 1, 1948, 171–199.
[57] Bush A.W., *Perturbation Methods for Engineers and Scientists.* Boca Raton, CRC Press, 1992.
[58] Cesari L., *Asymptotic Behavior and Stability Problems in Ordinary Differential Equations.* Berlin, Springer-Verlag, 1959.
[59] Champneys A.R., Kuznetsov Yu.A., Sandstede B., A numerical toolbox for homoclinic bifurcation analysis. *International Journal of Bifurcation and Chaos* 6(5), 1996, 867–887.
[60] Chen X., Lorenz equations Part I: Existence and nonexistence of homoclinic orbits. *SIAM Journal on Mathematical Analysis* 27(4), 1996, 1057–1069.
[61] Cheung Y.K., Chen S.H., Lau S.L., A modified Lindstedt - Poincaré method for certain strongly nonlinear oscillators. *International Journal of Nonlinear Mechanics* 26(3/4), 1991, 361–368.
[62] Crocco L., Coordinate perturbation and multiple scale in gasdynamics. *Philos. Transactions of Royal Society, Ser. A* 272(1222), 1972, 275–301.
[63] Cole J.D., *Perturbation Methods in Applied Mathematics.* Waltham, MA, Ginn-Blaisdell, 1968.
[64] Deimling K., *Multivalued Differential Equations.* Berlin, Walter de Gruyter, 1992.
[65] Eckhaus W., *Asymptotic Analysis of Singular Perturbations.* Amsterdam, North Holland, 1979.
[66] Erdeyn A., *Asymptotical Series.* Moscow, Fizmatgiz, 1962, in Russian.
[67] Fathi M.A., Salam A., The Melnikov technique for highly dissipative systems. *SIAM Journal of Mathematical Analysis* 47, 1987, 232–243.
[68] Farkas M., *Periodic Motions*, Appl. Math. Sci. 104., New York, Springer-Verlag, 1994.
[69] Fečkan M., Chaotic solutions in differential inclusions: chaos in dry friction problems. *Trans. American Mathathematical Society* 351, 1999, 2861–2873.
[70] Filippov A.F., *Differential Equations with Discontinuous Right-Hand Side.* Dordrecht, Kluwer, 1988.
[71] Friedrichs K.O., Asymptotic phenomena in mathematical physics. *Physics Bulletin American Mathematical Society* 61, 1955, 485–504.

[72] Friedrichs K.O., *Perturbation of Spectra in Hilbert Space.* Providence, RI, AMS, 1965.

[73] Ganiyev R.F., Ukrainskiy L.E., *Dynamics of Particles Subject to Vibrations.* Kiev, Naukova Dumka, 1975, in Russian.

[74] Gelfreich V.G., Melnikov method and exponentially small splitting of separatrices. *Physica D* 101, 1997, 227–248.

[75] Golidenveizer A.L., *Theory of Thin Elastic Shells.* Moscow, Gostekhizdat, 1953, in Russian.

[76] Golidenveizer A.L., Lidskaya V.A., Tovstik P.E., *Free Vibrations of Thin Elastic Shells.* Moscow, Nauka, 1979, in Russian.

[77] Govindan Potti P.K., Sarma M.S., Nageswara Rao B., On the exact periodic solution for $(\ddot{x} + \operatorname{sign} x) = 0$. *Journal of Sound and Vibration* 220, 1999, 378–381.

[78] Greben E.S., On the deformation and equilibrium of reinforced shells. *Izvestia AN SSSR MTT* 5, 1969, 106–114, in Russian.

[79] Grigoluk E.I., Tolkochev V.M., *Contacts Problems of Plates and Shells.* Moscow, Mashinostroyeniye, 1980, in Russian.

[80] Guckenheimer J., Holmes P.J., *Nonlinear Oscillations, Dynamical Systems and Bifurcations of Vector Fields.* Berlin, Springer, 1983.

[81] Gruendler J., The existence of homoclinic orbits and the method of Melnikov for systems in R^n. *SIAM Journal on Mathematical Analysis* 16, 1985, 907–931.

[82] Hassard B., Zhang J., Existence of a homoclinic orbit of the Lorenz system by precise shooting. *SIAM Journal on Mathematical Analysis* 25(1), 1994, 179–196.

[83] Hastings S.P., Troy W.C., A proof that the Lorenz equations have a homoclinic orbit. *Journal of Differential Equations* 113, 1994, 166–188.

[84] Hinch E.J., *Perturbation Methods.* Cambridge, Cambridge University Press, 1991.

[85] Kamke E., *Handbook on Ordinary Differential Equations.* Moscow, Nauka, 1971, in Russian.

[86] Kaplun S., Low Reynolds number flow past a circular cylinder. *Journal of Mathematics and Mechanics* 6, 1957, 595–603.

[87] Kassoy D.R., A note on asymptotic methods for jump phenomena. *SIAM Journal of Applied Mathematics* 42(4), 1982, 926–932.

[88] Keller J.B., *Perturbation Theory.* Lecture Notes, Michigan, Michigan State University, Departament of Mathematics, 1968.

[89] Kevorkian J., Cole J.D., *Multiple Scales and Singular Perturbation Methods.* Berlin, Springer-Verlag, 1996.

[90] Kluwick A., The analytical method of characteristic. *Prog. Aerospace Science* 19(2-4), 1981, 197–313.

[91] Krylov N.M., Bogulobov N.N., *Introduction into Nonlinear Mechanics.* Kiev, AN SSSR, 1937, in Russian.

[92] Kun P., *Strength Computations of Shells in Airplanes.* Moscow, Oborongiz, 1961, in Russian.

[93] Kunze M., *Nonsmooth Dynamical Systems.* Lecture Notes in Mathematics, Vol. 1744, Berlin, Springer-Verlag, 2000.

[94] Lagerstrom P.A., *Matched Asymptotic Expansions: Ideas and Techniques.* Berlin, Springer Verlag, 1988.

[95] Lagerstrom P.A., Casten R.G., Basic concepts underlying singular perturbation techniques. *SIAM Reviews* 14, 1972, 63–120.

[96] Lamarque C.-H., Bastien J., Numerical study of a forced pendulum with dry friction. *Nonlinear Dynamics* 23, 2000, 335–352.

[97] Lambert F., Padé approximants and closed form solutions of the KdV and MKdV equations. *Z Physik C, Particles & Fields* 5, 1980, 147–150.

[98] Lambert F., Musette M., Solitary waves, padeons and solitons. *Lecture Notes in Mathematics* 1071, 1984, 197–212.

[99] Lambert F., Musette M., Solitons from a direct point of view: padeons. *Journal of Computational and Applied Mathematics* 15, 1986, 235–249.

[100] Langer R.E., On the asymptotic solutions of ordinary differential equation with reference to Stoke's phenomenon about a singular point. *Trans. Mathematical Society* 37, 1935, 397–420.

[101] Lassoued L., Mathlouthi S., A numerical method for finding homoclinic orbits of Hamiltonian systems. *Numerical Functional Analysis and Optimization* 13(1–2), 1992, 155–172.

[102] Leine R.I., Van Campen D.H., Van de Vrande B.L, Bifurcations in nonlinear discontinuous systems. *Nonlinear Dynamics* 23, 2000, 105–164.

[103] Lighthill M.J., The position of the shock-wave in certain aerodynamics problems. *Quart. Journal of Mechanics and Applied Mathematics* 1(2), 1948, 309–318.

[104] Lighthill M.J., A technique for rendering approximate solutions to physical problems uniformly valid. *Philosophical Magazine* 40, 1949, 1179–1201.

[105] Lipscomb T., Mickens R.E., Exact solution to the antisymmetric, constant force oscillation equation. *Journal of Sound and Vibration* 169, 1994, 138–140.

[106] Lomov S.A., *Introduction in Theory of Singular Perturbation.* Moscow, Nauka, 1981, in Russian.

[107] Luke J.C., A perturbation method for nonlinear dispersive wave problems. *Proceedings of the Royal Society London Ser. A* 292, 1966, 403–412.

[108] Liverani C., Turchetti G., Existence and asymptotic behaviour of Padé approximants to the Korteveg - de Vries multisolitons solutions. *Journal of Mathematical Physics* 24(1), 1983, 53–64.

[109] Madelung E., *Mathematical Approaches in Physics.* Moscow, Nauka, 1968, in Russian.

[110] Malinetskiy G.G., *Chaos, Structures and Numerical Experiments.* Moscow, Nauka, 1997, in Russian.

[111] O'Malley Jr R.E., *Singular Perturbation Methods for Ordinary Differential Equations.* New York, Springer-Verlag, 1991.

[112] Marchuk G.I., Agoshkov V.I., Shutyaev V.P., *Adjoint Equations and Perturbation Algorithms in Nonlinear Problems.* Boca Raton, CRC Press, 1996.

[113] Martin P., Baker Jr G.A., Two-point quasifractional approximant in physics. Truncation error. *Journal of Mathematical Physics* 32(6), 1991, 313–328.

[114] Maslov V.P., Omel'yanov G.A., *Geometric Asymptotics for Nonlinear PDE. I.* Providence, Rhode Island, AMS, 2001.

[115] Melnikov V.K., On the stability of the center for time-periodic perturbations. *Trans. Moscow Mathematical Society* 12, 1963, 1–56.

[116] Mitropolskiy Yu.A., *Lectures of Averaging in Nonlinear Mechanics.* Kiev, Naukova Dumka, 1986, in Russian.

[117] Mitropolskiy Yu.A., Moseenkov B.N., *Lectures on Application of Asymptotical Methods for Solution of Partial Differential Equations.* Kiev, Naukova Dumka, 1968, in Russian.

[118] Moiseev N.N., *Asymptotical Methods of Nonlinear Mechanics.* Moscow, Nauka, 1969, in Russian.

[119] Nayfeh A.H., *Perturbation Methods.* New York, Wiley, 1973.

[120] Nayfeh A.H., Numerical-perturbation methods in mechanics. *Computations and Structures* 30(1-2), 1988, 185–204.

[121] Nayfeh A.H., *Introduction to Perturbation Techniques.* New York, John Wiley & Sons, 1981.

[122] Nayfeh A.H., Kluwick A., A comparison of three perturbation methods for nonlinear hyperbolic waves. *Journal of Sound and Vibration* 48(2), 1976, 293–299.

[123] Nayfeh A.H., Mook D., *Nonlinear Oscillations.* New York, Wiley Interscience, 1986.

[124] Nigmatulin R.I., *Dynamics of Multi-Phase Media.* Moscow, Nauka, 1987, in Russian.

[125] Obraztsov I.F., Nerubaylo B.V., Andrianov I.V., *Asymptotic Methods in Structural Mechanics of Thin-Walled Structures.* Moscow, Mashinostroyeniye, 1991, in Russian.

[126] Panchenkov N.S., *Asymptotical Methods of Solutions of Mathematical Physics Problems.* Applied Mathematics, Irkutsk, Irkutsk University Press, 1969, in Russian.

[127] Panchenkov A.W., Sigalov G.F., Method of full approximation in problems of transonic flow dynamics. *Applied Mathematics,* Vol. 2, Irkutsk, Irkutsk University Press, 1971, in Russian.

[128] Petrov A.G., Parametric method of Poincaré maps construction in hydrodynamic systems. *Prikladnaya Matematika i Mekhanika* 66(3), 2002, 948–967, in Russian.

[129] Petrov A.G., Modification of the method of invariant normalization of Hamiltonians using parametrization of canonical transformations. *Doklady Akademii Nauk, Mekhanika* 386(4), 2002, 482–486, in Russian.

[130] Petrov A.G., Zaripov M.N., Nonlinear oscillations of a swinging spring. *Doklady Akademii Nauk, Mekhanika* 399(3), 2004, 347–352, in Russian.

[131] Pilipchuk V.N., Analytical study of vibrating systems with strong nonlinearities by employing saw-tooth time transformations. *Journal of Sound and Vibration* 192(1), 1996, 43–64.

[132] Piskunov N.S., *Differential and Integral Computation for High Schools*, Vol. 2. Moscow, Science, 1978, in Russian.

[133] Poincaré A., *Collected Works in Three Volumes*, Volume II. Moscow, Nauka, 1972, in Russian.

[134] Prokopov V.K., Frame method of computation of ribbed cylindrical shells. *Scientific - Technical Information Bulletin, Series of Physical - Mathematical Science* 12, 1957, in Russian.

[135] Pug M., Saymon B., *Methods of Nowadays Mathematical Physics. Operator Analysis.* Moscow, Mir, 1982, in Russian.

[136] Pandey B. C., Study of cylindrical piston problems in water using PLK method. *ZAMP* 19, 1968, 962–963.

[137] Pfeiffer F., Hajek M., Stick-slip motions of turbine blade dampers. *Philosophical Transactions Royal Society London* 338, 1992, 503–517.

[138] McRae S.M., Vrscay E.R., Perturbation theory and the classical limit of quantum mechanics. *Journal of Mathematical Physics* 38(6), 1997, 2899–2921.

[139] Reiss E.L., A new asymptotic method for jump phenomena. *SIAM Journal of Applied Mathematics* 39(3), 1980, 440–455.

[140] Riabov V.M., Application of successive approximations while computing ribbed shells. *Izv. AN SSSR, Mechanika i Mashinostroyeniye* 6, 150–154, in Russian.

[141] Reed M., Simon B., *Methods of Modern Mathematical Physics. Analysis of Operations*, vol. 4. New York, Academic Press, 1978.

[142] Robinson J.C., All possible chaotic dynamics can be approximated in three dimensions. *Nonlinearity* 11, 1998, 529–545.

[143] Romakin A.G., Titarenko V.V., On improvement of approximation of weakly impacting waves. *Journal of Acoustics* 34(2), 1988, 303–305.

[144] Romakin A.G., Titarenko V.V., On damping of saw-like pressure wave in a pipe. *Journal of Acoustics* 76(1), 1990, 178–179.

[145] Romakin A.G., Titarenko V.V., Computation of nonstationary interactions of impacting waves using Padé approximation. Nonstationary flow with impacting waves. Leningrad, *FT I AN SSSR*, 1990, 216–221.

[146] Rosenberg R.M., On nonlinear vibrations of systems with many degrees of freedom. *Advances in Applied Mechanics* 9, 1996, 85–92.

[147] Rosenberg R.M., The Ateb(h)-functions and their properties. *Quartely of Applied Mathematics* 21(1), 1963, 37–47.

[148] Salenger G., Vakakis A.F., Gendelman O., Manevitch L.I., Andrianov I.V., Transition from strongly-to weakly-nonlinear motions of damped nonlinear oscillators. *Nonlinear Dynamics* 20, 1999, 99–114.

[149] Samarskii A.A., Galaktionov V.A., Kurdyumov S.P., Mikhaylov A.P., *Blow-up Quasilinear Parabolic Equations.* Berlin, New York, De Gruyter, 1995.

[150] Schester S., Rate of convergence of numerical approximations to homoclinic bifurcation points. *IMA Journal of Numerical Analysis* 15, 1995, 23–60.

[151] Sedov L.I., *Method of Similarities in Mechanics.* Moscow, Gostekhizdat, 1957, in Russian.

[152] Senik P.M., Inversions of Incomplete Beta-function. *Ukrainian Mathematical Journal* 21(3), 1969, 325–333, in Russian.

[153] Shamrovskiy A.D., Asymptotical integrations of static equations of elasticity in Descartes coordinates with automized search of parameters. *PMM* 45(5), 1979, 859–868.

[154] Sigalov G.F., *Method of Full Approximation of Theory of Transonic Flows.* Irkutsk, Irkutsk University Press, 1988.

[155] Smith P., The multiple scales method, homoclinic bifurcation and Melnikov's method for autonomous systems. *International Journal of Bifururcation and Chaos* 8(11), 1998, 2094–2105.

[156] Spanier J., Oldham K.B., *An Atlas of Functions.* Berlin, Springer Verlag, 1987.

[157] Starginsky, V.M., *Applied Methods of Nonlinear Oscillations.* Nauka, Moscow, 1977, in Russian.

[158] Stelter P., Nonlinear vibrations of structures induced by dry friction. *Nonlinear Dynamics* 3, 1992, 329–345.

[159] Sun J.H., Melnikov vector function for high-dimensional maps. *Physics Letters A* 216, 1996, 47–52.

[160] Trzaska Z., A simple approach for solving linear differential equations of the second order. *The Mathematical Gazette* 82(494), 1998, 294–297.

[161] Tsian S.S., Poincaré-Lighthill's method, Problems in Mechanics. Moscow, IL, 1959, 7–62, in Russian.

[162] Turchetti G., Padé approximants and soliton solutions of the KdV equation. *Lettres Nuovo Cimento* 27(4), 1980, 107–110.

[163] Tuwankotta J.M., Verhulst F., Symmetry and resonance in Hamiltonian systems. *SIAM Journal of Applied Mathematics and Mechanics* 61 (4), 2000, 1369–1385.

[164] Vakakis A.F., Manevitch L.I., Mikhlin Yu.V., Pilipchuk V.N, Zevin A.A., *Normal Modes and Localization in Nonlinear Systems.* New York, Wiley Interscience, 1996.

[165] Vakakis A.F., Azeez M.F.A., Analytic approximation of the homoclinic orbits of the Lorenz system at $\delta = 10$, $b = 8/3$ and $p = 13.926\ldots$. *Nonlinear Dynamics* 15, 1998, 245–257.

[166] Van Dyke M., *Perturbation Methods in Fluid Mechanics.* Stanford, California, The Parabolic Press, 1975.

[167] Van Dyke M.D., *Mathematical Approaches in Hydrodynamics.* Philadelphia, SIAM, 1991.

[168] Van Dyke M., Computer extension of perturbation series in fluid mechanics. *SIAM Journal of Applied Mathematics* 28, 1975, 72–734.

[169] Vasilieva A.B., Butuzov V.F., Kalachev L.V., *The Boundary Function Method for Singular Pertubation Problems.* Philadelphia, SIAM, 1995.

[170] Vasilieva A.B., Butuzov V.F., *Asymptotical Methods in Theory of Singular Perturbation.* Moscow, Vysshaia Shkola, 1990, in Russian.
[171] Verhulst, F., *Nonlinear Differential Equations and Dynamical Systems.* Berlin, Springer Verlag, 1996.
[172] Verhulst, F., *Methods and Applications of Singular Perturbations: Boundary Layers and Multiple Timescale Dynamics.* Berlin, Springer Verlag, 2005.
[173] Vishik M.I., Lyusternik L.A., The asymptotic behavior of solutions of differential equations with large or quickly changing coefficients and boundary conditions. *Russian Mathematical Surveys* 15(4), 1960, 23–91.
[174] Vishik M.I., Lusternik L.A., Regular singularity and boundary laser in linear differential equation with small parameter. *Uspekhi Matematicheskikh Nauk* 12(5), 1957, 3–122, in Russian.
[175] Weiyao Z., Jiaowan L., Exponential dichotomies and Melnikov functions for singularly perturbed systems. *Nonlinear Analysis* 36, 1999, 401–422.
[176] Whitham G.B., *Linear and Nonlinear Waves.* New York, Wiley, 1974.
[177] Wiggins S., *Global Bifurcations and Chaos.* Berlin, Springer, 1989.
[178] Zhimin P.A., Linear theory of ribbed shells. *Izvestia AN SSSR, MTT* 6, 1970, 150–162, in Russian.
[179] Zhuravlev V.F., *Introduction to Theoretical Mechanics.* Moscow, Nauka Fizmatlit, 1997, in Russian.
[180] Zhuravlev V.F., Invariant normalization of non-autonomous Hamiltonians. *Prikladnaya Matematika i Mekhanika* 66(3), 2002, in Russian.

Index

acceleration, 5
    of the convergence, 144
additive, xvii, 87, 88, 113
aerodynamics, 2, 48, 164
Agranowsky-Baglay-Smirnov method, 171
algebraic
    equations, 9, 37, 52, 69, 150, 174
    problems, 172
algorithm, 57, 221, 223, 224, 226
amplitude, 4–6, 8, 41, 44, 58, 73, 74, 92, 94, 95, 130, 146, 178, 226
    frequency characteristic, 94
    of vibrations, 4, 8
amplitude-frequency characteristics, 95, 226
analytical method of characteristics, 134, 136, 139
antisymmetric vibrations, 176
approximation
    first order, 40, 41, 45, 48, 49, 51, 76, 82, 84
area of nonuniformity, 25
asymptotic
    approaches, xvi, 9, 22, 35, 71, 108
    approximation with respect to parameter ε, 39
    decomposition, xvi, xvii, 23–30, 32, 33, 35–37, 39, 42, 43, 46, 47, 58, 59, 61, 62, 77–80, 82–85, 87, 88, 90–93, 112, 113, 146, 164
        direct, 93
        nonuniform, 9, 25
    determination, 128, 224
    matching decompositions, 77
    methods, xv, 149
    methods of motion separation, 74
    sequence, xvi, 9, 22, 27
    series, xvi, 9, 27–30, 35, 38–40, 43, 44, 47, 54–56, 62, 72, 79, 95, 113, 117, 121, 123–130, 132–137, 139, 157, 163
        infinite, 29, 31
        matched, 118, 121, 127, 129, 130
        nonuniform, 28
        two-term, 112
    solution, 224, 225
    splitting, 155
autonomous, xviii, 158, 192, 205, 217, 221, 223, 225
    Hamiltonian, 220
averaged
    equation, 168
    system, 177
    value within the period, 74
averaging, xvi, xvii, 35, 62, 72, 74–76, 97, 107, 113, 165, 167, 169, 181, 182, 226, 227
    method of, xvi, 35, 62, 72, 74, 75, 107, 113, 167, 169, 226, 227
    procedure, xvii, 75, 76, 165
axioms, 1

bending, 165, 173, 176, 177, 187, 190
Beta function, 105
    incomplete, 103, 105

biological models, xvii, 141, 149
Biot number, 2
blow-up
of the solution, 152
phenomenon, xvii, 141, 149, 151
boundaries, 176, 179, 184
boundary
condition, xv, 7, 16, 17, 26, 36, 47, 53, 79, 80, 82–84, 90, 93, 94, 115–117, 119, 120, 122, 123, 130, 132, 145, 151, 167–174, 176–181, 183–185, 187
effect, method of, 149
equation, 169
layer, xvi, 39, 77, 78, 82, 83, 92, 181, 183–185, 187
problem, 7, 145, 150, 169, 170, 176, 181, 182, 185
solution, xv
value problems, 16, 17, 57, 79–82, 113, 118
bounded, 11, 17, 26, 35, 47, 54, 56, 71, 73, 75
buckling of constructions, xvii, 141

canonical transformation, xviii, 217–222
Cauchy problem, 5, 14–16, 37, 80, 102, 112, 162
initial, 96
chaos, 191, 192, 199, 201
characteristic equation, 3, 58, 135
characteristics, xv–xvii, 1, 78, 115, 124, 133–137, 139, 141, 143, 164, 174
circumsonic
parameter of similarity, 145
regime, xvii, 141, 145
clamping, 174, 176
classical perturbation
approach, 40, 77, 92
method, xvi, 39, 40, 42, 49, 53, 58, 78, 92, 112, 116
coefficient α, 148
combinations, 2, 10, 14, 42, 59, 62, 73
combustion engine, xvii, 141, 149, 151
comparison functions, 19, 79
complex
decomposition, 79
decomposition additive, 87
form, 64
function, 29, 52, 67, 137
process, 8
singularity, 49
composite function differentiation rule, 4
compressibility, 2
conjugated quantities, 64
constant shock waves, 148, 149
constants variation, method of, 71, 73, 74
continuous
approximation, xviii, 203, 209, 212
functions, xviii, 68, 177, 203
convergence of iteration procedures, 158
convergent, 50, 59, 144, 150
series, 10, 11, 13, 29–31, 35, 143
corrections, 29, 52, 109, 167, 179–181, 184, 185, 187, 190
Cramer's
formulas, 49
rule, 52, 69
cusp, 199
cylindrical shell, 176

d'Alembert's principle, 10
damped linear oscillator analysis, 5
damping, 3, 5, 58, 98, 105–107, 145, 146
de l'Hospital
formula, 19
rule, 19
decomposition, xvi, xvii, 9, 17, 23–30, 32, 33, 35–37, 39, 42–

44, 46, 47, 58–63, 67, 68, 77–88, 92, 112, 113, 182, 186, 187
deficient, 19
definite integrals, 14
deflection of the infinite band, 187
deformation, 47, 48, 51–54, 56, 57, 125, 126, 129, 133, 134
  coefficient, 51, 53–56
  functions, 47
  of one independent variable, 53
  of two independent variables, 52, 55
deformed
  coordinates, method of, 52, 53, 57, 61, 90, 119, 123, 144, 145
  variable, 35, 50, 52, 54, 55, 57
  variables, method of, xvi, 47, 50, 52, 90
derivative, xvi, 52, 62, 63, 66, 82, 84, 92, 93, 137, 139, 183, 218
  decomposition, xvi, 62, 63
  function, 218
  highest, 92
  operator, 62
determination of eigenvalues, 109
differential equation, xv, xviii, 3–6, 14–16, 41, 43, 45, 48, 49, 61, 64, 68, 70, 73, 77, 80–82, 92, 93, 117, 134, 151, 156, 174, 185, 191, 203, 215
  approximated, xviii, 203
  first order, 14, 37, 49
  linear, 5, 15, 37, 63
  nonhomogeneous, 46, 67
  nonlinear, 40, 97, 115, 130
  normal, xviii
  ordinary, 14, 92, 170, 214, 215
  partial, 47, 52, 61, 92
  second order, 58, 68, 71, 72, 82, 105, 118, 135
  singular, xviii, 203
differential-integral equation, 7
dimension quantities, xvi, 1
dimensionless
  amplitude, 6
  displacement, 6
  form, 2, 3, 5, 6, 174
  model, 5
  number, 2
  parameter, 5, 7, 130
  quantity, xvi, 1, 2, 4, 5, 164
  variable, 6, 78
  viscosity coefficient, 4
direct decomposition along a coordinate, 17
Dirichlet series, 150
discontinuity, xviii, 177, 203
  level, 208, 212
discontinuous, xviii, 157, 190, 192, 201, 203, 205, 208, 212
  system, xviii, 203, 212
divergent, 29
  sequence, 31
  series, 30–32, 35, 143, 144, 163
Duffing
  equation, 6, 8, 41, 43, 54, 59, 66, 70, 94, 113
  oscillator, xvi, 44, 225
  problem, 7, 8, 40, 44, 45, 47, 58, 70, 75
    uniformly suitable solution, 46
  type stiffness, xvii
dynamic viscosity coefficient, 2

eigenfrequency, 3–5
eigenfunction, 109
eigenvalue, 109, 194, 211, 219
elastic deflection of a beam, 166
elementary functions, xvi, 9, 11, 14
elongated parameters, method of, xvi, 43, 44, 47, 48, 55, 68, 112, 113
entropy gain, 146
equation

linear differential homogeneous, 42
of characteristic, 131
of equilibrium, 187
of the motion of the supported plate, 175
equivalent averaging method, 107
errors' function, 30
estimation, xvi, 9, 13, 14, 18–21, 30, 31, 60, 98, 104, 105, 163, 201
Euler's formulas, 64
excited oscillators, 225
expansion, 59
asymptotic, 53, 207
coefficients, 91
external, 80
matched asymptotic, xvii, 87, 90
of elementary functions, 11
of functions, 9
of solution, 103
terms of, 39
experiment, 2, 7
exponential
approximation, 163
behavior, 194
dichotomies, 191
function, 22, 30, 59
exponentially small terms, 19, 86
external
decomposition, 78–80, 82, 85, 87–89
equations, 79
limit, 78, 85, 89, 90, 119
solution of k-th approximation, 79
space, 78, 80, 81, 89
variables, 78, 83, 86, 119
external and internal equations, 79

fast, 72, 74, 76, 137, 166–169, 178
variable, 72, 107
few perturbation parameters, xvii, 108, 110, 111
fixed point, 191, 208, 226
fluid, xvii, 92, 115, 191, 226, 227
mechanics, 48
Fourier series, 160
fragmentally continuous functions, 160
frequency, 8, 41, 43–46, 68, 73, 94, 95, 109, 148, 177, 178, 182, 191, 194, 226
friction, xvii, 2, 108, 191–193, 204, 205, 227
coefficient, 2, 204
frictional characteristics, 192
front of the impact wave, xvii, 115, 119
Froude number, 2
frozen coefficients, 107
full approximation
approach, 57
method of, xvi, 35, 52, 57, 58
fundamental matrix solution, 215

gases, xvii, 115
generating solution, xv
geometrical asymptotics, 149
Germain-Lagrange equation, 173
Gregory series, 144
guiding function, xviii, 217–220

Hamilton transformations, 217
Hamiltonian, 191, 194, 195, 217, 218, 220–227
system, 191, 218, 223
harmonic oscillations, 41
higher
approximations, 61, 109, 152
dimensional systems, xviii, 208
homoclinic orbits, xvii, 158, 191, 192, 194–196

idealization process, 1
imaginary, 154
impacting wave, xvii, 48, 116, 122, 123, 127, 130
processes, xvii, 115

independent variable, 3, 9, 16, 36, 44, 45, 47, 52, 53, 113, 121, 126, 133
inequality chain, 19
infinitely large functions, xvi, 9, 17, 18
infinitely small
  functions, xvi, 9, 17, 18
  terms, 22, 30
initial
  approximation, 171
  Cauchy problem, 96
  conditions, 4–7, 14–17, 36, 37, 40–43, 46, 49, 53, 71, 80, 81, 89, 93, 94, 98, 102, 106, 147, 148, 152, 153, 155–158, 160, 161, 205, 207, 210, 214–216
  coordinates, 103
  deflection, 96
  dislocation, 167, 178
  equation, 8, 14, 58, 74, 122, 167, 169, 178
  parameter, 3
  position, 43
  problem, 2, 5–7, 47, 49, 56, 84, 88, 94
  simplification, 221
  solution, 172
  system, 80, 223
  time instant, 6
  value problem, 16, 200, 212, 213, 215
  variable, 2, 183, 185, 186
  vectors, 194
integral, 7, 9, 15, 32, 50, 75, 102, 104, 106, 137, 191, 198, 220, 222, 223, 225, 227
internal
  boundary value problem, 79–81
  decomposition, 79, 81–86, 88, 89
  equation, 79, 80, 84, 89, 91
  equations, 79
  limit, 79, 85, 90, 119
  solution, 80, 81, 84, 86, 90, 119
    of l-th order approximation, 80
  space, 78, 80, 81, 89, 91
  variables, xvii, 78–80, 82, 84, 86, 89–91, 118, 119, 122, 123
internal equation and external boundary conditions, 80
invariant, 219, 221
  normalization, 217, 220, 221, 224, 225
inverse decomposition of a coordinate, 17
inversed transition, 57
irregular, xv, 25, 35, 82, 89, 91, 121, 128, 129

Jacobi matrix, 219
Jacobian, 218, 219

Kantorovich-Vlasov
  method, xvii, 165, 169, 171, 173, 184
  procedure, 174, 185
Kirchhoff-Klebsch theory, 165
Krylov-Bogolubov-Mitropolskiy method, xvi, 72

Lagrange's
  formula, 11
Laplace function, 14
Leibniz criterion, 13
Leray-Schauder degree theory, 212
Lighthill, 47, 49
  law, 125
  principle, 56
Lindstedt-Poincaré method, 43, 67, 76, 94, 97
linear oscillator, 3, 5, 58, 225
linearized system, 196
logarithmic function, 11, 22
Lorenz system, 158, 160

Mach number, 2, 93, 116, 126, 145
Maclaurin series, 5, 9, 10, 15
magnitude order, xvi, 9, 17, 19–23, 32, 50, 52, 88, 121, 160
manifold, 209, 212
maps, 191, 212
matching
  method, 80
  of asymptotic decompositions, method of, xvi, 35, 113
  rules, 80, 85
mathematical
  methods, xv, 16
  model, xvi, 1, 2, 7, 8, 116, 160, 165
Melnikov
  approach, 191, 192
  function, xviii, 191, 192, 195, 199, 200
  technique, 191, 192
Melnikov-Gruendler approach, xvii, 194
Melnikov-like
  approaches, 191
  techniques, 191
mulitiplicative, xvii
multiple scale method, xvi, 58, 59, 61–63, 76
multiplicative
  composite decomposition, 88

Newton's binomial, 9, 10
nonautonomous, xviii, 217, 221, 223, 225
nondefined coefficients, method of, 42
nondimensional form, 193
nonhomogeneity, 41
nonhomogeneous, 42, 43
  equation, 41, 42, 46
  problems, 17
noninvariance, 219
nonlinear
  force, 5
  phenomena, 48
  waves, 92
nonlinearities, xviii, 16, 150, 203
  discontinuous, xviii, 203
nonsigularity, 219
nonuniformity, xvi, xvii, 25, 26, 35, 37, 40, 49, 91–93, 112
nonuniformly exact, 25, 39
normal
  Birkhoff form, 220
  equation, xviii, 7, 203
  form, xviii, 217, 218, 220–223, 225–227
normalization, xviii, 217, 221
normalized
  Hamiltonian, 221, 222
  quantity, 14
number, 2, 7, 10, 11, 19, 20, 22, 23, 30, 31, 59, 60, 75, 85, 86, 93, 109, 116, 126, 145, 148, 153, 154, 170, 175, 182, 186, 187, 192, 226
numerical simulations, 98, 191, 201
Nusselt number, 2

one-degree-of-freedom, 72
  system, 3, 63, 192, 203, 204
one-dimensional nonstationary nonlinear waves, xvii, 115, 130
one-parametric families, xviii
operator, xvi, 9, 16, 36, 62, 159, 170, 174, 181
oscillator, 5, 8, 203, 217
overlapping space, 85
  of decompositions, 79

Padé approximation, xvii, 141–144, 148, 149, 151, 154, 157, 158, 160, 161, 163
  diagonal, 159
  two-point, 107
padeon, 149
parametric
  form, xviii, 217, 218
  transformations, 221
  variables, 217

partial
differential equation, 93
equation, 7, 36, 47, 48, 52, 61, 92, 130, 170
particular solution, 41, 42, 46, 67, 69, 180
periodic solution, 102, 103, 203, 205, 207, 208, 210–212, 215, 226
stability of, xviii, 203
periodically perturbed, 191
persistence, xviii, 203, 212
perturbation
along a parameter, 17
along a parameter and coordinates, 9, 16
along coordinates, 17
analysis, xv
parameter, xvi, 1, 25, 42, 93, 94, 109
one, 110
small, xv
regular and singular, xvi, 7, 8, 35, 36
singular, xv–xvii, 8, 29, 35–39, 77, 90, 92, 112, 115, 132, 139
phase portraits, 200, 201
physics, xv, 16, 22, 25, 35, 44, 48, 63, 72, 102, 131, 144, 226
piston, 116, 118, 119, 130
cylinder-like, xvii, 115, 116
plate, 172, 174, 175, 182, 184
ortothropic, 177, 178
rib-supported, 184
structural-ortothropic, 182, 185
Poincaré maps, xviii, 200, 201, 203
Poincaré-Birkhoff normal form, 217
Poisson brackets, 220
polynomials, 28, 142
Prandtl number, 148
precised characteristics, method of, 136
projection, 170, 172, 195

quasi-fractional approximations, 158
quasi-linear method, 98
quasi-periodic functions, 222, 223
quick-changing components, 181

random quantity, 14
rational function, 141, 152, 159
real, 4, 17, 30, 41, 46, 73, 83, 94, 154
rectangular plates, 173, 174
deflections, 187
regular, 201
asymptotics, 151
behaviour, 199
dynamics, 191, 199
outer solution, 191
perturbation, 8, 36
perturbations, xvi, 7, 36
problems, xvi, 35, 37
series, 151
renormalization method, xvii, 115, 124, 127, 129, 130, 133, 134, 136, 139
rescaling, 16, 35, 67, 68, 76, 90, 112, 119
Reynolds number, 2
ribbed
plates, xvii, 165
shell, 165
rigidity, 145, 165, 166, 168, 174, 177, 182, 184
roots, 4, 150
complex conjugated, 4, 58
Runge-Kutta method, 151, 162, 163

saw-shaped functions, 160
scalar product, 109, 212
scaling, 193
function, 17, 19, 20, 22–24, 78–80, 90
method of, xvi, 52, 53
multiple, 53, 55, 56
scheme, 159, 165, 171, 175, 182, 187, 190
second approximation, 17, 42, 131, 135, 137, 225

secular terms, 43, 44, 59, 61, 66–68
self-excited oscillations, 192
series of, 10, 17, 29, 30, 32, 50, 62, 70, 78, 80, 82, 84, 86, 89, 90, 113, 169, 224
shock waves, 146, 148
sigularity, 26, 49, 51, 56, 57, 79, 82, 84, 88, 91, 98, 125, 126, 129, 150
similar method, 169
simple support, 173, 176, 178
singular, 8, 25, 26, 35, 47, 53, 125, 203
    equations, xviii
    excitations, xvi
    forcing, 8
    layers, 39
    perturbation, xv–xvii, 7, 8, 29, 35–39, 49, 77, 87, 90, 92, 112, 115, 132, 139
    perturbations, theory of, xvii, 29, 35, 36, 39, 77, 115, 139
    points, 87, 150
    problems, xv, xvi, 35, 36, 38, 39, 132
    subspaces, 35
singularities, xvii, 56, 61, 84, 91–93, 115, 133
sinusoidal wave of pressure, 146
slow, 31, 66, 72–74, 76, 137, 167–169, 180–182
slow-changing components, 181
small δ method, 102, 160
small parameter, xv, xvi, 2, 8, 9, 16, 17, 35, 38, 42–45, 47, 53, 54, 68, 72, 73, 77, 82, 87, 92, 93, 98, 102, 107–109, 112, 113, 116, 118, 120, 121, 124, 127, 130, 131, 136, 145, 155, 159, 182
    method of, 35
smooth, xviii, 97, 102, 103, 188, 191, 192, 199, 201, 203, 209, 214, 216
space, 26, 36, 37, 39, 57, 72, 77–80, 85, 91, 92, 112, 146, 170, 171, 174, 194, 218, 219
spectrum, 77, 214, 215
spring stiffness, 2
stable, 195, 205, 208, 211, 216, 226
    solution, 212
stick-slip, xviii, 192, 199, 201
strained coordinates, method of, 124
strongly nonlinear dynamical problems, 97
Strouhal number, 2
structural-ortothropic theory, 178, 187
subspace of nonuniformity, 37
successive approximations, 46, 126
    method of, 169
successive differentiation, method of, 14, 16
summing of divergent series, 144
supersonic flow, xvii, 93, 141, 144, 145
swinging, xviii, 217, 218
    oscillator, xviii, 218
system
    linear, 43
    nonlinear, 218

tangencies, 191
tangent space, 212
Taylor
    formulas, 53, 54
    series, 9, 10, 55, 141, 145
theory of elasticity, 48, 77
    three-dimensional, 165
thin cone, xvii, 141, 145
threshold curve, 199, 201
Tichonov theorem, xviii, 203, 207, 209, 210
timescale, 6, 62, 68, 77
transcendental equation, 107
transcendentally small terms, 19

transformation, xviii, 28, 44, 47, 49, 57, 64, 79, 96, 97, 138, 217–223, 225–227
transversal, xviii, 191, 203, 210, 211
transverse vibrations, 174
trigonometric
 functions, 42, 75
 identities, 45, 75
 transformations, 41
truncated equations, 74
two families of characteristics, 135
two scaling, method of, 107
two timescales, 62
two-degrees-of-freedom, 205, 217

unary external series, 122
undetermined coefficients, 59, 67
uniformly
 exact, 25, 26, 39
 suitable solutions, xvii, 115
unit matrix, 219
unperturbed solution, 16
unstable, 167, 195, 200, 216
 solution, 213

value of a jump, 151
Van der Pol
 equations, 74
 method, xvi, 72, 74
variation of arbitrary constants, xvi, 70
variational iterations, method of, 171, 173, 174
variations of arbitrary constants, 68
vector, 16, 24, 25, 36, 72, 170, 191, 192, 194, 195, 205, 206, 212
velocity potential, 93, 116, 130–132, 134, 139, 145
vibrations, 5, 35, 43, 48, 72, 73, 92, 95, 96, 163, 176–178, 182, 185, 217
 free, 3, 177
 low-frequency, 177
vibro-impact
 model, 97
 process, 98, 102
 systems, 160

wave equation, 131, 132
 nonhomogeneous, 131
wave-impact processes, 115
waves, 130, 132, 135, 138, 146
Winkler's type, 167
wire explosion, 116
Witham method, 149
Wronski function, 69
Wronskian, 69

zero order approximation, 41, 45, 47, 70, 93, 97
Zhuravlev's theorem, 224, 226